DESCRIPTIVE INORGANIC CHEMISTRY RESEARCHES OF METAL COMPOUNDS

Edited by **Takashiro Akitsu**

Descriptive Inorganic Chemistry Researches of Metal Compounds
http://dx.doi.org/10.5772/65855
Edited by Takashiro Akitsu

Contributors

Yuichi Negishi, Tadeusz Michalowski, Anna Maria Michałowska-Kaczmarczyk, Aneta Spórna-Kucab, Valentina Uivarosi, Rodica Olar, Mihaela Badea, Messai A. Mamo, Francis Opoku, Ephraim Kiarii, Penny Govender, Miloš René, Aleksey Kuznetsov, Wafaa Hosny, Ming-Der Su, Hsin-Yi Liao, Jia-Syun Lu, Elena Pahontu

Published by InTech
Janeza Trdine 9, 51000 Rijeka, Croatia

Publishing Process Manager Nina Kalinic
Technical Editor SPi Global
Cover InTech Design team

First published July, 2017
Printed in Croatia
Legal deposit, Croatia: National and University Library in Zagreb

Additional hard copies can be obtained from orders@intechopen.com

Descriptive Inorganic Chemistry Researches of Metal Compounds, Edited by Takashiro Akitsu
p. cm.
Print ISBN 978-953-51-3397-1
Online ISBN 978-953-51-3398-8

For EU product safety concerns:
IN TECH d.o.o., Prolaz Marije Krucifikse Kozulić 3, 51000 Rijeka, Croatia,
info@intechopen.com or visit our website at intechopen.com.

Meet the editor

Dr. Takashiro Akitsu now is a professor of Chemistry, at the Department of Chemistry, Faculty of Science Division II, Tokyo University of Science. He got his BSc degree in Chemistry, MSc degree in Inorganic and Physical Chemistry, and PhD degree in Chemistry at Osaka University in 2000. He became a postdoctoral fellow at the Institute for Protein Research at Osaka University in 2000, and he was appointed as an assistant professor at Keio University in 2002 (he stayed at Stanford University in 2007–2008). He moved to Tokyo University of Science as a junior associate professor in 2008 and was promoted to the present position in 2012. His research field is coordination chemistry, physical inorganic chemistry, and crystallography (organic/inorganic hybrid materials composed of chiral metal complexes).

Contents

Preface

Metal ions play an important role in analytical chemistry, organometallic chemistry, bioinorganic chemistry, and materials chemistry. This book, *Descriptive Inorganic Chemistry Researches of Metal Compounds*, collects research articles, review articles, and tutorial description about metal compounds. To perspective contemporary researches of inorganic chemistry widely, the kinds of metal elements (typical and transition metals including rare earth; p, d, f-blocks) and compounds (molecular coordination compounds, ionic solid materials, or natural metalloenzyme) or simple substance (bulk, clusters, or alloys) to be focused are not limited.

In undergraduate level lectures on inorganic chemistry, as shown below, descriptive facts are introduced in typical fashions with important viewpoints, e.g., periodic table, electron (some principles), abundance, simple substance, basic reactions, complexes in addition applications or new topics, and so on. As you can notice, new developing researches provide and add more important information (applications or new topics) to inorganic chemistry, which will be written in this book.

For example, the 12 group elements (Zn, Cd, Hg, and Cn) are sometimes not classified into transition elements but classified into d-block elements. They are easy to be divalent cations like two group elements, which have filled d-shells as electron configuration of Zn and Zn^{2+} indicates:

$$_{30}Zn: (1s)^2(2s)^2(2p)^6(3s)^2(3p)^6(3d)^{10}(4s)^2(4p)^0$$

$$Zn^{2+}: (1s)^2(2s)^2(2p)^6(3s)^2(3p)^6(3d)^{10}(4s)^0(4p)^0$$

Typical similarity of 2 and 12 group elements may be structures and properties of solutions, while differences are distortion due to filled d-shell (d^{10} configuration), which results in no color (no d-d transition), and diamagnetic and four-coordinated complexes afford sterically favored tetrahedral geometry.

Natural sphalerite (both zinc blend and wurtzite are ZnS) contains Zn, which coexists in galena (PbS). Simple substance Zn can be dissolved by water ($Zn^{2+}(aq) + H_2O \rightarrow ZnH^+(aq) + H^+$), acid, and base ($Zn + 2OH^- \rightarrow ZnO_2^{2-} + H_2$). $ZnSO_4.7H_2O$, ZnO (photocatalysts, pigments), ZnS (emitting phosphor, semiconductor), and ZnF_2 (rutile-type crystal) are known commonly. As Irving-Williams series Ni < Cu > Zn exhibits, formation constants of Zn^{2+} complexes are relatively small.

In many cases, properties of 3d metals are different from that of 4d and 5d ones. Naturally, isostructural replacement of Cd to Zn is found in minerals. Simple substance Cd can be dissolved by acid and base, and Zn is more ionized than Cd ($Zn + Cd^{2+} = Zn^{2+} + Cd$ $E^0 = +0.36V$).

CdO, CdS, and CdF_2 (fluorite-type crystal) are known commonly, and Cd^{2+} reacts with water ($2Cd^{2+}(aq) + H_2O = CdOH^{3+}(aq) + H^+$).

Metal Hg is known as a sole liquid simple substance at room temperature or component of amalgam alloy, which is obtained by reduction of slightly soluble cinnabar HgS (solubility product = 10^{-54}) or $Hg_2^{2+} + 2OH^- \rightarrow Hg + HgO(s) + H_2O$ and red $HgO(s) \rightarrow Hg(l) + 1/2\ O_2$ $\Delta H_{diss} = 90.4\ kJmol^{-1}$. $Hg(OH)_2$ is a weak acid ($K=[Hg^{2+}][OH^-]^2 / [Hg(OH)_2] = 1.8 \times 10^{-22}$), and salts are easy to be hydrolyzed (covalent $HgCl_2 + H_2O = Hg(OH)Cl + H^+ + Cl^-$, but only HgF_2 is an ionic halide). Complexes afford two-coordinated linear or four-coordinated tetrahedral geometries ($[Hg(CN)_4]^{2-}$). Because of Hg^{2+}/Hg_2^{2+} equilibrium, disproportionation ($Hg_2^{2+} = Hg + Hg^{2+}$, $E^0 = -0.131V$; $K = [Hg^{2+}]/[Hg_2^{2+}] = 6.0 \times 10^{-3}$) can occur because of these redox potentials ($Hg_2^{2+} + 2e^- = 2Hg$, $E^0 = 0.789V$; $2Hg^{2+} + 2e^- = 2Hg_2^{2+}$, $E^0 = 0.920V$; $Hg^{2+} + 2e^- = Hg$, $E^0 = 0.854V$).

As for solid-state chemistry, magnetic moment of spinel in Fe oxides that is increased by doping of diamagnetic Zn^{2+} is a famous theme of students' experiments. Basically, spin orientation of solid-state magnetic oxides is paramagnetism (random), ferromagnetism (parallel), antiferromagnetism (antiparallel), and ferrimagnetism (antiparallel of strong and weak spins). Evaluating m_{eff} for inverse spinel ferrite Fe_3O_4 (per formula) is to be $[Fe^{3+}]_{tet}[Fe^{2+},Fe^{3+}]_{oct}O_4$, $Fe^{3+}(3d^5) = 5\mu_{BM}$, and $Fe^{2+}(3d^6) = 4\mu_{BM}$; therefore, (5+4)-5 = $4\mu_{BM}$. The reason why increasing magnetic moment after doping $ZnFe_2O_4$ in Fe_3O_4 is explained to be composition $Fe^{2+}{}_{1-x}+Zn^{2+}{}_xFe^{3+}{}_2O_4$, tetrahedral site: $Fe^{3+}{}_{1-x}+Zn^{2+}{}_x$, octahedral site: $Fe^{3+}{}_{1+x}+Fe^{2+}{}_{1-x}$, hence, $[5\mu_{BM}(1+x) + 4\mu BM(1-x)] - 5\mu_{BM}(1-x) = (4+6x)\mu_{BM}$. As doped Zn^{2+} (x) increased, magnetic moment is increased.

As for biochemical application, some Zn^{2+} complexes are used as fluorescence probes. Recently, Zn^{2+} ion has a wide range of different roles in immunity such as the second messenger in signal transduction. Zn^{2+} ion is an inhibited phosphatase PTEN, which enhances the phosphorylation of AKT by interleukin 2 (IL2). Therefore, went and put a stimulus to the cells in a state in which a certain amount of fluorescence by the addition of Zn^{2+} complex was out fluorescence disappears, and external field stimulation sex fluorescent probe has been sought as a medical research tool. This is either $Z^{n2}+$ complex that has been made to work in the same way as zinc, range of applications will change depending on whether there is an inhibitory effect as of pyrithione.

Takashiro Akitsu
Department of Chemistry,
Faculty of Science,
Tokyo University of Science,
Tokyo, Japan

Chapter 1

Nanoformulation as a Tool for Improve the Pharmacological Profile of Platinum and Ruthenium Anticancer Drugs

Valentina Uivarosi, Rodica Olar and Mihaela Badea

Additional information is available at the end of the chapter

http://dx.doi.org/10.5772/intechopen.68306

Abstract

Cisplatin and analogs are used for the treatment of some type of cancers in combination with organic cytostatics. Also, two ruthenium (III) complexes are in clinical trials as anticancer drugs. In order to overcome toxicity and resistance associated with this therapy and/or enhance stability, a large variety of formulations based on organic, inorganic, or hybrid matrix were developed and tested both *in vivo* and *in vitro*. The best results were obtained for systems properly functionalized in order to enhance the metal content and/or to specific target the tumor tissue through overexpressed receptors.

Keywords: platinum, ruthenium, anticancer metal-based drugs, nanoformulations, conjugation

1. Introduction

Despite the use of metal compounds in empirical medicines since the ancient civilization time of Mesopotamia, Egypt, India, and China, the pharmacological bases of their therapeutic action were just began to be understood in the last 50 years [1].

A milestone in the development of inorganic medicinal chemistry was represented by the serendipitously discovery of the anticancer agent cisplatin (Platinol) [2], which opened the gate of extensive and rigorous research for anticancer metal-based drugs. Cisplatin quickly became a successful antitumor agent, but over time, its severe side effects and installation of resistance led to the orientation of research toward finding new cisplatin analogs. Thus, "the second-generation platinum drugs" (e.g., carboplatin) with improved toxicological profiles and "the third-generation drugs" (e.g., oxaliplatin) overcoming cisplatin resistance have been developed [3].

Having in view the systemic administration, the patients experienced severe symptoms since cisplatin and its analogs, carboplatin and oxaliplatin, were introduced in cancer therapy. Moreover, the intrinsic or acquired resistance and the fact that many cancers are insensitive to platinum-based drug therapy started an assiduous search for formulations that are able to deliver these drugs with reduced toxicity but with a similar or even enhanced cytotoxic profile [4–9].

A promising strategy able to overcome most of the above limitations consists in embedding either the original drug or a precursor in a proper matrix that is able to release a high amount of active species at target site. As result, several formulations based on organic, inorganic, or hybrid materials were designed. Among organic-based materials, a large variety of lipids, polymers, or mixed species were developed as platinum- and ruthenium-based drug carriers while magnetite, gold, graphene, and silica were studied as inorganic-based materials for the same purpose. Moreover, hybrid materials based on functionalized graphene, gold, iron oxides, silica, or polinuclear complexes and polysilsesquioxanes were studied in order to facilitate the delivery of these drugs [6–9].

Beyond improving solubility and reducing toxicity, a main challenge of these formulations was to increase their selectivity for tumor cells in order to achieve an optimum pharmacological profile. The first formulation developed by platinum-based drugs embedding through noncovalent interactions generated systems with a low loading capacity. A proper functionalization of the embedding matrix with Pt(II) drugs or Pt(IV)/Ru(III) prodrugs and/or with a responsive stimulus or a targeting moiety provided species with an increased cytotoxicity [6–9].

A large variety of encapsulation matrices and conjugations were developed, and formulations exhibit a promising cytotoxicity against either multidrug resistant or platinum insensitive cancer cells.

2. Anticancer metallodrugs

Apart from extensive research undertaken in the field of platinum complexes, other metals or other therapeutic strategies have attracted attention in order to reduce the side effects, to mitigate the resistance, and to achieve the oral administration.

The anticancer metallodrugs known at this time belong to three main classes:

- anticancer therapeutics
- therapeutic radiopharmaceuticals
- photochemotherapeutic metallodrugs [10].

http://dx.doi.org/10.5772/intechopen.68306

Numerous *chemotherapeutic metallodrugs* developed in the last 4 decades are based on a large variety of metals: Pt, Ru, Au, Sn, Al, Ga, In, Ti [11–16]. Among the metal-based compounds, complexes of platinum (Pt(II) and Pt(IV)), ruthenium (Ru(II) and Ru(III)), gold (Au(I) and Au(III)), and titanium (Ti(IV)) are the most studied [13].

Therapeutic radiopharmaceuticals include a β-emitting radionuclide (^{89}Sr, ^{90}Y, ^{153}Sm, ^{213}Bi) or a α-emitting radionuclide (^{223}Ra). In general, α- and β-(electrons) emitters are used in radiotherapy, while β+ (positrons) and γ-emitters are used in radiodiagnosis [14].

Utilization of *photochemotherapeutic metallodrugs* is based on the photodynamic therapy (PDT). In PDT, a photosensitizing agent is delivered in tumor cells, which are activated with light, generating cytotoxic singlet oxygen. Starting to observation that Photofrin, a haematoporphyrin derivative is a strong chelator, forming a complex with Zn^{II} *in vivo*, some photochemotherapeutic metallodrugs have been developed [15].

The main platinum-based anticancer drugs currently used in clinic are presented in **Table 1**, while the emerging platinum- and ruthenium-based anticancer agents are listed in **Table 2**.

Metal	Compound	Indications	Commercial names
Chemotherapeutic metallodrugs			
Pt	**Cisplatin** (*cis*-diamminedichloroplatinum (II)) 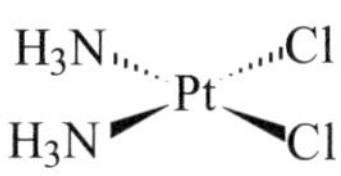	Testicular, ovarian and colorectal cancer	Cisplatin Platosin Sinplatin Platinol
	Carboplatin (*cis*-diammine (1,1-cyclobutanedicarboxylatoplatinum (II)) 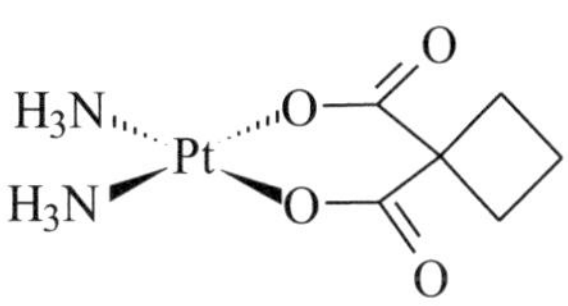		Carboplatin Paraplatin
	Oxaliplatin ((1*R*, 2*R*)-(N, N'-1,2 diamminecyclohexan)-(O-O')-etandioatoplatinum (II)) 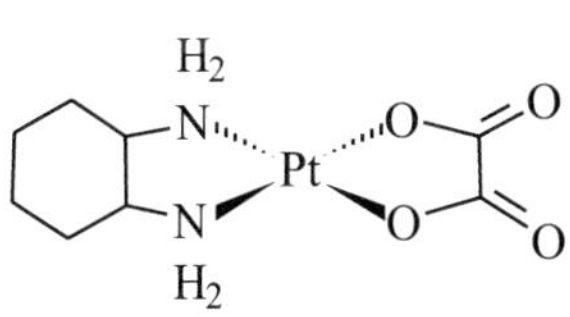		Oxaliplatin Eloxatin

Table 1. Platinum-based anticancer drugs currently used in clinic.

Metal	Compound	Uses/Comments
Chemotherapeutic metallodrugs		
Pt	**Pt^{IV}: Satraplatin (JM 216)** **Pt^{II}: BBR3464**	Satraplatin: first orally bioavailable platinum drug; extended activity spectrum; reduced resistance. Investigated in phase III clinical trials for hormone-refractory prostate cancer [14]. This compound are not only multinuclear but also polycationic, breaking the traditional design rules of platinum complexes. Has undergone phase II clinical trials for metastatic small cell lung cancer [15].
Ru	**Ru^{III}**: [H_2im][***trans***-RuCl_4(DMSO-S)(Him)] **NAMI-A** Imidazolium *trans*-[tetrachloro (dimethylsulfoxide) (imidazole) ruthenate(III)] (a); **Ru^{III}**: [H_2ind][***trans***-RuCl_4(Hind)$_2$] **KP1019** Indazolium *trans*-[tetrachlorobis(1Hindazole) ruthenate(III)] (b)	NAMI-A (Ru) in combination with gemcitabine as antimetastatic agent accomplished phase I/II [16] KP1019 antimetastatic agent; completed phase I clinical trials [16]

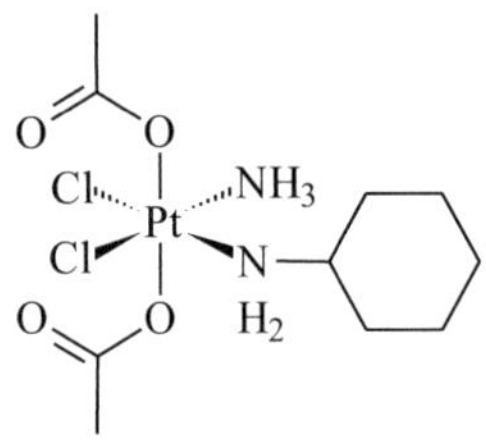

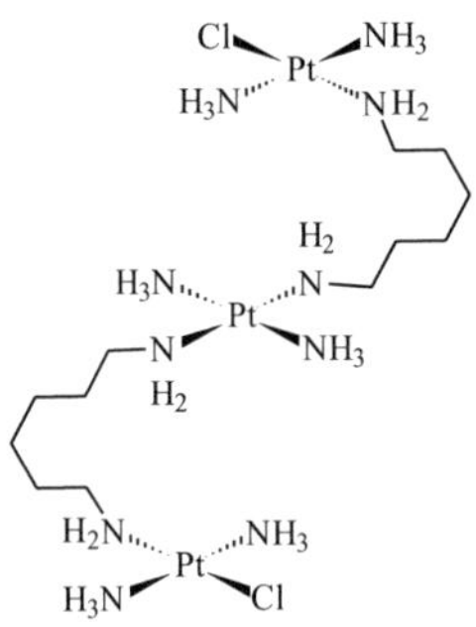

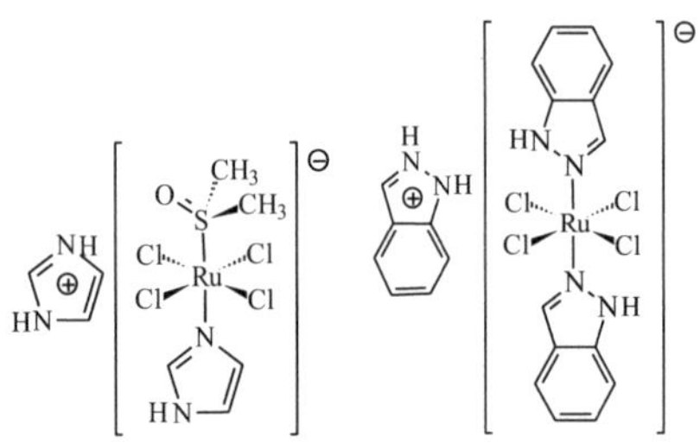

(a) (b)

Table 2. Platinum and ruthenium-based anticancer drugs subjected to clinical trials.

3. Platinum-based drugs nanoformulations

The clinical use of cisplatin and its analogs evidenced pharmacological deficiencies such as poor water solubility, low bioavailability, and short circulating time, besides toxicity and resistance. Moreover, a few types of cancers are sensitive to platinum-based drugs treatment.

Therefore, in the last decades, the researches were focused in designing drug delivery systems that are able to overcome these issues, but with preserving or even enhancing the drug efficacy. A brief overview concerning nanoscale drug delivery systems based on worldwide approved platinum-based cytostatic drugs cisplatin, carboplatin, and oxaliplatin is presented with focus on systems that advanced in clinical trials or exhibited promising pharmacological profile *in vitro* or *in vivo* preclinical assays.

3.1. Cisplatin-based nanoformulations

Cisplatin was the pioneering metallodrug introduced for the cancer treatment with the best result obtained in testicular cancer cure, for which a rate of 90% survival was achieved.

An impressive work was directed in the last time to overcome the severe side effects and intrinsic or acquired resistance by its inclusion in a proper matrix. This approach provided a way to extend its curative effect to other types of cancer proved so far to be insensitive to platinum-based drugs alone or in combination with other organic antineoplastic drugs. Its encapsulation into liposomes or polymeric species seems to provide the most promising formulations so far, since some of these formulations are currently in clinical trials.

Many formulations were developed by cisplatin encapsulation in the aqueous core of liposomes, with differences that consist in the composition of lipid bilayer, platinum content, and release profile. These attempts to incorporate cisplatin into liposomes were limited by its low both hydrophilicity and lipophilicity that resulted in a very low drug-lipid ratio and unstable systems, especially when injected into the blood stream [6–9].

In order to increase the liposomes stability, these systems were coated with a biocompatible hydrophilic polymer such as polyethylene glycol (PEG). Among these, lipoplatin was developed by cisplatin incorporation in a mixture of lipids from vegetable and animal sources, some being PEGylated [17]. An optimum pharmacological profile was observed in phase I clinical trial and significant improvements in patients with acquired resistance in phase II, in combination with gemcitabine, advanced this formulation in phase III clinical trials for both nonsmall-cell lung and pancreatic carcinoma [18]. Moreover, a preclinical study evidenced the potential of lipoplatin for cisplatin-resistant cervical cancer treatment [19].

A modest pharmacological profile was evidenced in clinical trials for a similar formulation SPI-77 as a result of low amount of cisplatin released [20, 21], while for LiPlaCis, a significant renal nephrotoxicity and infusion reactions were observed during phase I clinical trial [22].

As a result, the studies were directed to increase the amount of platinum species embedded either by using negatively charged phospholipids to entrap electrostatic $[Pt(NH_3)_2(H_2O)_2]^{2+}$ species or by lipid bilayer functionalization and conjugation with platinum(II) or platinum(IV) species [7]. In this respect, some formulations with a high loading capacity were developed by *cis*-$Pt(NH_3)_2$and *cis*-$Pt(NH_3)_2Cl$ moieties coordinated to carboxylate groups of lipids, and some exhibited a significant antitumor activity both *in vitro* and *in vivo* assays [23, 24].

Moreover, systems with a pendant group having selectivity for an overexpressed receptor in the cancer cells have been exploited to enhance the platinum species accumulation through receptor-mediated endocytosis. Such liposomal system targeting epidermal growth factor receptor (EGFR)-expressing tumors was developed by conjugation with sodium alginate and indeed exhibited enhanced delivery ability into ovarian tumor tissues and a reduced nephrotoxicity in mice [25].

A variety of polymeric formulations designed as micelle, hydrogels, nanoparticles, and nanocapsules were also studied as cisplatin carrier. The noncovalent encapsulation provided systems with similar or even lower efficacy in comparison with free cisplatin and as a result polymer conjugates were developed by Pt(II) or Pt(IV) species in reversible coordination to a functional group from the polymer backbone or its branches [6–9, 26].

Among these, nanoplatin (NC-6004) was obtained as micellar formulation by cisplatin entrapping in the core of polyethylene glycol-poly(glutamic acid) copolymer. The *in vitro* and *in vivo* preclinical assays evidenced a complete tumor regression as well as a low nephrotoxicity and neurotoxicity in C26 murine colon carcinoma cell [27]. The phase I trial evidenced a better tolerability and reduced side effects in comparison with cisplatin [28] and thus advanced nanoplatin in phase II trials for nonsmall-cell lung cancer, bladder cancer, and bile duct cancer, respectively [8].

The conjugated polymer (AP5280) was developed as nanoparticles by *N*-(2-hydroxypropyl) methacrylamide copolymer conjugation by *cis*-$Pt(NH_3)_2$ moiety to the peptidyl side chains (Gly-Phe-Leu-Gly) ended with amidomalonate group. This formulation exhibited an increased cytotoxicity in murine tumor models [29] and, moreover, evidenced reduced side effects in a phase I clinical trial conducted by intravenous infusion administration [30].

The conjugate designed by cis-$Pt(NH_3)_2$ moiety coordination to polyethylene glycol branched with citric acid exhibited an enhanced cytotoxicity in both sensitive and resistant HT1080 human fibro sarcoma cells, CT26 fibroblasts, and SKOV3 human ovarian cells [31], while another one based on poly(ethylene glycol)-poly(acrylic acid) copolymer and encapsulated in calcium phosphate evidenced its cytotoxicity against a lung cancer cisplatin-resistant cell line [32].

A good antitumor activity was also achieved by *cis*-$Pt(NH_3)_2$ moiety coordination to the carboxyl groups of poly(γ,L-glutamic acid)-based polymer [33], while by conjugation with polyamidoamines dendrimers developed nanocarriers that inhibit the subcutaneous B16F10 murine melanoma, a cisplatin insensitive tumor [34].

On the other hand, the conjugation and/or encapsulation of an organic cytostatic or a sensitive trigger together with platinum species were exploited to enhance the cytotoxicity of these formulations.

As a result, micellar carriers developed by poly(ethyleneglycol)-b-poly(L-glutamic acid)-b-poly(L-phenylalanine) tri-block copolymer conjugation with paclitaxel- and cisplatin-derived moieties exhibited an enhanced activity against A549 human lung tumor cells both *in vitro* and *in vivo* [35], while conjugates of both paclitaxel and *cis,cis,trans*-$[Pt(NH_3)_2Cl_2(OH)(HSucc)]$ (H_2Succ: succinic acid) (**Figure 1a**) prodrug with poly(ethylene glycol)-b-poly(ε-caprolactone)-b-poly(l-lysine) tri-block amphiphilic biodegradable copolymer exhibited an enhanced efficacy in U14 cervical tumor line xenograft in mice as a result of the synergistic effect [36].

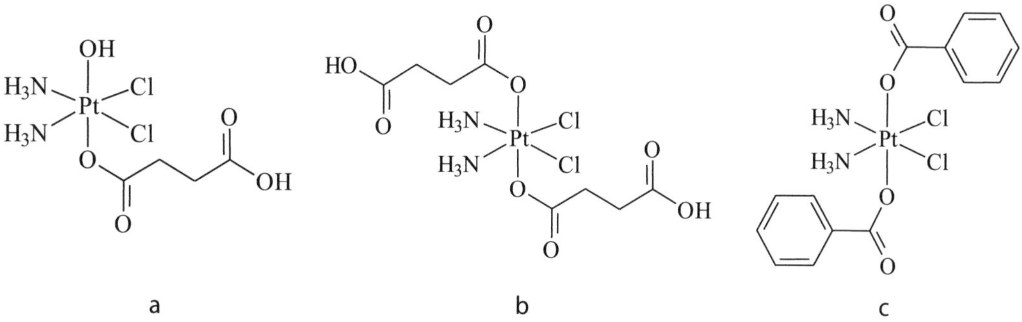

Figure 1. Platinum (IV) complexes embedded into cisplatin based formulations: cis,cis,trans-[$Pt(NH_3)_2Cl_2(OH)(HSucc)$] (a), *cis,cis,trans*-[$Pt(NH_3)_2Cl_2(HSucc)_2$] (b), and *cis,cis,trans*-[$Pt(NH_3)_2Cl_2(Bz)_2$] (c).

A combination of doxorubicin- and peptide-modified with *cis*-$Pt(NH_3)_2$ moiety loaded in positively charged mucoadhesive chitosan-polymethacrylic acid-based nanocapsules demonstrated an enhanced cytotoxicity against UMUC3 human urothelial carcinoma cell line [37]. Likewise, glutathione-sensitive micelles based on carboxymethyl chitosan crosslinked with 3,3′-dithiobis-N-hydroxysuccinimidyl propionate modified with folic acid exhibited synergistic cisplatin-doxorubicin effect against HeLa tumor cell line [38].

Another co-delivery system was developed by self-assembly of the anionic polyglutamic polymer *cis*-$Pt(NH_3)_2$ conjugated with an cationic metformin polymer. This formulation suppressed tumor growth for H460 human NSCLC xenografts in mice by a synergistic effect related to protein kinase α pathway activation and mammalian target rapamycin inhibition [39].

In order to achieve a high selectivity in targeting tumor cells, peptide and glycoside residues were inserted in the polymer backbone as groups that can be specifically recognized by the tumor tissue. This strategy resulted in thermosensitive nanoparticles obtained by cisplatin and indocyanine green loading in a complex matrix of poly(lactic-co-glycolic acid) copolymer and lipids functionalised with Gly-Cys-Gly-Ala-Ala-Asn-Leu heptapeptide. This formulation was designed to target MGC803 gastric tumor cells that overexpress the legumain and as a result exhibited a good activity *in vitro* [40].

Another formulation was developed as lyophilized system by *cis*-$Pt(NH_3)_2(OH_2)$ moiety coordination to carboxyl groups of hyaluronan, a naturally occurring glycosaminoglycan polysaccharide that targets tumor cells through specific interactions with CD44 receptor highly overexpressed in many cancers tissues. This conjugate demonstrated a suppressed cancer progression through intratracheal administration in Lewis lung carcinoma allografts in mice [41]. A platform targeting the same receptor was prepared by cisplatin incorporation in calcium phosphate and then embedded in hyaluronan-chitosan cross-linked polymer shell. These nanoparticles demonstrated target specific delivery in A549 human lung cancer cells confirmed by an eightfold increase of drug efficacy [42].

Some inorganic materials such as magnetite, graphene, gold, and silica were also studied in order to develop proper formulations for cisplatin delivery. The attempts to obtain nanoparticles based on these species have been discouraged by the low amount of cisplatin that can be noncovalent-retained and consequently promote an early release of the active species in the plasma. This problem has been solved either by coating the inorganic species-cisplatin assembly with an organic shell or by its surface functionalization with groups that are able to coordinate platinum species [6–9].

Following these strategies, an enhanced therapeutic effect in A549 human lung cancer xenograft model was obtained by magnetite-cisplatin assembly encapsulated in poly(vinyl alcohol) and poly(acrylic acid) [43]. Another formulation designed by *cis*-$Pt(NH_3)_2$ conjugation and magnetite embedded in (methacrylic acid)-*g*-poly(ethylene glycol methacrylate) polymer exhibited an enhanced anticancer efficacy in cisplatin-resistant HT-29 human colon adenocarcinoma model, particularly when a magnetic field gradient was applied at the tumor site [44].

An improved antitumor effect was also obtained either for gold nanoparticles PEGylated and *cis,cis,trans*-$[Pt(NH_3)_2Cl_2(HSucc)_2]$ (**Figure 1b**) conjugated [45] or for that functionalized with oligonucleotide and *cis,cis,trans*-$[Pt(NH_3)_2Cl_2(OH)(HSucc))]$ conjugated [46]. It is to be pointed the higher cytotoxicity against cisplatin-resistant line exhibited by such formulations.

Nanoparticles developed by *cis*-$Pt(NH_3)_2(OH)$ moiety coordination to functionalized mesoporous silica exhibited also an enhanced cytotoxicity on HT-29 colon cancer cell line [47].

Concerning graphene-based materials, a cisplatin nanotube conjugate modified with epidermal growth factor (EGF) proved an enhanced activity against EGF overexpressing head and neck squamous carcinoma cells [48], while functionalized multi-walled carbon nanotubes (MWCNTs) conjugated with *cis,cis,trans*-$[Pt(NH_3)_2Cl_2(Bz)_2]$ (HBz: benzoyc acid) (**Figure 1c**) exhibited a selective accumulation in mice lungs [49].

Some hybrid materials based on coordination polymers were also developed as cisplatin carriers. Such supramolecular assembly was developed by $[Tb_2\{Pt(NH_3)_2Cl_2(Succ)_2\}_3]_n$ encapsulation in amorphous silica (**Figure 2**) as cytotoxic agent against HT-29 human colon carcinoma cell line [50].

Another platform was designed by hetero-metallic coordination polymer $[Zn_2\{Pt(NH_3)_2Cl_2(Ncp)_2\}]_n$ (Ncp: N-carbamoyl phosphate) embedding in an asymmetric lipid layer modified with polyethylene glycol. This assembly, with a high amount of cisplatin incorporated, exhibited an

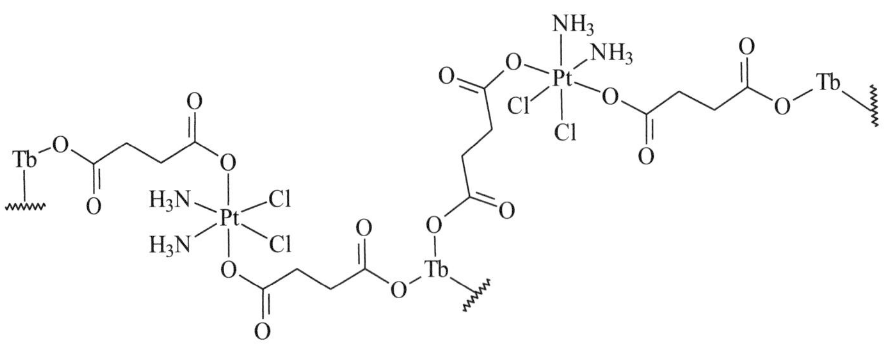

Figure 2. Hybrid nanoformulation developed by $[Tb_2\{Pt(NH_3)_2Cl_2(Succ)_2\}_3]$ encapsulation.

enhanced efficacy in comparison with free cisplatin in H460 human nonsmall cell lung cancer and AsPC-1 human pancreatic cancer xenograft in mice [51].

A carrier system based on the same hetero-metallic coordination polymer and pyrolipid as photosensitizer exhibited a synergistic effect in cisplatin-resistant human head and neck cancer SQ20B xenograft in mice [52], while another formulation with small interfering RNA (siRNA) in addition and coated with a cationic lipid layer exhibited cytotoxicity both *in vitro* and *in vivo* against SKOV-3cisplatin-resistant ovarian cancer [53].

On the other hand, polysilsesquioxane-based hybrid nanomaterials developed by *cis,cis,trans*-$[Pt(NH_3)_2Cl_2(HptsSucc)_2]$ (H_2ptsSucc: propyltriethoxysilane succinic acid) polymerization (**Figure 3**) and coated with polyethylene glycol demonstrated an enhanced efficacy in combination with radiotherapy against A549 and H460 human lung cancer cells xenograft in mice [54].

These formulations can be internalized into the cancer tissues through passive or active transport. The passive transport is based on the ability of nanosystems to accumulate better in tumor tissue as a result of its increased permeability and poor lymphatic clearance, phenomenon known as enhanced permeability and retention (EPR) effect [55]. Moreover, the intratumoral nanoparticles content can be enhanced through an active transport facilitated by an overexpressed receptor.

Upon endo- or phagocytosis, the platinum species release is triggered in cytosol or other cellular compartments by several processes that can be acid, redox, and/or enzymatic assisted. For conjugated formulations, the cisplatin structure is restored either by reaction of Pt(II) species with chloride anions or by Pt(IV) species reduction with glutathione or ascorbic acid [6–9].

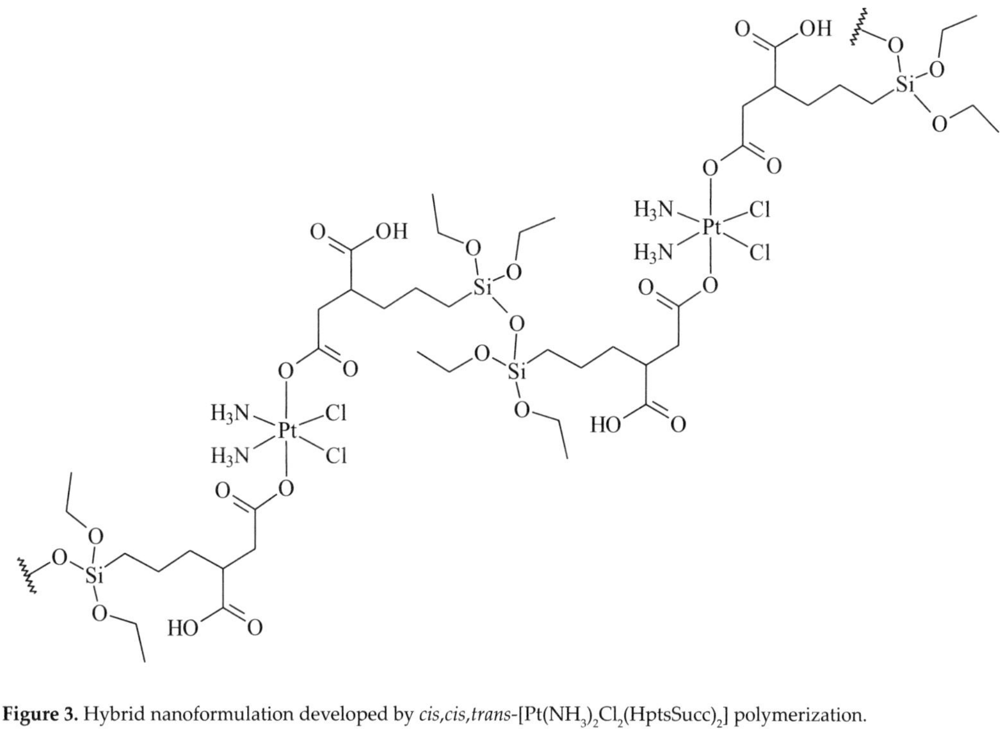

Figure 3. Hybrid nanoformulation developed by *cis,cis,trans*-$[Pt(NH_3)_2Cl_2(HptsSucc)_2]$ polymerization.

3.2. Carboplatin-based nanoformulations

The structural difference between cisplatin and carboplatin consists in replacing the chloride leaving groups by 1,1-ciclobutandicarboxylate as chelate ligand. Although carboplatin is often preferred over cisplatin in cancer therapy based on a lower nephrotoxicity, this exhibits a limited therapeutic efficacy related to its reduced uptake by the tumor cells. Moreover, the treatment induces myelosuppression and cross-resistance [14].

As a result, few studies were concerned on developing carboplatin-based formulations. Based on experience accumulated in cisplatin-based formulation development, carboplatin was embedded through noncovalent interactions especially in polymeric or hybrid materials, some proper functionalized in order to achieve either a targeted delivery or an enhanced efficacy, especially against multidrug resistant cancer cell lines.

Such a polymeric formulation was developed by loading in poly(D-L-lactide-*co*-glycolide) polymer. This nanocarrier exhibited an enhanced cellular uptake in both A549 lung and MA148 ovarian tumor cells [56], while that based on poly(ε-caprolactone) was also efficient uptakes and displayed a significant cytotoxicity in U-87 human glioma cell line, without inducing haemolysis [57]. Moreover, carboplatin-loaded apotransferrin and lactoferrin nanoparticles with high encapsulation efficacy exhibited a significantly cellular uptake and sustained intracellular drug retention in retinoblastoma cells [58], while a chitosan-based formulation demonstrated an enhanced antiproliferative effect against MCF-7 breast cancer cell line [59].

The hybrid materials were also studied in order to improve the pharmacological profile of carboplatin. Such supramolecular assembly based on multiple functionalizations of MWCNTs with amino groups resulted in a dramatic decrease of the MDA-MB-231 human mammary adenocarcinoma derived epithelial cells viability, which was related to superoxide anions production. This study also evidenced that expression of some proteins was inhibited, while the Beclin1 was overexpressed. As a result, most probably this system triggers the cell death through autophagy [60]. Another nanohybrid formulation developed by carboplatin loading in the nanographene oxide-gelatine material exhibited an enhanced efficacy in IMR-32 human neuroblastoma cell line [61].

3.3. Oxaliplatin-based nanoformulations

Oxaliplatin was introduced as first-line chemotherapeutic for the treatment of advanced colorectal cancer based on a different antineoplastic spectrum in comparison with cisplatin. However, the peripheral neuropathy and a moderate myelotoxicity in cumulative dose dependence were observed in many patients [62].

As a result, the attempts to improve its pharmacological profile and reduce the side effects resulted in several valuable formulations for this antineoplastic drug. Similar with cisplatin, a variety of organic, inorganic, and hybrid materials were studied for embedding either the original species [Pt(dach)(C_2O_4)] (dach: (1*R*,2*R*)-1,2-diaminocyclohexane) or another Pt(II) or Pt(IV) complex bearing dach as chelate ligand.

Among these, lipoxal developed as liposomal PEGylated formulation exhibited an acceptable pharmacological profile in a phase I clinical study for advanced gastrointestinal cancer [63]. By this formulation injected directly in F98 glioma implanted in rats, a reduced toxicity with preservation of the antitumor potential of oxaliplatin was achieved as well [64].

Similar with cisplatin, the efficacy of oxaliplatin-based formulation has been improved by surface of the liposomes modification with moieties that are able to assure either a specific targeting or a rapid release after the internalization of delivery system in tumor tissue.

These strategies resulted in developing a transferrin target sensitive liposomal formulation, which demonstrated increased tumor suppression in C-26 colon cancer cell line xenograft in mice as a result of transferrin receptor overexpression in this line [65]. This transferrin-targeted liposomal formulation is currently under phase II clinical investigation for the treatment of gastric cancer and gastroesophageal junction cancer [66].

By oxaliplatin encapsulation in PEGylated cationic liposomes, a formulation with a selective delivery in tumor vasculature was developed. The assays evidenced a complete suppressing tumor-induced angiogenesis and antitumor efficacy in mouse dorsal air sac as a result of dual-targeting both tumor cells and its vascular endothelial structure [67]. Moreover, the efficacy of this assembly can be improved by a sequential administration of oxaliplatin containing PEG-coated cationic liposomes [68].

Several oxaliplatin-based polymeric systems were also developed in order to enhance its cytotoxicity. Such oxaliplatin containing micelles (NC4016), in addition, proved the ability to overcome the oxaliplatin resistance *in vivo* are currently in clinical trials in patients with advanced solid tumors or lymphoma [69].

Another micellar formulation was developed by $[Pt_2(dach)_2(dah)_2](NO_3)_2$ (dah: 1,2-diaminohexane) complex (**Figure 4a**) embedding into methoxylpoly(ethylene glycol)-b-poly(lactide-co-2-methyl-2-carboxylpropylene carbonate) (mPEG-b-P(LA-co-MCC)) copolymer. This pH sensitive assembly exhibited a significant cytotoxicity against H22 liver cancer cell line xenograft in mice [70].

The polymeric systems were exploited not only to enhance the drug cytotoxicity through conjugation with Pt(dach) moieties, but for a combined delivery as well.

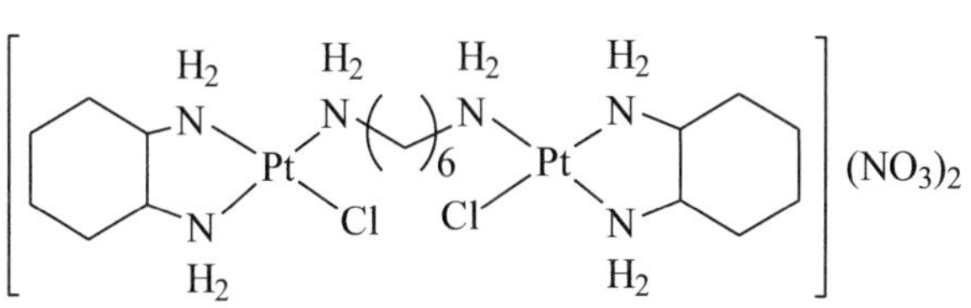

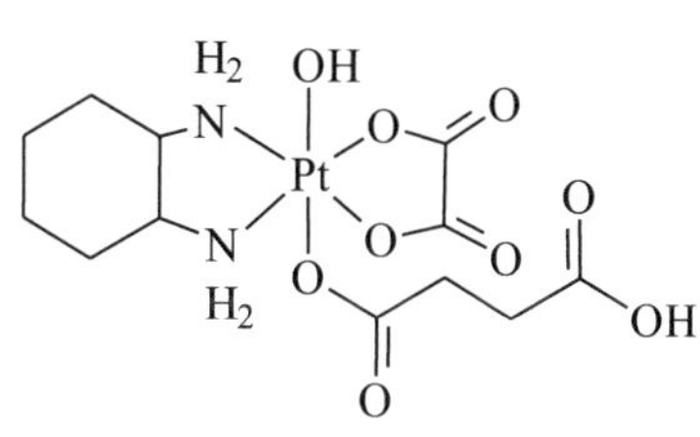

Figure 4. Platinum (IV) complexes embedded into oxaliplatin based formulations: $[Pt_2(dach)_2(dah)_2](NO_3)_2$ (a), and $[Pt(dach)(C_2O_4)(OH)(HSucc)]$ (b).

In this respect, micelles based on poly(ethylene glycol)-b-poly(glutamic acid) copolymer conjugated with Pt(dach) moiety demonstrated a potent tumor growth inhibition after an intraperitoneal injection in HeLa tumor cell xenograft in mice [71], while a similar micellar formulation inhibited the tumor growth in OCUM-2MLN scirrhous gastric cancer cell line and their lymphatic metastases in mice [72].

The polymer conjugate AP5346 was developed by Pt(dach) moiety coordination to the pH-sensitive amidomalonato chelating group from a *N*-(2-hydroxypropyl) methacrylamide-based copolymer structure. This conjugate exhibited an improved cytotoxicity in comparison with oxaliplatin in some colon tumor cell line xenograft in mice [73]. Based on pharmacological profile observed in patients with advanced solid tumors in phase I trial [69], this formulation advanced in phase II trial in recurrent ovarian cancer was initiated, but the results are so far disappointing [73].

Hybrid micelles containing mPEG-b-P(LA-co-MCC) copolymer conjugated with both Pt(dach) moiety and gemcitabine showed a low systemic toxicity and a synergic efficacy against MCF7 human breast cancer cell line xenograft in mice [74], while a similar system based on this copolymer conjugates with both [Pt(dach)(C_2O_4)(OH)(HSucc)] (**Figure 4b**) and daunorubicin showed reduced systemic toxicity and a synergistic effect in H22 hepatocarcinoma xenograft in mice [75].

In order to enhance the concentration of active species released in tumor tissue through a targeted delivery, some oxaliplatin-based polymer formulations were functionalized with glycoside residues and antibodies. Such polymeric nanoparticles were designed by carboplatin embedding in the supramolecular assembly of chitosan conjugated with hyaluronan and additionally coated with Eudragit S100. The oral administration of this formulation resulted in an enhanced activity in HT-29 cell line xenograft in mice [76].

Nanoparticles with a high amount of oxaliplatin embedded in a hybrid material consisting in a polymeric chitosan layer [77, 78] and a mixture of phospholipids conjugated with a thiolated antibody for tumour necrosis factor induced protein were developed as well [77]. Such formulations exhibited an increased cytotoxicity in comparison with oxaliplatin in HT-29 [77] and MCF7 cell lines [78].

Moreover, the functionalization allowed extending the cytotoxic effect to oxaliplatin insensitive tumors such as breast and gastric cancer. Thus, a pH-responsive nanocarrier was constructed by Pt(dach) moiety conjugation in citrate cross-linked chitosan matrix. The enhanced cytotoxicity of these nanoparticles in MCF-7 human breast cancer cell line was related to apoptosis induced in a caspase-dependent manner [67]. The nanogel system developed by embedding oxaliplatin in hydroxypropylcellulose-poly(acrylic acid) exhibited cytotoxicity against BGC823 human gastric cancer cell line [79].

Several systems based on hybrid materials were also developed for achieving an oxaliplatin enhanced delivery. Among these, superparamagnetic iron oxide nanoparticles encapsulated in pectin Ca^{2+} cross-linked exhibited 10-fold enhanced cytoxicity in comparison with free drug in MIA-PaCa-2 pancreas cancer cell line [80].

Another formulation developed by oxaliplatin incorporation into the inner cavity of PEGylated MWCNTs demonstrated a significantly improved cytotoxicity against HT-29 colorectal cell

http://dx.doi.org/10.5772/intechopen.68306

line [81], while similar nanocomposites additionally decorated with magnetite exhibited antitumor effect and low toxicity in HCT116 human colon cancer cell line xenograft in mice [82].

Naked gold nanoparticles functionalized with a thiolated poly(ethylene glycol) monolayer capped with a carboxylate group and conjugated with $[Pt(dach)(H_2O)_2](NO_3)_2$ yielded a supramolecular complex with about 280 Pt(dach) moieties per nanoparticle. This formulation demonstrated a similar or significant enhanced cytotoxicity in comparison with free oxaliplatin in A549 lung epithelial cancer cell line and HCT116, HCT15, HT29, and RKO colon cancer cell lines. Moreover, an unusual ability to penetrate the nucleus in the lung cancer cells was observed in these assays [83].

Mesoporous silica nanoparticles functionalised with carboxyl groups and conjugated with Pt(dach) moiety were also obtained with an improved cytotoxicity against HepG-2 human liver cell line [84].

Data concerning a platform constructed by $[Zn_2\{Pt(dach)Cl_2(Ncp)_2\}]_n$ hetero-metallic coordination polymer conjugation to an asymmetric lipid bilayer modified with polyethylene glycol (**Figure 5**) were reported. This assembly with a high amount of platinum species incorporated exhibited cytotoxicity in H460 human nonsmall cell lung and AsPC-1 human pancreatic cancer cell lines xenograft in mice [53].

Hybrid nanoparticles were also obtained by $[Pt(dach)Cl_2(triethoxysilylpropylsuccinate)_2]$ base-catalyzed sol-gel polymerization similar to cisplatin derivative. Moreover, the silanol and carboxyl groups were functionalised with cyclic arginine-glycine-aspartate peptide and anisamide and then the surface was PEGylated. The cytotoxicity assay clearly indicated an increased uptake of this assembly by DLD-1 and HT-29 human adenocarcinoma cancer cells through integrin receptor and by AsPC-1 human pancreatic cancer cells through sigma receptor together with the tumor growth inhibition efficacy in pancreatic cancer xenograft in mice [85].

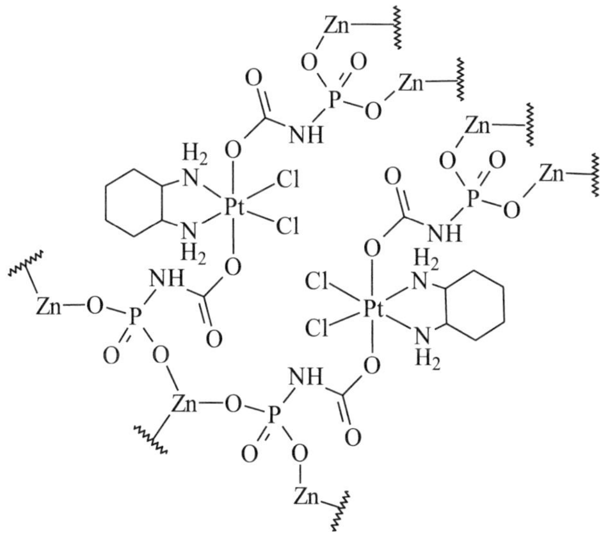

Figure 5. Hybrid nanoformulation developed by $[Zn_2\{Pt(dach)Cl_2(Ncp)_2\}]_n$ encapsulation.

4. Ruthenium (III)-based drugs nanoformulations

The studies regarding ruthenium complexes as anticancer agents were developed as an alternative of platinum complexes, especially for their reduced toxicity, large spectrum of activities (including against cisplatin-resistant tumors) and selectivity [86–88]. Among the various compounds of ruthenium investigated for their anticancer activity, two are in phase II clinical trials, namely NAMI-A (**Table 2**) as antimetastatic agent and KP1019 (**Table 2**) as antitumor for primary tumor site [89–93].

Both are pseudo-octahedral complexes having four chloride ions in the equatorial plane. The axial ligands are imidazole and DMSO molecules in NAMI-A complex, while for KP1019 are two indazole molecules. Both complexes undergo hydrolysis in aqueous solutions (chloride ions being replaced by water and/or hydroxide ions) and interact with biological reductants leading to ruthenium (II) species. These two processes seem to provide the active species in the body [94–96].

In order to improve the stability in aqueous systems, especially at physiological pH, and the delivery of drugs to the solid tumors, various drug delivery carriers have been designed and investigated. Two major ways were followed namely chemical conjugation and physical encapsulation [97].

4.1. Physical encapsulation of ruthenium-based drugs

Physical encapsulation is based on the capacity of carriers to retain the drug by physical bonds in a matrix. Different solid nanoparticles were used in order to encapsulate ruthenium complexes [97] such as poly(lactic acid) [98], mesoporous silica nanoparticles [99], or metal-organic frameworks [100]. The promising ruthenium (III) drug KP1019 was co-precipitated with poly(lactic acid) in a single oil-in-water emulsion with two different surfactants [98]. The obtained nanoparticles have an improved cytotoxicity comparing with KP1019.

4.2. Chemical conjugation of ruthenium-based drugs

4.2.1. Polymer conjugates

The main idea of this approach is to obtain a polymer, which contains a moiety that can act as ligand for ruthenium. In case of NAMI-A, this moiety can be an imidazole group. Thus, the Stenzel group [101] reports the polymerization of 4-vinil imidazole followed by addition of adequate ruthenium precursor complex. They obtained an amphiphilic co-polymer capable of self-assembly into micelles (**Figure 6**).

The tests on ovarian and pancreatic cancer cells revealed a 1.5 times increased cytotoxicity for polymeric micelles. Furthermore, these were tested for antimetastatic activity on breast cancer cells proving a higher activity comparing to NAMI-A complex.

4.2.2. Lipid base conjutates/liposomes

The Paduano group focused on developing drug carriers for NANI-A analog, named AZIRu (**Figure 7**) [102–108] and investigating their anticancer activity. Unlike NAMI-A, AZIRu contains a pyridine ligand instead of imidazole and sodium as counterion.

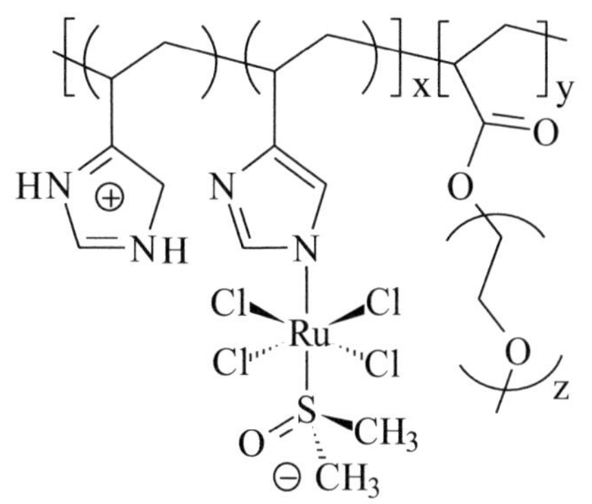

Figure 6. NAMI-A conjugated to polymer.

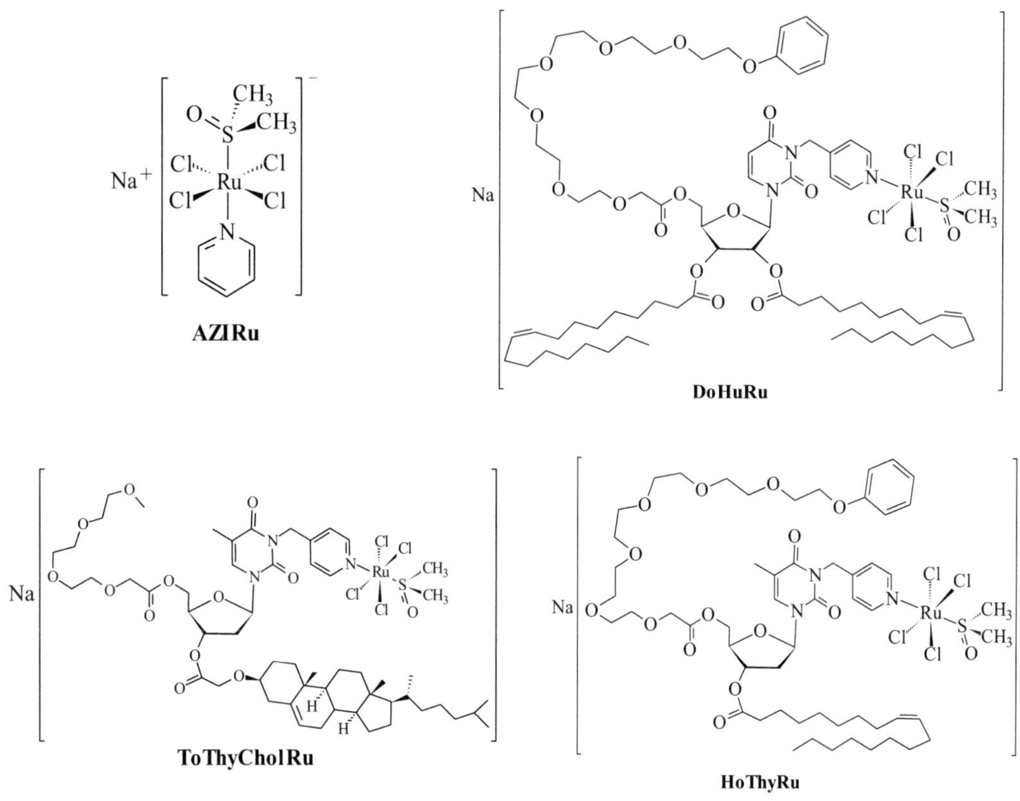

Figure 7. AZIRu and selected amphiphilic nucleolipid-based AZIRu.

New amphiphilic derivatives of nucleosides have been developed in order to act as drug carriers for AZIRu complex. In detail, a nucleobase (thymidine or uridine), which was attached with a pyrimidilmethyl group at the N-3 position (in order to act as ligand for ruthenium) was selected as starting material. The resulted compounds were further bonded to one or two lipid residues (oleoyl or cholesteroxyacetyl) and one hydrophilic oligo(ethylene glycol) chain of variable lengths. There were thus obtained amphiphilic supramolecular aggregates, essentially liposomes [102–105].

The nucleolipidic compounds proved to have similar instability in aqueous systems as NAMI-A and AZIRu, forming insoluble precipitates in few hours. In order to reduce the hydrolysis processes, the nucleolipidic compounds were formulated with biocompatible phospholipids, POPC (1-palmitoyl-2-oleoyl-sn-glycero-3-phosphocholine) [103–105] and DOTAP (1,2-dioleoyl-3-trimethylammoniumpropane) [106, 107]. The bioactivity of these Ru^{III}-containing nucleolipids was tested on human and nonhuman cancer cells proving higher anticancer activity, higher stability in aqueous systems, and lower toxicity than AZIRu [108].

4.2.3. Dendrimers

The interest in dendrimers as drug carriers comes from their characteristics namely highly branched three-dimensional molecules containing functional groups at periphery, which can react with drug molecules. So far, only one potential anticancer ruthenium (III) drug, RAPTA-C, was incorporated into dendrimer (**Figure 8**) [109], but there is no study regarding the anticancer activity.

Interactions of ruthenium (II) complexes with dendrimers and the anticancer activity of the resulted compounds, which are described in some reviews, have also attracted much interest [110, 111].

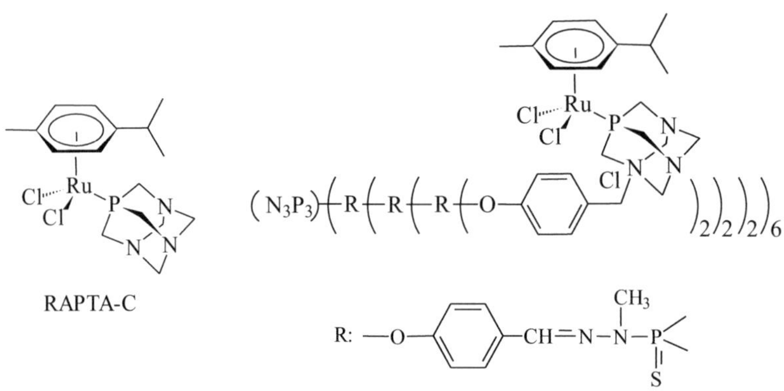

Figure 8. Ruthenium (III) drug RAPTA-C and its dendrimeric nanoformulation.

Author details

Valentina Uivarosi[1†], Rodica Olar[2†*] and Mihaela Badea[2†]

*Address all correspondence to: rodica_m_olar@yahoo.com

1 Carol Davila University of Medicine and Pharmacy, Bucharest, Romania

2 University of Bucharest, Bucharest, Romania

† The authors contributed equally to this work

References

[1] Sadler PJ, Muncie C, Shipman MA. Metals in Medicine in Biological Inorganic Chemistry, Structure and Reactivity. In: Bertini I, Gray HB, Stiefel EI, Selverstone Valentine J, editors. Sausalito, California: University Science Books; 2007. pp. 95-135

[2] Rosenberg B, Vancamp L, Krigas T. Inhibition of cell division in *Escherichia coli* by electrolysis products from a platinum electrode. Nature. 1965;**205**:698-699. DOI:10.1038/205698a0

[3] Kostova I. Platinum complexes as anticancer agents. Recent Patents on Anti-Cancer Drug Discovery. 2006;**1**:1-22. DOI:10.2174/157489206775246458

[4] Mjos KD, Orvig C. Metallodrugs in medicinal inorganic chemistry. Chemical Reviews. 2014;**114**:4540-4563. DOI: dx.doi.org/10.1021/cr400460s

[5] Kozarich JW. Medicinal Inorganic Chemistry: Promises and Challenges in Medicinal Inorganic Chemistry. In: Sessler JL, Doctrow SR, McMurry TJ, Lippard SJ, editors. Washington, DC: American Chemical Society;2005. pp. 4-14

[6] Johnstone TC, Suntharalingam K, Lippard SJ. The next generation of platinum drugs: Targeted Pt(II) agents, nanoparticle delivery, and Pt(IV) prodrugs. Chemical Reviews. 2016;**116**:3436-3486. DOI: 10.1021/acs.chemrev.5b00597

[7] Wani WA, Prashar S, Shreaz S, Gómez-Ruiz S. Nanostructured materials functionalized with metal complexes: In search of alternatives for administering anticancer metallodrugs. Coordination Chemistry Reviews. 2016;**312**:67-98. DOI: 10.1016/j.ccr.2016.01.001

[8] Duan X, He C, Kron SJ, Lin W. Nanoparticle formulations of cisplatin for cancer therapy. WIREs Nanomedicine and Nanobiotechnolology. 2016;**8**:776-791. DOI: 10.1002/wnan.1390

[9] Cheng Q, Liu Y. Multifunctional platinum-based nanoparticles for biomedical applications. WIREs Nanomedicine and Nanobiotechnolology. 2017;**9**:e1410. DOI: 10.1002/wnan.1410

[10] Farrell N. Metal complexes as drugs and chemotherapeutic agents in volume 9: Applications of coordination chemistry. In: Ward MD, editor: from Comprehensive Coordination Chemistry II–From Biology to Nanotechnology. 2nd ed. editors: McCleverty JA, Meyer TJ. Amsterdam: Elsevier 2003, pp. 809-840

[11] Muhammad N, Guo Z. Metal-based anticancer chemotherapeutic agents. Current Opinion in Chemical Biology. 2014;**19**:144-153. DOI: 10.1016/j.cbpa.2014.02.003

[12] Gaynor D, Griffith DM. The prevalence of metal-based drugs as therapeutic or diagnostic agents: Beyond platinum. Dalton Transactions. 2012;**41**:13239-. DOI: 10.1039/c2dt31601c

[13] Farrer NJ, Sadler PJ. Medicinal inorganic chemistry: State of the art, new trends, and a vision of the future. In: Alessio E, editors. Bioinorganic Medicinal Chemistry. Weinheim, Germany: Wiley-VCH Verlag & Co.;2011. pp. 1-48

[14] Fricker SP. Metal based drugs: From serendipity to design. Dalton Transactions. 2007;**43**:4903-4917. DOI: 10.1039/b705551j

[15] Hannon MJ. Metal-based anticancer drugs: From a past anchored in platinum chemistry to a post-genomic future of diverse chemistry and biology. Pure and Applied Chemistry. 2007;**79**:2243-2261. DOI:10.1351/pac200779122243

[16] Barry NPE, Sadler PJ. Exploration of the medical periodic table: Towards new targets. Chemical Communications. 2013;**49**:5106-5131. DOI: 10.1039/c3cc41143e

[17] Newman MS, Colbern GT, Working PK, Engbers C, Amantea MA. Comparative pharmacokinetics, tissue distribution, and therapeutic effectiveness of cisplatin encapsulated in long-circulating, pegylated liposomes (SPI-077) in tumor-bearing mice. Cancer Chemotherapy and Pharmacology. 1999;**43**:1-7. DOI: 10.1007/s002800050855

[18] Boulikas T. Clinical overview on Lipoplatin™: A successful liposomal formulation of cisplatin. Expert Opinion on Investigational Drugs. 2009;**18**:1197-1218. DOI: 10.1517/13543780903114168.

[19] Casagrande N, De Paoli M, Celegato M, Borghese C, Mongiat M, Colombatti A, Aldinucci D. Preclinical evaluation of a new liposomal formulation of cisplatin, lipoplatin, to treat cisplatin-resistant cervical cancer. Gynecologic Oncology 2013;**131**:744-752. DOI: 10.1016/j.ygyno.2013.08.041

[20] White SC, Lorigan P, Margison GP, Margison JM, Martin F, Thatcher N, Anderson H, Ranson M. Phase II study of SPI-77 (sterically stabilized liposomal cisplatin) in advanced non-small-cell lung cancer. British Journal of Cancer. 2006;**95**:822-828. DOI:10.1038/sj.bjc.6603345

[21] Seetharamu N, Kim E, Hochster H, Martin F, Muggia F. Phase II study of liposomal cisplatin (SPI-77) in platinum-sensitive recurrences of ovarian cancer. Anticancer Research. 2010;**30**:541-545

[22] de Jonge MJ, Slingerland M, Loos WJ, Wiemer EA, Burger H, Mathijssen RH, Kroep JR, den Hollander MA, van der Biessen D, Lam M-H. Early cessation of the clinical development of LiPlaCis, a liposomal cisplatin formulation. European Journal of Cancer. 2010;**46**:3016-3021. DOI: 10.1016/j.ejca.2010.07.015

[23] Guo S, Miao L, Wang Y, Huang L. Unmodified drug used as a material to construct nanoparticles: Delivery of cisplatin for enhanced anti-cancer therapy. Journal of Controlled Release. 2014;**174**:137-142. DOI: 10.1016/j.jconrel.2013.11.019

[24] Li Q, Tian Y, Li D, Sun J, Shi D, Fang L, Gao Y, Liu H. The effect of lipocisplatin on cisplatin efficacy and nephrotoxicity in malignant breast cancer treatment. Biomaterials. 2014;**35**:6462-6472. DOI: 10.1016/j.biomaterials.2014.04.023

[25] Wang Y, Zhou J, Qiu L, Wang X, Chen L, Liu T, Di W. Cisplatine-alginate conjugate liposomes for targeted delivery to EGFR-positive ovarian cancer cells. Biomaterials 2014;**35**:4297-4309. DOI: 10.1016/j.biomaterials.2014.01.035

[26] Callaria M, Aldrich-Wright JR, de Souza PL, Stenzel MH. Polymers with platinum drugs and other macromolecular metal complexes for cancer treatment. Progress in Polymer Science2014;**39**:1614-1643. DOI: 10.1016/j.progpolymsci.2014.05.002

[27] Uchino H, Matsumura Y, Negishi T, Koizumi F, Hayashi T, Honda T, Nishiyama N, Kataoka K, Naito S, Kakizoe T. Cisplatin-incorporating polymeric micelles (NC-6004) can reduce nephrotoxicity and neurotoxicity of cisplatin in rats. British Journal of Cancer 2005;**93**:678-687. DOI:10.1038/sj.bjc.6602772

[28] Plummer R, Wilson R, Calvert H, Boddy A, Griffin M, Sludden J, Tilby M, Eatock M, Pearson D, Ottley C. A phase I clinical study of cisplatin incorporated polymeric micelles (NC-6004) in patients with solid tumours. British Journal of Cancer. 2011;**104**:593-598. DOI: 10.1038/bjc.2011.6

[29] Lin X, Zhang Q, Rice J, Stewart D, Nowotnik D, Howell S. Improved targeting of platinum chemotherapeutics: The antitumour activity of the HPMA copolymer platinum agent AP5280 in murine tumour models. European Journal of Cancer. 2004;**40**:291-297. DOI: 10.1016/j.ejca.2003.09.022

[30] Rademaker-Lakhai JM, Terret C, Howell SB,Baud CM, de Boer RF, Pluim D, Beijnen JH, Droz J-P. A phase I and pharmacological study of the platinum polymer AP5280 given as an intravenous infusion once every 3 weeks in patients with solid tumors. Clinical Cancer Research. 2004;**10**:3386-3395. DOI: 10.1158/1078-0432.CCR-03-0315

[31] Haririan I, Alavidjeh MS, Khorramizadeh MR, Ardestani MS, Ghane ZZ, Namazi H. Anionic linear-globular dendrimer-cis-platinum(II) conjugates promote cytotoxicity in vitro against different cancer cell lines. International Journal of Nanomedicine. 2010;**5**:63-75. DOI: 10.2147/IJN.S8595

[32] Ding Y, Zhai K, Pei P, Lin Y, Ma Y, Zhu H, Shao M, Yang X, Tao W. Encapsulation of cisplatin in a pegylated calcium phosphate nanoparticle (CPNP) for enhanced cytotoxicity to cancerous cells. Journal of Colloid and Interface Science. 2017;**493**:181-189. DOI: 10.1016/j.jcis.2017.01.032.

[33] Xiong Y, Jiang W, Shen Y, Li H, Sun C, Ouahab A, Tu J. A poly(γ, L-glutamic acid)-citric acid based nanoconjugate for cisplatin delivery. Biomaterials. 2012;**33**:7182-7193. DOI: 10.1016/j.biomaterials.2012.06.071

[34] Malik N, Evagorou EG, Duncan R. Dendrimer-platinate: A novel approach to cancer chemotherapy. Anti-Cancer Drugs. 1999;**10**:767-776

[35] Song W, Tang Z, Li M, Lv S, Sun H, Deng M, Liu H, Chen X. Polypeptide-based combination of paclitaxel and cisplatin for enhanced chemotherapy efficacy and reduced side-effects. Acta Biomaterialia. 2014;**10**:1392-1402. DOI: 10.1016/j.actbio.2013.11.026

[36] Xiao H, Song H, Yang Q, Cai H, Qi R, Yan L, Liu S, Zheng Y, Huang Y, Liu T, Jing X. A prodrug strategy to deliver cisplatin(IV) and paclitaxel in nanomicelles to improve efficacy and tolerance. Biomaterials. 2012;**33**:6507-6519. DOI: 10.1016/j.biomaterials.2012.05.049

[37] Lu S, Xu L, Kang ET, Mahendran R, Chiong E, Neoh KG. Co-delivery of peptide-modified cisplatin and doxorubicin via mucoadhesive nanocapsules for potential synergistic intravesical chemotherapy of non-muscle-invasive bladder cancer. European Journal of Pharmaceutical Sciences. 2016;**10**;84:103-115. DOI: 10.1016/j.ejps.2016.01.013

[38] Zhang X, Li L, Li C, Zheng H, Song H, Xiong F, Qiu T, Yang J. Cisplatin-crosslinked glutathione-sensitive micelles loaded with doxorubicin for combination and targeted therapy of tumors. Carbohydrate Polymers. 2017; **155**:407-415. DOI: 10.1016/j.carbpol.2016.08.072

[39] Xiong Y, Zhao Y, Miao L, Lin CM, Huang L. Co-delivery of polymeric metformin and cisplatin by self-assembled core-membrane nanoparticles to treat non-small cell lung cancer. Journal of Controlled Release. 2016;**244**(Pt A):63-73. DOI: 10.1016/j.jconrel.2016.11.005

[40] Shi T, Gu L, Sun Y, Wang S, You C, Zhang X, Zhu J, Sun B. Enhanced legumain-recognition and NIR controlled released of cisplatin-indocyanine nanosphere against gastric carcinoma. European Journal of Pharmacology. 2017;**5**;794:184-192. DOI: 10.1016/j.ejphar.2016.11.039

[41] Ishiguro S, Cai S, Uppalapati D, Turner K, Zhang T, Forrest WC, Forrest ML, Tamura M. Intratracheal administration of hyaluronan-cisplatin conjugate nanoparticles significantly attenuates lung cancer growth in mice. Pharmaceutical Research. 2016;**33**:2517-2529. DOI: 10.1007/s11095-016-1976-3

[42] Suh MS, Shen J, Kuhn LT, Burgess DJ. Layer-by-layer nanoparticle platform for cancer active targeting. International Journal of Pharmaceutics. 2017;**517**:58-66. DOI: 10.1016/j.ijpharm.2016.12.006

[43] Chiang C-S, Tseng Y-H, Liao B-J, Chen SY. Magnetically targeted nanocapsules for PAA-cisplatin-conjugated cores in PVA/SPIO shells via surfactant-free emulsion for reduced nephrotoxicity and enhanced lung cancer therapy. Advanced Healthcare Materials. 2015;**4**:1066-1075. DOI: 10.1002/adhm.201400794

[44] Voulgari E, Bakandritsos A, Galtsidis S, Zoumpourlis V, Burke BP, Clemente GS, Cawthorne C, Archibald SJ, Tuček J, Zbořil R, Kantarelou V, Karydas AG, Avgoustakis K. Synthesis, characterization and in vivo evaluation of a magnetic cisplatin delivery nanosystem based on PMAA-graft-PEG copolymers. Journal of Controlled Release. 2016;**243**:342-356. DOI: 10.1016/j.jconrel.2016.10.021

[45] Min Y, Mao C-Q, Chen S, Ma G, Wang J, Liu Y. Combating the drug resistance of cisplatin using a platinum prodrug based delivery system. Angewandte Chemie, International Edition. 2012;**51**:6742-6747. DOI: 10.1002/anie.201201562

[46] Dhar S, Daniel WL, Giljohann DA, Mirkin CA, Lippard SJ. Polyvalent oligonucleotide gold nanoparticle conjugates as delivery vehicles for platinum(IV) warheads. Journal of the American Chemical Society. 2009;**131**:14652-14653. DOI: 10.1021/ja9071282

[47] Gu J, Liu J, Li Y, Zhao W, Shi J. One-pot synthesis of mesoporous silica nanocarriers with tunable particle sizes and pendent carboxylic groups for cisplatin delivery. Langmuir. 2012;**29**:403-410. DOI.org/10.1016/j.ijpharm.2012.03.026

[48] Bhirde AA, Patel V, Gavard J, Zhang G, Sousa AA, Masedunskas A, Leapman RD, Weigert R, Gutkind JS, Rusling JF. Targeted killing of cancer cells in vivo and in vitro with EGF-directed carbon nanotube-based drug delivery. ACS Nano. 2009;**3**:307-316. DOI: 10.1021/nn800551s

[49] Li J, Pant A, Chin CF, Ang WH, Ménard-Moyon C, Nayak TR, Gibson D, Ramaprabhu S, Panczyk T, Bianco A, Pastorin G. In vivo biodistribution of platinum-based drugs encapsulated into multi-walled carbon nanotubes. Nanomedicine: Nanotechnology, Biology and Medicine. 2014;**10**:1465-1475. DOI: 10.1016/j.nano.2014.01.004

[50] Rieter WJ, Pott KM, Taylor KML, Lin W. Nanoscale coordination polymers for platinum-based anticancer drug delivery. Journal of the American Chemical Society. 2008;**130**:11584-11585. DOI: 10.1021/ja803383k

[51] Liu D, Poon C, Lu K, He C, Lin W. Self-assembled nanoscale coordination polymers with trigger release properties for effective anticancer therapy. Nature Communications. 2014;**5**. DOI:10.1038/ncomms5182

[52] He C, Liu D, Lin W. Self-assembled core-shell nanoparticles for combined chemotherapy and photodynamic therapy of resistant head and neck cancers. ACS Nano. 2015;**9**:991-1003. DOI: 10.1021/nn506963h

[53] He C, Liu D, Lin W. Self-assembled nanoscale coordination polymers carrying siRNAs and cisplatin for effective treatment of resistant ovarian cancer. Biomaterials. 2015;**36**:124-133. DOI:10.1016/j.biomaterials.2014.09.017

[54] Rocca JD, Werner ME, Kramer SA, Huxford-Phillips RC, Sukumar R, Cummings ND, Vivero-Escoto JL, Wang AZ, Lin W. Polysilsesquioxane nanoparticles for triggered release of cisplatin and effective cancer chemoradiotherapy. Nanomedicine: Nanotechnology, Biology and Medicine. 2014;**11**:31-38. DOI: 10.1016/j.nano.2014.07.004

[55] Maeda H, Bharate GY, Daruwalla J. Polymeric drugs for efficient tumor-targeted drug delivery based on EPR-effect. European Journal of Pharmaceutics and Biopharmaceutics. 2009;**71**:409-419. DOI: 10.1016/j.ejpb.2008.11.010

[56] Sadhukha T, Prabha S. Encapsulation in nanoparticles improves anti-cancer efficacy of carboplatin. AAPS Pharmaceutical Science Technology. 2014;**5**:1029-1038. DOI: 10.1208/s12249-014-0139-2

[57] Karanam V, Marslin G, Krishnamoorthy B, Chellan V, Siram K, Natarajan T, Bhaskar B, Franklin G. Poly(ε-caprolactone) nanoparticles of carboplatin: Preparation, characterization and in vitro cytotoxicity evaluation in U-87 MG cell lines. Colloids and Surfaces B: Biointerfaces. 2015;**130**:48-52. DOI: 10.1016/j.colsurfb.2015.04.005

[58] Ahmeda F, Alib MJ, Kondapi AK. Carboplatin loaded protein nanoparticles exhibit improve anti-proliferative activity in retinoblastoma cells. International Journal of Biological Macromolecules. 2014;**70**:572-582. DOI: 10.1016/j.ijbiomac.2014.07.041

[59] Khan MA, Zafaryab M, Mehdi SH, Quadri J, Rizvi MM. Characterization and carboplatin loaded chitosan nanoparticles for the chemotherapy against breast cancer in

vitro studies. International Journal of Biological Macromolecules. 2017;**97**:115-122. DOI: 10.1016/j.ijbiomac.2016.12.090

[60] Balas M, Constanda S, Duma-Voiculet A, Prodana M, Hermenean A, Pop S, Demetrescu I, Dinischiotu A. Fabrication and toxicity characterization of a hybrid material based on oxidized and aminated MWCNT loaded with carboplatin. Toxicology in Vitro. 2016;**37**:189-200. DOI: 10.1016/j.tiv.2016.09.011

[61] Makharza S, Vittorio O, Cirillo G, Oswald S, Hinde E, Kavallaris M, Büchner B, Mertig M, Hampel S. Graphene oxide–Gelatin nanohybrids as functional tools for enhanced carboplatin activity in neuroblastoma cells. Pharmaceutical Research. 2015;**32**:2132-2143. DOI: 10.1007/s11095-014-1604-z

[62] Yang C, Fu ZX. Liposomal delivery and polyethylene glycol-liposomal oxaliplatin for the treatment of colorectal cancer. Biomedical Reports. 2014;**2**:335-339. DOI:10.3892/br.2014.249

[63] Stathopoulos GP, Boulikas T, Kourvetaris A, Stathopoulos J. Liposomal oxaliplatin in the treatment of advanced cancer: A phase I study. Anticancer Research. 2006;**26**:1489-1493

[64] Shi M, Fortin D, Paquette B, Sanche L. Convection-enhancement delivery of liposomal formulation of oxaliplatin shows less toxicity than oxaliplatin yet maintains a similar median survival time in F98 glioma-bearing rat model. Investigational New Drugs. 2016;**34**:269-276. DOI: 10.1007/s10637-016-0340-0

[65] Suzuki R, Takizawa T, Kuwata Y, Mutoh M, Ishiguro N, Utoguchi N, Shinohara A, Eriguchi M, Yanagie H, Maruyama K. Effective anti-tumor activity of oxaliplatin encapsulated in transferrin–PEG–liposome. International Journal of Pharmaceutics. 2008;**346**:143-150. DOI: 10.1016/j.ijpharm.2007.06.010

[66] Abu-Lila AS, Kiwada H, Ishida T. Selective delivery of oxaliplatin to tumor tissue by nanocarrier system enhances overall therapeutic efficacy of the encapsulated oxaliplatin. Biologicaland Pharmaceutical Bulletin. 2014:**37**;206-211. DOI:10.1248/bpb.b13-00540

[67] Abu-Lila A, Suzuki T, Doi Y, Ishida T, Kiwada H. Oxaliplatin targeting to angiogenic vessels by PEGylated cationic liposomes suppresses the angiogenesis in a dorsal air sac mouse model. Journal of Controlled Release. 2009;**134**:18-25. DOI:10.1016/j.jconrel.2008.10.018

[68] Oberoi HS, Nukolova NV, Kabanov AV, Bronich TK. Nanocarriers for delivery of platinum anticancer drugs. Advanced Drug Delivery Reviews. 2013;**65**:1667-1685. DOI: 10.1016/j.addr.2013.09.014

[69] Wang R, Hu X, Xiao H, Xie Z, Huang Y, Jing X. Polymeric dinulcear platinum(II) complex micelles for enhanced antitumor activity. Journal of Materials Chemistry B. 2013;**1**:744-748. DOI: 10.1039/C2TB00240J

[70] Cabral H, Nishiyama N, Kataoka K. Optimization of (1,2-diaminocyclohexane)platinum (II)-loaded polymeric micelles directed to improved tumor targeting and enhanced anti-tumor activity. Journal of Controlled Release. 2007;**121**:146-155. DOI:10.1016/j.jconrel.2007.05.024

[71] Rafi M, Cabral H, Kano MR, Mi P, Iwata C, Yashiro M, Hirakawa K, Miyazono K, Nishiyama N, Kataoka K. Polymeric micelles incorporating (1,2-diaminocyclohexane)platinum(II) suppress the growth of orthotopicscirrhous gastric tumors and their lymph node metastasis. Journal of Controlled Release. 2012;**159**:189-196. DOI: 10.1016/j.jconrel.2012.01.038

[72] Nowotnik DP, Cvitkovic E. ProLindac (AP5346): A review of the development of an HPMA DACH platinum polymer therapeutic. Advanced Drug Delivery Reviews. 2009;**61**:1214-1219. DOI:10.1016/j.addr.2009.06.004

[73] Apps MG, Choi EH, Wheate NJ. The state of-play and future of platinum drugs. Endocrine-Related Cancer. 2015;**22**:R219–R233. DOI: 10.1530/ERC-15-0237

[74] Song H, Xiao H, Zheng M, Qi R, Yan L, Jing X. A biodegradable polymer platform for co-delivery of clinically relevant oxaliplatin and gemcitabine. Journal of Materials Chemistry B. 2014;**2**:6560-6570. DOI: 10.1039/C4TB00678J

[75] Xiao H, Li W, Qi R, Yan L, Wang R, Liu S, Zheng Y, Xie Z, Huang Y, Jing X. Co-delivery of daunomycin and oxaliplatin by biodegradable polymers for safer and more efficacious combination therapy. Journal of Controlled Release. 2012;**163**:304-314. DOI: 10.1016/j.jconrel.2012.06.004

[76] Jain A, Jain SK, Ganesh N, Barve J, Beg AM. Design and development of ligand-appended polysaccharidic nanoparticles for the delivery of oxaliplatin in colorectal cancer. Nanomedicine. 2010;**6**:179-190. DOI: 10.1016/j.nano.2009.03.002

[77] Tummala S, Gowthamarajan K, Satish Kumar MN, Wadhwani A. Oxaliplatin immuno hybrid nanoparticles for active targeting: An approach for enhanced apoptotic activity and drug delivery to colorectal tumors. Drug Delivery. 2016;**23**:1773-1787. DOI: 10.3109/10717544.2015.1084400

[78] Vivek R, Thangam R, Nipunbabu V, Ponraj T, Kannan S. Oxaliplatin-chitosan nanoparticles induced intrinsic apoptoticsignaling pathway: A "smart" drug delivery system to breastcancer cell therapy. International Journal of Biological Macromolecules. 2014;**65**:289-297. DOI: 10.1016/j.ijbiomac.2014.01.054

[79] Chen Y, Ding D, Mao Z, He Y, Hu Y, Wu W, Jiang X. Synthesis of hydroxypropylcellulose-poly(acrylic acid) particles with semi-interpenetrating polymer network structure. Biomacromolecules. 2008;**9**:2609-2614.DOI: 10.1021/bm800484e

[80] Dutta RK, Sahu S. Development of oxaliplatin encapsulated in magnetic nanocarriers of pectin as a potential targeted drug delivery for cancer therapy. Results in Pharma Sciences. 2012;**2**:38-45. DOI: 10.1016/j.rinphs.2012.05.001

[81] Wu L, Man C, Wang H, Lu X, Ma Q, Cai Y, Ma W. PEGylated multi-walled carbon nanotubes for encapsulation and sustained release of oxaliplatin. Pharmaceutical Research. 2013;**30**:412-423. DOI: 10.1007/s11095-012-0883-5

[82] Lee PC, Lin CY, Peng CL, Shieh MJ. Development of a controlled-release drug delivery system by encapsulating oxaliplatin into SPIO/MWNT nanoparticles for effective colon

cancer therapy and magnetic resonance imaging. Biomaterials Science. 2016;**4**:1742-1753. DOI:10.1039/c6bm00444j

[83] Brown SD, Nativo P, Smith J-A, Stirling D, Edwards PR, Venugopal B, Flint DJ, Plumb JA, Graham D, Wheate NJ. Gold nanoparticles for the improved anticancer drug delivery of the active component of oxaliplatin. Journal of the American Chemical Society. 2010; **132**:4678-4684. DOI: 10.1021/ja908117a

[84] He H, Xiao H, Kuang H, Xie Z, Chen X, Jing X, Huang Y. Synthesis of mesoporous silica nanoparticle–oxaliplatin conjugates for improved anticancer drug delivery. Colloids and Surfaces B: Biointerfaces. 2014;**117**:75-81. DOI: 10.1016/j.colsurfb.2014.02.014

[85] Della Rocca J, Huxford RC, Comstock-Duggan E, Lin W. Polysilsesquioxane nanoparticles for targeted platin-based cancer chemotherapy by triggered release. Angewandte Chemie, International Edition. 2011;**50**:10330-10334. DOI: 10.1002/anie.201104510

[86] Bergamo A, Gagliardi R, Scarcia V, Furlani A, Alessio E, Mestroni G, Sava G. In vitro cell cycle arrest, in vivo action on solid metastasizing tumors, and host toxicity of the antimetastatic drug NAMI-A and cisplatin. Journal of Pharmacology and Experimental Therapeutics. 1999;**289**:559-564

[87] Dragutan I, Dragutan V, Demonceau A. Editorial of special issue ruthenium complex: The expanding chemistry of the ruthenium complexes. Molecules. 2015;**20**:17244-17274. DOI: 10.3390/molecules200917244

[88] Bergamo A, Gaiddon C, Schellens JHM, Beijnen JH, Sava G. Approaching tumour therapy beyond platinum drugs: Status of the art and perspectives of ruthenium drug candidates. Journal of Inorganic Biochemistry. 2012;**106**:90-99. DOI: 10.1016/j.jinorgbio.2011.09.030

[89] Rademaker-Lakhai JM, Van Den Bongard D, Pluim D, Beijnen JH, Schellens JHM. A phase I and pharmacological study with imidazolium-trans-DMSO-imidazole-tetrachlororuthenate, a novel ruthenium anticancer agent. Clinical Cancer Research. 2004;**10**: 3717-3727. DOI: 10.1158/1078-0432.CCR-03-0746

[90] Leijen S, Burgers SA, Baas P, Pluim D, Tibben M, van Werkhoven E, Alessio E, Sava G, Beijnen JH, Schellens JH. Phase I/II study with ruthenium compound NAMI-A and gemcitabine in patients with non-small cell lung cancer after first line therapy. Investigational New Drugs. 2015;**33**:201-214. DOI: 10.1007/s10637-014-0179-1

[91] Hartinger CG, Zorbas-Seifried S, Jakupec MA, Kynast B, Zorbas H, Keppler BK. From bench to bedside–preclinical and early clinical development of the anticancer agent indazolium trans-[tetrachlorobis(1H-indazole)ruthenate(III)] (KP1019 or FFC14A). Journal of Inorganic Biochemistry. 2006;**100**:891-904. DOI: 10.1016/j.jinorgbio.2006.02.013

[92] Hartinger CG, Jakupec MA, Zorbas-Seifried S, Groessl M, Egger A, Berger W, Zorbas H, Dyson PJ, Keppler BK. KP1019, a new redox-active anticancer agent-preclinical development and results of a clinical phase I study in tumor patients. Chemistry & Biodiversity. 2008;**5**:2140-2155. DOI: 10.1002/cbdv.200890195

[93] Lentz F, Drescher A, Lindauer A, Henke M, Hilger RA, Hartinger CG, Scheulen ME, Dittrich C, Keppler BK, Jaehde U. Pharmacokinetics of a novel anticancer ruthenium complex (KP1019, FFC14A) in a phase I dose-escalation study. Anti-Cancer Drugs. 2009;**20**:97-103. DOI: 10.1097/CAD.0b013e328322fbc5

[94] Mestroni G, Alessio E, Sava G, Pacor S, Coluccia M, Boccarelli A. Water-soluble ruthenium(III)-dimethyl sulfoxide complexes: Chemical behaviour and pharmaceutical properties Met. Based Drugs. 1994;**1**:41-63. DOI: 10.1155/MBD.1994.41

[95] Ott I, Gust R. Non platinum metal complexes as anti-cancer drugs. Archiv der Pharmazie. 2007;**340**:117-126. DOI: 10.1002/ardp.200600151

[96] Levina A, Aitken JB, Gwee YY, Lim ZJ, Liu M, Singharay AM, Wong PF, Lay PA. Biotransformations of anticancer ruthenium(III) complexes: An X-ray absorption spectroscopic study. Chemistry--A European Journal. 2013;**19**:3609-3619. DOI: 10.1002/chem.201203127

[97] Blunden BM, Stenzel MH. Incorporating ruthenium into advanced drug delivery carriers – An innovative generation of chemotherapeutics. Journal of Chemical Technology and Biotechnology. 2015;**90**: 1177-1195. DOI: 10.1002/jctb.4507

[98] Fischer B, Heffeter P, Kryeziu K, Gille L, Meier SM, Berger W, Kowol CR, Keppler BK. Poly(lactic acid) nanoparticles of the lead anticancer ruthenium compound KP1019 and its surfactant-mediated activation. Dalton Transactions. 2014;**43**:1096-1104. DOI: 10.1039/c3dt52388h.

[99] He L, Huang Y, Zhu H, Pang G, Zheng W, Wong Y-S, Chen T. Cancer-targeted monodisperse mesoporous silica nanoparticles as Carrier of ruthenium polypyridyl complexes to enhance theranostic effects. Advanced Functional Materials. 2014;**24**:2754-2763. DOI: 10.1002/adfm.201303533

[100] Rojas S, Quartapelle-Procopio E, Carmona FJ, Romero MA, Navarro JAR, Barea E. Biophysical characterisation, antitumor activity and MOF encapsulation of a half-sandwich ruthenium(II) mitoxantronato system. Journal of Materials Chemistry B. 2014;**2**:2473-2477. DOI: 10.1039/C3TB21455A

[101] Blunden BM, Rawal A, Lu H, Stenzel MH. Superior chemotherapeutic benefits from the ruthenium-based anti-metastatic drug NAMI-A through conjugation to polymeric micelles. Macromolecules. 2014;**47**:1646-1655. DOI: 10.1021/ma402078d

[102] Vaccaro M, Del Litto R, Mangiapia G, Carnerup AM, D'Errico G, Ruffo F, Paduano L. Lipid based nanovectors containing ruthenium complexes: A potential route in cancer therapy. Chemical Communications. 2009;1404-1406. DOI: 10.1039/B820368G

[103] Mangiapia G, D'Errico G, Simeone L, Irace C, Radulescu A, Di Pascale A, Colonna A, Montesarchio D, Paduano L. Ruthenium-based complex nanocarriers for cancer therapy. Biomaterials. 2012;**33**:3770-3782. DOI: 10.1016/j.biomaterials.2012.01.057

[104] Simeone L, Mangiapia G, Vitiello G, Irace C, Colonna A, Ortona O, Montesarchio D, Paduano L. Cholesterol-based nucleolipid-ruthenium complex stabilized by lipid aggregates for antineoplastic therapy. Bioconjugate Chemistry. 2012;**23**:758-770. DOI: 10.1021/bc200565v

[105] Montesarchio D, Mangiapia G, Vitiello G, Musumeci D, Irace C, Santamaria R, D'Erricoa G, Paduano L. A new design for nucleolipid-based Ru(III) complexes as anticancer agents. Dalton Transactions. 2013;**42**:16697-16708. DOI: 10.1039/c3dt52320a.

[106] Mangiapia G, Vitiello G, Irace C, Santamaria R, Colonna A, Angelico R, Radulescu A, D'Errico G, Montesarchio D, Paduano L. Anticancer cationic ruthenium nanovectors: From rational molecular design to cellular uptake and bioactivity. Biomacromolecules. 2013;**14**:2549-2560. DOI: 10.1021/bm400104b

[107] Vitiello G, Luchini A, D'Errico G, Santamaria R, Capuozzo A, Irace C, Montesarchio D, Paduano L. Cationic liposomes as efficient nanocarriers for the drug delivery of an anticancer cholesterol-based ruthenium complex. Journal of Materials Chemistry B. 2015;**3**:3011-3023. DOI: 10.1039/C4TB01807A

[108] Riccardi C, Musumeci D, Irace C, Paduano L, Montesarchio D. Ru^{III} complexes for anticancer therapy: The importance of being nucleolipidic. European Journal of Organic Chemistry. 2016. DOI: 10.1002/ejoc.201600943

[109] Servin P, Laurent R, Gonsalvi L, Tristany M, Peruzzini M, Majoral J-P, Caminade A-M. Grafting of water-soluble phosphines to dendrimers and their use in catalysis: positive dendritic effects in aqueous media. Dalton Transactions. 2009;**38**:4432-4434. DOI: 10.1039/B906393P

[110] Valente A, Garcia MH. Syntheses of macromolecular ruthenium compounds: A new approach for the search of anticancer drugs. Inorganics. 2014;**2**:96-114; DOI: 10.3390/inorganics2010096

[111] Govender P, Therrien B, Smith GS. Bio-metallodendrimers–Emerging strategies in metal-based drug. European Journal of Inorganic Chemistry. 2012;**2012**:2853-2862. DOI: 10.1002/ejic.20120016

Chapter 2

Formation of Fe(III) Ternary Complexes with Related Bio-relevant Ligands

Wafaa Mahmoud Hosny

Additional information is available at the end of the chapter

http://dx.doi.org/10.5772/intechopen.69158

Abstract

Ternary complexes of iron(III)-glycine(Gly)-nitrilotriacetate (NTA) system determined by electrochemical measurements of the dissolved iron(III)-Gly-NTA mixed ligand system in the 0.1 mol·dm^{-3} $NaClO_4$ aqueous solution at pH = 8.0 ± 0.1and 25 ± 1°C. The coordination number of Fe in Fe(EDTA)-L is seven in coordinate complex, where L can be a DNA constituent like uracil, uridine, thymine, thymidine, and inosine. The nonlinear least-squares program MINIQUAD-75 is used to deduce the hydrolysis constants of $[Fe(EDTA)(H_2O)]^-$ and its formation constant in solution. The antimicrobial activity of Fe(III) complexes of salicylhydroxamic acid (SHAM) and 1,10-phenanthroline (PHEN) studied against representative pathogenic bacteria and fungi.

Keywords: iron(III), MINIQUAD-75, speciation, stability constant, glycine, sulfamethoxazole, salicylhydroxamic acid

1. Introduction

Iron is consider as an essential element; its chemical and biological functions evolved from its oxidation and reduction processes and interactions with oxygen [1]. These are very important biogeochemical in natural aquatic systems [2, 3]. It is one of the most abundant metals in the Earth's crust [4]. However, very low concentrations (<10^{-9} mol·dm^{-3}) of dissolved, mostly iron (III) organic complexes are present in natural waters due to the low solubility of its thermodynamically stable 3+ ionic form [5, 6]. Iron is used to treat chlorosis (green disease), which often resulting from deficiency of the iron concentration [7]. However, 80 years ago, we did not have any information about the importance of inorganic iron in synthesis of hemoglobin [8]. Many years ago, the nutritional experts became interested in the role of iron in oxygen transport and hemoglobin formation [9]. Most anemia diseases in industrialized countries result from low iron intake and bioavailability. On the other side, they are responsible for only about half of the anemia in developing countries [10]. There are other important causes [11] like infectious and inflammatory diseases (especially malaria), blood loss from parasitic infections, and other nutrient deficiencies (vitamin A, riboflavin, folic acid, and vitamin B12).

2. Analysis of aqueous solution complexes by different methods

1. The first and most common method is ion-selective electrode. It defines the position of dynamic equilibrium. The most important electrode is the glass one. There is also hydrogen gas electrode that can be used in hydrogen ion calculations.
2. Metal amalgam electrodes are a second choice. It can be used for some metal ions, but they are not as precise as the hydrogen ion electrode. We prefer the ion-selective electrode in-calculation because the results are collected from series of data taken through a titration procedure. A good method to check for this prerequisite is to make repeated high-resolution electrode readings at predetermined time intervals, since this will make sluggish attainments of equilibrium clearly visible.
3. Spectrophotometry can be used if the metal ion or the ligand is colored, so that the color will change (in intensity and/or frequency) upon complexation.

There are other experimental techniques that we are going to give some examples for them in the following lines. The specific method for diamagnetic metal ion is the nuclear magnetic resonance (NMR). It gives one separated signal for each unique chemical surrounding. In other words, it can inform us about concentration of the ligand, the free metal ion, the number of species, and their concentration for a given chemical composition. The important feature of this technique is that the positions of these selective signals are responsible for the protonation and deprotonation ones. In the case of fast reactions, we can use a stopped-flow technique.

2.1. The addition of water to iron(III)

It is observed that Fe^{3+} hydrolyses in water goes as follows:

$$Fe^{3+} + H_2O \leftrightharpoons FeOH^{2+} + H^+ \tag{1}$$

The addition of water to Fe^{3+} is carried out through a series of deprotonation reactions, resulting in formation of ferric hydroxides and oxyhydroxides [12, 13] as in Eq. (1), the equilibrium constant for this reaction was calculated to be 6.78×10^{-3} at 298 K (total iron(III) concentration of 0.5 $mol{\cdot}dm^{-3}$ and an ionic strength of 0.1 $mol{\cdot}dm^{-3}$). **Figure 1** shows the speciation of the two iron-containing species of Eq. (1) as a function of pH, calculated at $T = 298$ K. It seen that $FeOH^{2+}$ will become the common species above pH = 2.17. At lower pH, there are small amount of it; at pH value 1.2, more than 9% of all iron (III) is present as the hydroxide. As a result of that, the calculations for the spectroscopic measurements were carried out at pH values up to 1. The equilibrium constant for the reaction shown in Eq. (1) has been studied at temperature range from

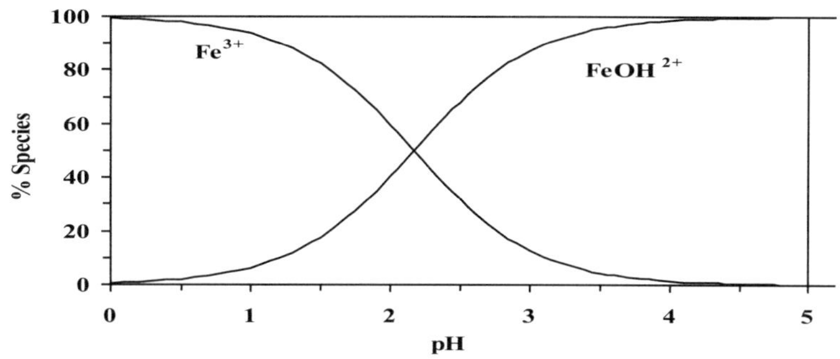

Figure 1. The speciation of the two species of iron as in Eq. (1) as a function of pH at $T = 298$ K for all the iron (III) concentration of 0.5 mol·dm^{-3} and an ionic strength of 0.1 mol·dm^{-3}.

$T = 298$ to 353 K and ionic strength range from $I = 0.1$ until 2.67 mol·dm^{-3} in perchlorate media [14].

For Eq. (1) to be done we must prevent the formation of hydrolyzed iron(III). This can be accomplished under the conditions of $T = 293$–323 K, and $I \sim 0.1$ mol·dm^{-3}.

2.2. How can you determine the stability constants of mixed ligand complexes

For inorganic chemistry, it is very important to determine the stability constant, or the equilibria constant or we can refer to it as the formation constant, for the reaction [15]. It is not easy to get the solution equilibria constants between the ligands and the metal ions. Proton ions and a range of metal ions fight for a range of donor sites. There are many factors that decide who will be the winner whether the proton ions or the metal ions. These factors are the concentration and pH. Potentiometry and spectrophotometry are used to determine the stability constants of metal complexes. Legget [16] and Meloum et al. [17] calculated the equilibrium constants from experimental data for the first time. Nowadays, many programs were published for these calculations using microcomputers. **Table 1** presents some of these programs [15, 19–30]. These programs are very helpful as they quickly present the best fit. They use the least-square method to reduce the differences between calculated and experimental data. The sum of square of residuals between calculated and experimental values is very small; it is nearly between 10^{-6} and 10^{-9}. Potentiometry is used to determine the stability constants of metal complexes. It is based on pH-metric titration of the ligand, and the availability of metal ions. Data obtained from potentiometry are analyzed by the least-square method to derive the formation constant. This later can describe the solution equilibria. For the measurements, there must be two conditions: the first one is a constant ionic strength of the solution and the second condition concerns the ionic strength that have to be higher than the concentration of metal ion. The reaction of all mixed complex:

$$L(M) + p(L_1) + q(L_2) + r(H) \leftrightarrow \left[(M)_1(L_1)_p(L_2)_q(H)_r\right] \quad (2)$$

System	Data type[a]	Reference
MINIQUAD	V	[19]
MINIQUAD75	V	[20]
TITAN	V	[25]
SCOGS2a	V	[21]
D SCOGS2ba	V	[26]
MINIQUAD	V, A	[22]
PSEQUAD	V, A	[23]
SPECFIT	A(E)	[18]
PKAS	V	[15]
HYDROUAD	V	[24]
STAR	A	[29]
HYPNMR	N	[30]

V, potentiometric experiments; A, spectrophotometric experiments; E, ESR; N, NMR.
[a]Additional data used in calculations are taken from different sources.

Table 1. The programs commonly used for calculating formation equilibrium constants.

The total stability constant, β_{lpqr}, can be calculated from the equation by:

$$\beta_1 pqr = [(M)l(L_1)p(L_2)q(H)r]/[M]^1[L_1]p[L_2]q[H]_r (\text{for simplicity charges are omitted}) \quad (3)$$

where M, L_1, L_2, and H stand for metal ion, ligand (1), ligand (2), and proton, respectively. For OH^-, the coefficient (r) for H = −1.

2.3. Calculation of speciation

Pettit program computes speciation based on the concentrations of metal ions and the complexing species. This program specifies a certain pH range. Then the former calculates the species distribution of a certain series of complexes and plots it. We enter some data such as the total concentrations of metal and ligand ions and pH range. After that the best-fit set of β values will be used later to compute the equilibrium concentrations of those complex species over the pH range which we have specified before. We can use this program for all types of complexes: mixed complexes, protonated, hydroxo, and polynuclear species. The program produces a graphical recording of the most predominant complex species at any pH and the physiological pH range. In this chapter, we reviewed several iron complexes.

3. Different complexes between Fe(III) and biological active ligands

3.1. Studies of binary and ternary complexes of sulfamethoxazole (SMZ) and glycine with metal ions

Sulfamethoxazole (4-amino-N-(5-methyl-3-isoxazolyl)-benzenesulfonamide (SMZ)) is the most predominant sulfonamide in human medicine. Sulfonamides are synthetic antimicrobial agents derived from sulfanilamide, whose antibacterial activity was discovered in the early 1930s by Domagk and Tréfouel [31–33].

3.1.1. Stability constants of ternary complexes (metal-SMZ-Gly)

Different metal ions Ti(II), Zr(IV), Sr(II), Al(III), Cr(III), Fe(III), Th(IV), Pb(II), La(III), and Co(II) were selected to make further investigation to elucidate the interaction of these metal ions with solution of SMZ and Gly (mixed ligand complexes). The potentiometric equilibrium measurements were made, at constant ionic strength $I = 0.1$ M $NaClO_4$ at 25 ± 0.1°C, for the interaction of SMZ and the selected 10 metal ions, with biologically important secondary ligand glycine (Gly) in a (1:1:1) molar ratio (1×10^{-3} M for each). The solutions were titrated pH-metrically against standard carbonate-free NaOH solution, as illustrated in **Figure 2**.

Figure 2 represents the titration curves for the metal-SMZ-Gly system studied. It is observed that the metal ion-SMZ titration curve (c) diverges from the SMZ curve (b) at variable pH values (pH ≈ 2.8 for Fe(III), pH ≈ 3.5 for La(III), pH ≈ 4.2 for Th(IV), pH ≈ 5.5 for Zr(IV), pH ≈ 4.5 for Al(III), and pH ≈ 6.06 for Co(II)) denoting the formation of metal ions-SMZ binary complexes. For the titration curves of the ternary systems studied, it can be observed that the curves (c) and (f), however, overlap with each other at lower pH values in the case of Fe(III) and La (III), whereas that for Sr(II), Pb(II), Cr(III), and Ti(II) are well separated. This indicates the formation of metal ions-SMZ-Gly ternary complexes at lower pH values, which can be considered as an evidence for the formation of protonated SMZ mixed ligand complex. The stability constants of the ternary metal ion complexes containing SMZ and Gly were calculated from Eqs. (4) and (5),

$$M + SMZ \leftrightarrow M\ (SMZ) \tag{4}$$

$$K^{M(\mathbf{SMZ})}_{M(\mathbf{SMZ})(Gly)} = \frac{[M(SMZ)(Gly)]}{[M(SMZ)(Gly)]} \tag{5}$$

using the data obtained from potentiometric titrations ($I = 0.1$M $NaClO_4$ at 25 ± 0.1°C).

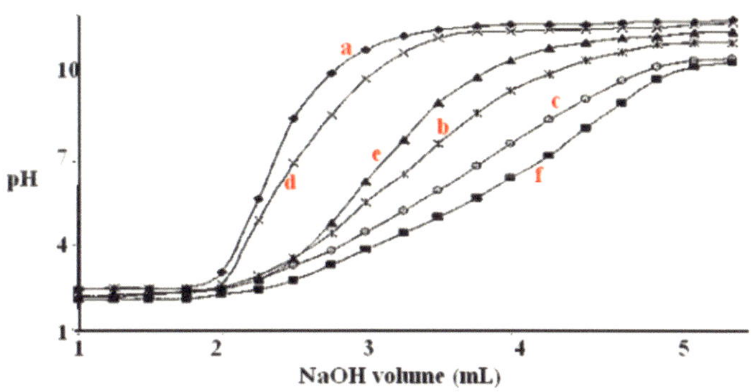

Figure 2. Potentiometric curves of SMZ in 0.1 M $NaClO_4$ at 25 ± 0.1°C: (a) 0.01 M $HClO_4$, (b) a + 0.001 M SMZ, (c) b + 0.001 M Sr (II), (d) b + 0.001 M Pb(II), (e) b+ 0.001 M Co(II), (f) b + 0.001 M Fe(III), and (g) b + 0.001 M Al (III).

3.2. Ternary complexes of iron(III)-glycine(Gly)-nitrilotriacetate (NTA) system

Electrochemical measurements of the dissolved iron(III)-Gly-NTA mixed ligand system in the 0.1 mol·dm^{-3} $NaClO_4$ aqueous solution were performed at pH = 8.0 ± 0.1and 25 ± 1°C, using differential pulse cathodic voltammetry (DPCV), cyclic voltammetry (CV), and direct current (d. c.) polarography. Iron(III) concentrations were varied from 2.5 × 10^{-5} to 6 × 10^{-4} mol·dm^{-3}, NTA total concentrations varied from 2 × 10^{-5} to 1 × 10^{-3} mol·dm^{-3} and glycine total concentrations were 0.2, 0.02, and 0.002 mol·dm^{-3}. **Figure 3** shows the differential pulse voltammograms of iron (III) in a mixture of glycine (0.2 mol·dm^{-3}) and NTA (5 × 10^{-4} mol·dm^{-3}). Reduction peak currents of mixed ligand complex depend on iron(III) concentrations, as shown in **Figure 3**. Basic line (voltammogram) represents the solution with both ligands present, without iron(III). It does not contain any reduction peak. This later implies electrochemical inactivity of these two ligands under the applied experimental conditions. When iron(III) is added, the reduction peak potentials remain constant at –0.112 V, indicating stability of the formed species. These peaks are the response to iron(III) reduction in mixed ligand complexes.

3.3. Determination of formation equilibria of seven-coordinate Fe(EDTA) complexes with DNA and related biorelevant ligands

Fe(EDTA)-L is a seven-coordinate complex as the coordination number of Fe is seven. L can be a DNA constituent like uracil, uridine, thymine, thymidine, and inosine. To understand the chemistry of this seven-coordinate complex, we did some investigations using methylamine, ammonium chloride, or imidazole. The complexes produced are 1:1 with DNA constituents and other ligands. This complex indicates that the total coordination number of Fe(III) ion is seven. Potentiometric titration is carried out at 25°C and ionic strength 0.1 mol·L^{-1} using $NaNO_3$ to measure the stability constant. Besides, the nonlinear least-squares program MINIQUAD-75 is used to deduce the hydrolysis constants of $[Fe(EDTA)(H_2O)]^-$ and its formation constant in solution. The concentration distributions of the different species formed in solution were evaluated as a pH dependent.

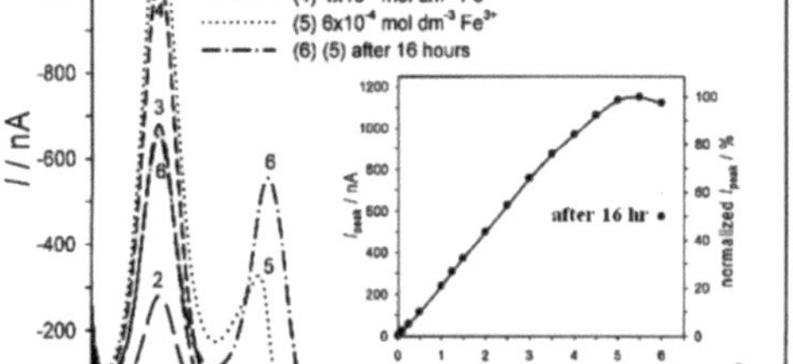

Figure 3. DPC voltammograms; iron(III)-Gly-NTA peak currents on added iron(III). 0.2 molċdm^{-3} glycine, 5 ± 10^{-4} molċdm^{-3} NTA, 0.1 molċdm^{-3} $NaClO_4$; pH = 8.0 ± 0.1, E_{inc} = 2 mV, a = 25 mV, t_p = 0.05 s, t_{int} = 0.2 s.

3.3.1. *Calculated equilibria of the [Fe(EDTA)(H_2O)]$^-$ ion*

Different equilibrium models were tested [34] to fit the experimental potentiometric data for the hydrolysis of [Fe(EDTA)(H_2O)]$^-$ ion. The best-selected model consists of the formation of the 10^{-1} species, as given in Eq. (6). This supports the presence of one water molecule coordinated in the [Fe(EDTA)(H_2O)]$^-$ ion:

$$\underset{(100)}{[\mathrm{Fe(EDTA)(H_2O)}]^-} \overset{\mathrm{K_{al}}}{\rightleftharpoons} \underset{(10-1)}{[\mathrm{Fe(EDTA)(OH)}]^{2-}} + \mathrm{H}^+ \tag{6}$$

The pH-meter readings (B) recorded in dioxane-water mixtures were converted to hydrogen ion concentrations [H^+] with the widely used relation given by the Van Uitert and Haas equation [35]:

$$-\log_{10}[\mathrm{H}^+] = B + \log_{10} U_\mathrm{H} \tag{7}$$

where $\log_{10} U_\mathrm{H}$ is the correction factor for the solvent composition and ionic strength at which β was determined. Values of pK_w in dioxane-water mixtures were determined as described previously [36, 37]. Different amounts from NaOH of the known concentration were added to a solution of ionic strength 0.1. The amount of base added determines the [OH^-], unlike [H^+], which is calculated from the pH value. The product of ([OH^-]. [H^+]) is used to calculate the mean values ($pK_{w)}$ which is $-\log_{10}$ [H^+][OH^-]. The mean values at 25°C are 14.17, 14.37, 14.50, and 15.44 for 12.5, 25, 37.5, and 50% dioxane, respectively. These percentages are the mass percentage of dioxane in water solution. The equilibrium constants obtained from the titration data (summarized in **Table 2**) are defined by Eqs. (8) and (9), where M, L, and H stand for [Fe(EDTA)(H_2O)]$^-$, ligand, and proton, respectively

$$p\mathrm{M} + q\mathrm{L} + r\mathrm{H}\ \mathrm{M}_p\mathrm{L}_q\mathrm{H}_r \tag{8}$$

$$\beta_{pqr} = [\mathrm{M}_p\mathrm{L}_q\mathrm{H}_r]/[\mathrm{M}]^p[\mathrm{L}]^q[\mathrm{H}]^r \tag{9}$$

The speciation distribution diagram for the hydrolysis of [Fe(EDTA)(H_2O)]$^-$is given in **Figure 4**. The fraction of the monohydroxo species increases with increasing pH, attaining a maximum of 99.9% at a pH = 10.6.

3.3.2. *Complex formation equilibria of the [Fe(EDTA)(H_2O)]$^-$ion*

The potentiometric titration curve, given in **Figure 5**, illustrates the result where imidazole is taken as an example. This curve has two plots one for the [Fe(EDTA)(H_2O)]$^-$-imidazole system and the other for imidazole. The complex formation curve for the [Fe(EDTA)(H_2O)]$^{--}$-imidazole system is lower than the imidazole's one. This is because of the hydrogen ion evolved during the formation of a complex species. This potentiometric data are products for an experiment composed of the species 110.

There are many examples for the pyrimidinic species like uridine, uracil, thymine, and thymidine. The dissociable proton of the pyrimidinic species lies in the N3–C4O group. The acid dissociation constants for pyrimidinic species and the N1 proton of inosine are compared. The

System	p	Q	r^a	$\log_{10}\beta^b$	S^c
$[Fe(EDTA)(H_2O)]^-$	1	0	−1	−7.60(0.008)	4.7E–8
Uracil	0	1	1	9.35(0,002)	4.5E–7
	1	1	0	5.12(0.1)	8.8E–6
Thymine	0	1	1	9.50(0.01)	8.1E–8
	1	1	0	5.98(0.1)	2.5E–6
Thymidine	0	1	1	9.06(0.01)	8.7E–8
	1	1	0	5.89(0.1)	3.0E–5
Uridine	0	1	1	9.01(0.02)	1.1E–7
	1	1	0	4.93(0.05)	2.0E–5
Methylamine·HCL	0	1	1	10.03 (0.04)	4.4E–7
	1	1	0	6.13(0.1)	2.2E–5
Ammonium chloride	0	1	1	9.32(0.01)	7.2E–5
	1	1	0	3.91(0.03)	1.2E–6
Imidazole	0	1	1	7.04(0.01)	2.6E–9
	1	1	0	2.23(0.04)	5.6E–7
Inosine	0	1	1	8.43(0.01)	5.0E–9
	1	1	0	5.96(0.02)	2.9E–5
	1	1	1	13.14(0.06)	

[a] p, q, and r are the stoichiometric coefficients corresponding to $[Fe(EDTA)(H_2O)]^-$, L, and H^+, respectively.
[b] Standard deviations are given in parentheses.
[c] Sum of the squares of residuals.

Table 2. Stability constant of mixed complexes in water at 25 ± 0.1 °C and 0.1 ionic strength.

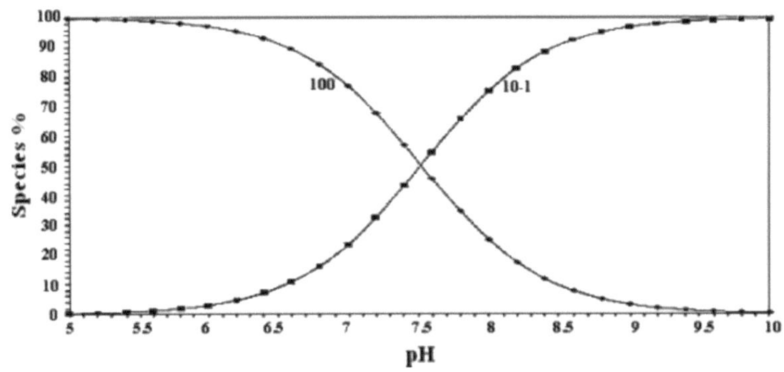

Figure 4. Speciation distribution of different species as pH dependence in the Fe(EDTA)-OH system at 1.25 mmol·L^{-1} $[Fe(EDTA)(H_2O)]^-$, in aqueous solution at 25°C and ionic strength $I = 0.1$.

latter is slightly more acidic. The anionic form of purinic derivatives is the reason for that as it occurred in a large number of resonance forms. These resonance forms are created by the two condensed rings in the inosine ligand, as shown in **Scheme 1**. We can conclude that uracil,

http://dx.doi.org/10.5772/intechopen.69158

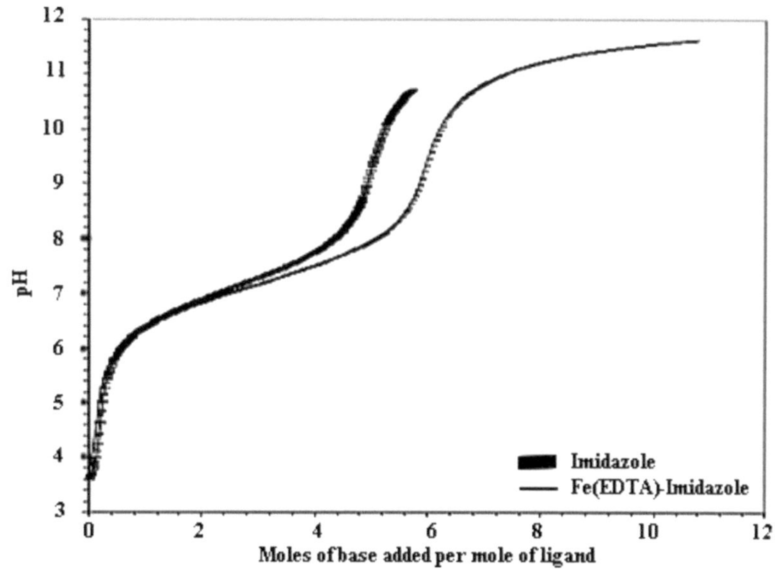

Figure 5. Titration curves of the Fe(EDTA)-imidazole system in aqueous medium.

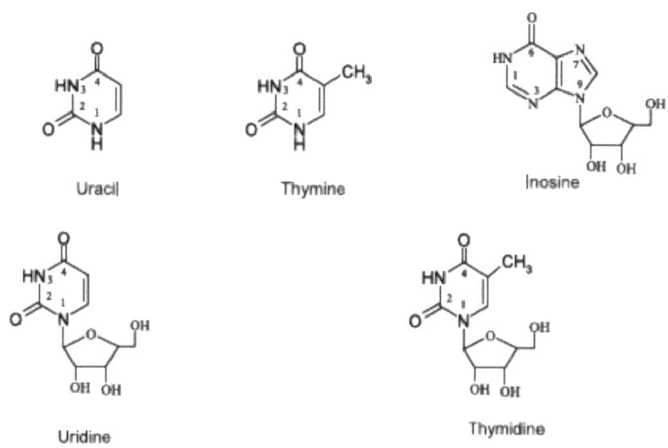

Scheme 1. Structural formulae of the DNA used.

uridine, thymine, and thymidine do not form protonated complexes. They coordinates through N_3 in their deprotonated form; the monoanion one.

Thymine and thymidine have a methyl group that donates an extra electron of an inductive effect. This increases the basicity of the N3 site of thymine and thymidine complexes and stabilizes them more than uracil and uridine ones. Pyrimidines are monodentate ligands of $pKa \approx 9$, so their complexes are absent below pH = 6. This indicates that in the neutral or nearly basic pH media, the negatively charged nitrogen donors of pyrimidines bases are vital binding sites.

Inosine complex chelates as a monodentate has two sites of chelation N(1) and N(7). In the acidic medium, N1 is protonated and N7 is attached to the metal ion. When pH increases, the metal ion moves from N7 to N1. This motion was recorded by nuclear magnetic resonance (NMR) spectroscopy [38, 39]. Chelation depends on the pH range; in basic medium N(1) is a coordination site in the complex formation [40]. The data show the formation of the ternary complexes with stoichiometric coefficients 110 and 111. To know the main features observed in the species distribution in these systems, the speciation diagram obtained for the Fe(EDTA)-uracil and Fe(EDTA)-inosine complexes, as shown in **Figure 6**, as examples of DNA constituents. The pK_a value of the N1H group of the protonated complex (log10 β111 – log10 β110) amounts to 7.18. This indicates the acidification of the N1H site by 1.25 log units through coordination with the $[Fe(EDTA)(H_2O)]^-$ complex, which is in agreement with previous results for similar systems [41]. Detection of the concentration distribution of the various species in solution provides a useful picture of metal ion binding. At pH 4.0, the mixed complex of Fe(EDTA) with uracil, species 110, occurs. This occurrence increases with increasing the pH of the medium, up to it reaches 84% at pH 8.5. After 8.5 the concentration of Fe(EDTA)-uracil system

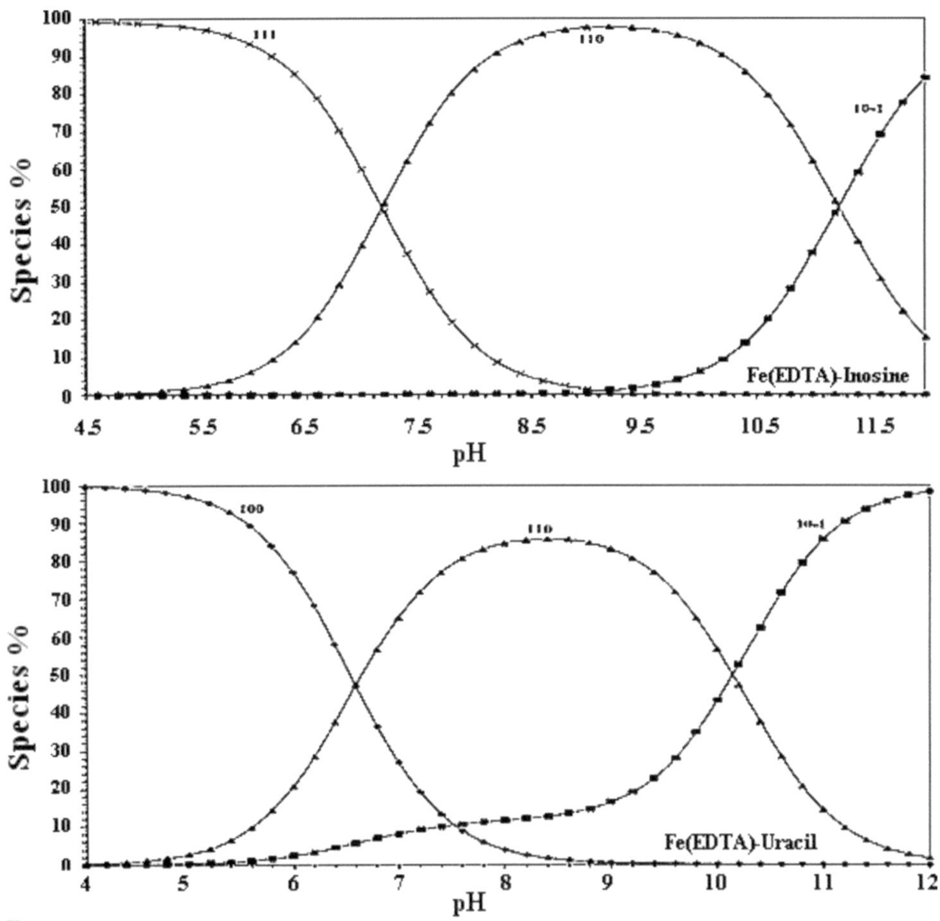

Figure 6. Speciation distribution curve of different species as pH dependence in the Fe(EDTA)-inosine and Fe(EDTA)-uracil systems (at 1.25 mmol·L^{-1} for Fe(EDTA) and 6.25 mmol·L^{-1} for inosine and uracil), in water at room temperature and 0.1 ionic strength.

http://dx.doi.org/10.5772/intechopen.69158

drops and $[Fe(EDTA)(OH)]^{2-}$ develop. This result concludes that the Fe(EDTA) complex can interact with bioligands as DNA units. Not only does the pH affects the appearance of Fe (EDTA)-uracil system, it also affects the Fe(EDTA)-inosine system. At low pH mediums, species 111 is present; N7 coordinates to the complex and the N1 nitrogen is protonated. Whereas at high pH like 8.8, species 110 occurs where the concentration of N1 coordinated is 98%. This is the maximum concentration obtained for N1 coordinated. As the pH plays an important role before in the absence and presence of some systems, it has its role with cytosine and cytidine chelates. At low pH, they have their N3 protonated. It was recorded by the NMR spectroscopy in its solution state and with X-ray crystallography in its solid state.

3.3.3. The influence of thermodynamic parameters

Thermodynamic parameters are useful tools for studying the interactions with DNA constituents and understanding the relative stability of the complexes formed. The thermodynamic parameters ΔG^0, ΔH^0, and ΔS^0 were easily determined using Van't Hoff relation (Eq. (10)). If we take the protonation of uracil and its complex formation with [Fe(EDTA)], as representative example. Since, we have a known value for protonation constant, the stability formation constant (*K*) and gas constant (*R)* in this reaction. So, we can apply the Van't Hoff equation to obtain the value of those parameters at the required temperature (*T*) in Kelvin. Then we can use the results to draw a graph of ln *K* versus $1/T$ and the intercept will be $\Delta S^0/R$.

$$lnK = -\Delta H^0/RT + \Delta S^0/R \tag{10}$$

and a slope parameters ΔH^0. The formation constants and the thermodynamic parameters values are presented in **Tables 3** and **4** and can be interpreted as follows:

1. The protonation reaction of uracil can be represented as:

$$\mathbf{L^+ + H^+ \; LH} \tag{11}$$

   The thermodynamic processes accompanying the protonation reactions are as follows:

   i. The neutralization reaction is considered as an exothermic equation

   ii. Desolvation of ions is considered as an endothermic process; and

   iii. Changes in the configurations and arrangements of the hydrogen bonds around the free and protonated ligands.

2. The $\log_{10} K^H$ values decrease with increasing temperature, revealing that their acidity increases with increasing temperature, as shown in **Figure 7**.

3. The protonation reaction of uracil has a positive entropy change, which may be due to increased disorder as a result of desolvation processes and the breaking of hydrogen bonds.

4. The negative value of ΔH^0 for the protonation process of uracil ligand indicates that its association process is accompanied by a release of energy and the process is exothermic.

System	T (°C)	p	q	r^{a}	$\log_{10}\beta^{b}$	S^{c}
Uracil	15	0	1	1	9.55(0.002)	1.6E–8
[Fe(EDTA)]		1	0	−1	−7.78(0.009)	1.6E–7
[Fe(EDTA)-uracil]		1	1	0	5.20(0.08)	1.3E–5
Uracil	20	0	1	1	9.46(0.002)	1.1E–8
[Fe(EDTA)]		1	0	−1	−7.69 (007)	1.0E–7
[Fe(EDTA)-uracil]		1	1	0	5.07(0.06)	9.7E–6
Uracil	25	0	1	1	9.35(0.002)	4.5E–7
[Fe(EDTA)]		1	0	−1	−7.60(0.008)	3.7E–8
[Fe(EDTA)-uracil]		1	1	0	4.93(0.05)	8.8E–6
Uracil	30	0	1	1	9.25(0.003)	8.6E–9
[Fe(EDTA)]		1	0	−1	−7.53(0.008)	1.2E–7
[Fe(EDTA)-uracil]		1	1	0	4.80(0.09)	1.0E–5
Uracil	35	0	1	1	9.15(0.003)	3.1E–7
[Fe(EDTA)]		1	0	−1	−7.46(0.008)	1.4E–7
[Fe(EDTA)-uracil]		1	1	0	4.69(0.14)	1.0E–5

[a] p, q, and r are the stoichiometric coefficient corresponding to $[Fe(EDTA)(H_2O)]^-$, uracil, and H^+, respectively.
[b] Standard deviations are given in parentheses.
[c] Sum of the squares of residuals.

Table 3. Protonation constants of uracil and the formation constants of the Fe(EDTA)-uracil complex in aqueous solution and different temperatures and 0.1 ionic strength.

Equilibrium	ΔH^0 (kJ·mol^{-1})	ΔS^0 (J·k1^{-1}·mol^{-1})	ΔG^0 (kJ·mol^{-1})
Fe-EDTA hydrolysis			
(1) $[Fe(EDTA)(H_2O)]$ ($[Fe(EDTA)(OH)]^-$ + H^+	−54.40 ± 0.75	27.20 ± 2.5	43.4 ± 1.5
Uracil			
(2) L^- + H^+ (LH	−34.29 ± 0.75	63.91 ± 2.5	−53.39 ± 1.5
Fe-EDTA-uracil			
(3) $[Fe(EDTA)(H_2O)]$ + L ([Fe-EDTA-L] + H_2O	−43.84 ± 0.68	−52.64 ± 2.3	−28.15 ± 1.4

L denotes uracil.

Table 4. Thermodynamic parameters (ΔH^0, ΔS^0 and ΔG^0) for the interaction n of Fe–EDTA with uracil in aqueous solution.

The values of formation constants of the complexes are plotted in **Figure 8** at different temperatures. The plotted line shows that the formation constants of the complexes are inversely proportion to increasing temperature. Therefore, the complexation process requires low temperatures. At the end, we can say that

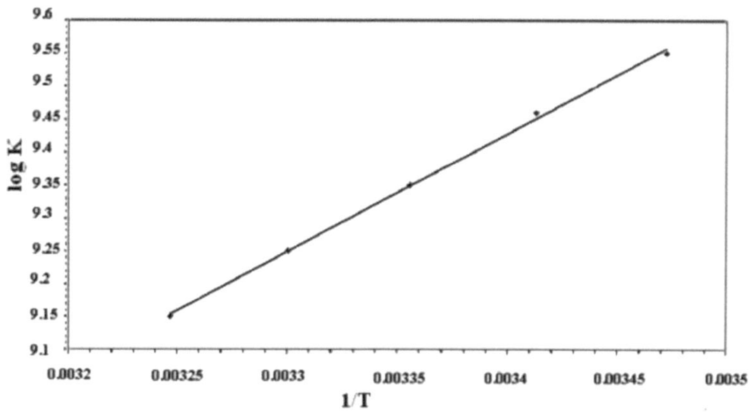

Figure 7. Effect of temperature on the protonation constant of uracil.

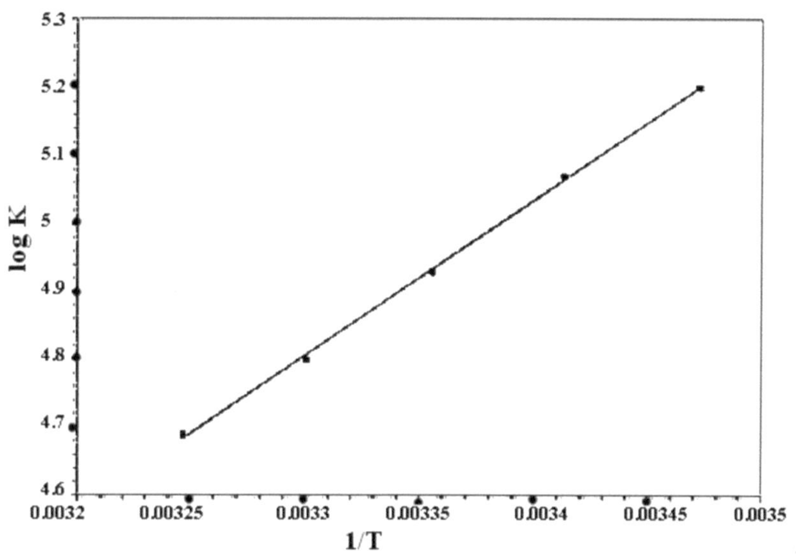

Figure 8. Effect of temperature on the stability constant of $[Fe(EDTA)(uracil)]^{2-}$.

1. The values of ΔG^0 for the coordination process are negative, which indicates that the reaction is spontaneous.

2. The positive value of ΔS^0 upon complexation is because of the increase in entropy. Within complexation a bounded solvent molecule was released. The release of the molecule produces increased the entropy, however, the complexation process causes a small decreased in the entropy. Because the solvent molecules are arranged around the ligand in a certain way. Unlike, the metal ions are configured randomly upon complexation.

3. The negative values of ΔH^0 indicate that the coordination processes are exothermic, so that the complexation reactions are favored at low temperatures.

3.3.4. How can you expect the effect of solvent composition

It is well known that the "effective" or "equivalent solution" dielectric constants in a protein [42, 43], or active site cavities of enzymes [44] are small compared to that in bulk water. The dielectric constants detecting in such locations range from 30 to 70 [43, 44]. Therefore, by using aqueous solutions containing ~10–50% dioxane, one may expect to simulate to some degree the situation in active site cavities [45], and hence to extrapolate the data to physiological conditions. We asked what the relation between the solvent occurs in media and formation constant of complexes on the equilibrium constants (**Table 4**) reveals the following points:

1. The value of pK_a of uracil (N_3-site) increases directly with increasing dioxane in the medium as shown in **Figure 9**. Dioxane has a low-dielectric constant, which increased the electrostatic forces between the proton and the ligand. Finally, the pK_a increases.

2. The stability constant ($\log_{10} K_1$) of the Fe(EDTA)-uracil complex increases with increase of the dioxane concentration (**Figure 10**). This is due to that lowering the dielectric constant of

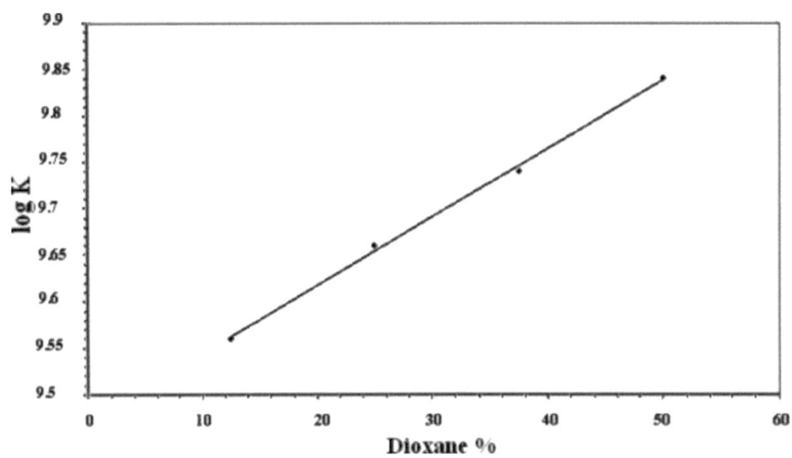

Figure 9. Effect of dioxane on the protonation constant of uracil.

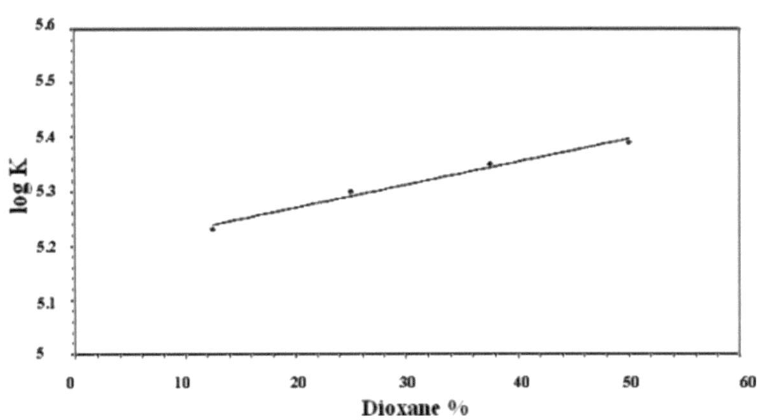

Figure 10. Effect of dioxane on the stability constant of the $[Fe(EDTA)uracil)]^{2-}$ species.

System	% dioxane	p	q	r^a	$\log_{10}\beta^b$	S^c
[Fe(EDTA)]	12.5%	1	0	−1	−7.68(0.002)	2.5E−9
Uracil		0	1	1	9.56(0.004)	2.3E−8
[Fe(EDTA)-uracil]		1	1	0	5.23(0.16)	1.0E−5
[Fe(EDTA)]	25%	1	0	−1	−7.83(0.002)	4.1E−9
Uracil		0	1	1	9.66(0.009)	2.3E−7
[Fe(EDTA)-uracil]		1	1	0	5.30(0.06)	3.1E−6
[Fe(EDTA)]	35%	1	0	−1	−7.98(0.003)	6.9E−9
Uracil		0	1	1	9.74(0.01)	2.6E−7
[Fe(EDTA)-uracil]		1	1	0	5.35(0.19)	1.2E−5
[Fe(EDTA)]	50%	1	0	−1	−8.14(0.006)	2.0E−8
Uracil		0	1	1	9.84(0.01)	3.0E−7
[Fe(EDTA)-uracil]		1	1	0	5.39(0.13)	1.2E−5

[a] p, q, and r are the stoichiometric coefficient corresponding to $[Fe(EDTA)(H_2O)]^-$, uracil, and H^+, respectively.
[b] Standard deviations are given in parentheses.
[c] Sum of the squares of residuals.

Table 5. Effect of solvent (dioxane) on the stability constant of [Fe(EDTA)(uracil)] at 25°C.

the medium (by increasing the dioxane content) favors the interaction between Fe(EDTA) and uracil, and consequently the stability constant of the complex increases. These finding, shown in **Table 5**, is in agreement with literature data [46].

4. Biological activity of mixed ligand Fe(III) complexes with bioactive ligands

4.1. The study of antibacterial activity of Fe Complex

4.1.1. The effect of microorganisms and media

Salicylhydroxamic acid (SHAM) and its binary and ternary complexes (I–VI) were screened [47] to do an experiment to test its antibacterial activity on six bacterial strains: *Escherichia coli* (E.c.), *Staphylococcus aureus* (S. a.), *Enterobacter cloacae* (E. c.), *Salmonella gallinarum* (S. g.), *Bacillus subtilis* (B. s.), and *Pseudomonas aeruginosa* (P. a.). This experiment is illustrated in **Table 6**. These strains were obtained from the Microbial Centre of Ain Shams University, Egypt. Some other microorganisms like (*Aspergillus fumigatus* (A. f.), *Candida albicans* (C. a.), *Alternaria alternata* (A. a.), *Penicillium italicum* (P. i.), *Saccharomyces cerevisiae* (S. c.), and *Microsporum canis* (M. c.) were used for the fungistatic evaluation. These were later provided by the National Research Centre and the Microbial Centre of A in Shams University, Egypt. The media used were Mueller Hinton agar medium, tryptic soy broth (TSB), (ICN, biochemical Co., USA).

	Antibacterial bioassay						Antifungal bioassay					
	Bacterial strain						Fungi strain					
Compd.	***S.a.***	***S.g.***	***B.s.***	***P.a.***	***E.c.***	***E.c.***	***A.f.***	***C.a.***	***A.a.***	***P.i.***	***S.c.***	***M.c.***
SHAM	++++	++++	++++	++++	+++	++++	++++	+++	++++	++++	++++	++++
I	+++	+++	+++	+++	++	++	++++	+++	+++	+++	++++	+++
II	++	++	++	++	+	+	++	+++	++	++	+	+
III	+++	+++	+++	+++	++	++	+++	++	+++	+++	++	+++
IV	++++	++++	++++	++++	+++	++++	++++	++++	+++	++	++++	+++
V	+++	+++	+++	+++	++	+++	++++	++	+++	++++	++	++
VI	++++	+++	++++	++++	++++	+++	++++	+++	+++	+++	++	+++
A	+++	+++	++	++	+++	++++	+++	++	++++	++	++	++

I, II, III, IV, V, VI are complexes of SHAM (I-VI) $Cu(SHAM)_2{\cdot}2H_2O$ (I), $Ni(SHAM)_2{\cdot}2H_2O$ (II), $Fe(SHAM)_2{\cdot}2H_2O$ (III), [Cu(Phen)(SHAM)] (IV), [Ni(Phen)(SHAM)] (V), [Fe(Phen)(SHAM)] (VI), and the reference drug (A) against six bacterial species and six fungi species in the agar disc diffusion method measured by diameter of inhibition zones (DIZ, mm).

Table 6. Antimicrobial activities of SHAM and its complexes.

4.1.2. The calculation of minimum inhibitory concentration (MIC)

The apparatus used for the experiments are 0.5 mL volume 96-well microplates (ICN Biochem. Co., USA). Fill the first row of wells in each of 96-well microdilution plates with 100 μL of double strength TSB and fill the others with single strength TSB. Dilute the test compound and pour 100 μL in the first row. Every well filled must have its content mixed. The mixing technique based on suctioning and dispensing its content back five times. Take 100 μL from the first row and fill the second one. Then take from the second and fill the third. Repeat this procedure until the 23 rows. Each organism was experimented twice and the experiment as a whole was repeated three times. Before inoculating the 96-well plates with 96-prong inoculator, this later was in the inoculums. The microdilution plates were sealed with tape to prevent the drying of samples. At the end, plates were incubated at suitable temperatures to allow the growth to start. The growth was observed after 1 day.

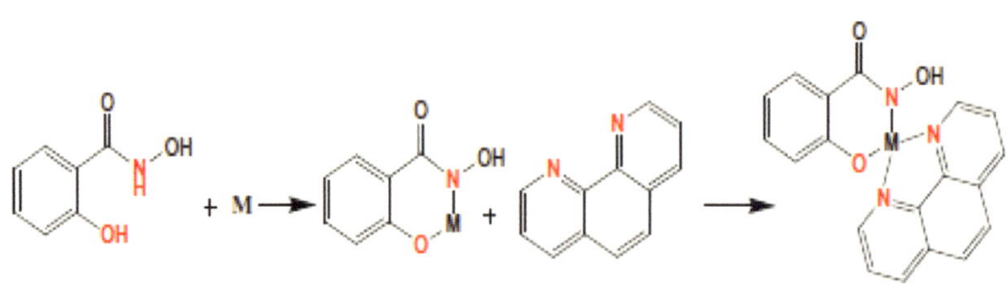

Metal complexes of salicylhydroxamic acid (SHAM) and 1,10-phenanthroline (PHEN)

This stable can be summarized in the following points:

- DIZ 7–9 mm get + sign which indicates a weak activity.
- DIZ 10–14 mm get ++ sign which indicates a moderate activity.
- DIZ 15–18 mm get +++ sign which indicates a good activity.
- DIZ >18 mm get +++ sign which indicates a significant activity.

The bacteria used were (*S.a.*), (*S.g.*), (*B.s.*), (*P.a.*), (*E.c.*), and (*E.c.*). While the fungi used were (*A.f.*), (*C.a.*), (*A.a.*), (*P.i.*), (*S.c.*), and (*M.c.*). The reference drugs: (A) Ampicillin (H_2O).

5. Conclusion

In this chapter, a detailed survey of the formation equilibria of Fe^{3+} with ligands of biological significance is presented. The main conclusions may be summarized as follows:

- Stability constant of the ternary complexes formed between glycine combines with the binary complex (metal:SMZ) (1:1) in similar manner was calculated with respect to the binary complexes.
- Mixed ligand complex formed between iron(III) and glycine(Gly) as the first ligand and nitroacetic acid as the second one have been characterized using differential pulse cathodic voltammetry (DPCV), cyclic voltammetry (CV), and direct current (d.c.) polarography, where iron(III) concentrations varied from 5×10^{-6} to 6×10^{-4} mol·dm^{-3}, (nitroacetic acid), its concentrations varied from 2×10^{-5} to 1×10^{-3} mol·dm^{-3} and glycine's concentrations were 0.2, 0.02, and 0.002 mol·dm^{-3}, in a 0.1 mol·dm^{-3} $NaClO_4$ at $pH = 8.0 \pm 0.1$ and 289 K.
- The concentration distribution curves of the complexes are plotted against the pH. The concentration of $[Fe(EDTA)(uracil)]^{2-}$ complex starts to increase from pH = 4.0 and continues up to pH 8.5, where it reaches 84% which represents its highest value. After that the concentration decreases and the concentration of the hydrolyzed species $[Fe(EDTA)(OH)]^{2-}$ develops.
- The antimicrobial activities of SHAM and PHEN as bioligands and their synthesized metal complexes (I–VI) against representative pathogenic bacteria and fungi. The minimum inhibitory concentration (MIC) value was defined as the lowest concentration of the antibacterial, antifungal agents at which there showed optically clear. Quality control was performed using ampicillin as a standard antibiotic.

Acknowledgements

As the chairman of the Chemistry Department, Faculty of Science, Cairo University, Egypt, the author like to thank all members of her department, especially his colleagues in the laboratory, for their efforts and co-operation. The author is also grateful to Prof. Dr. Said Faheem, Dean of Faculty of Science, for the financial support and his constant encouragement for this work.

Abbreviations

A. a	*Alternaria alternata*
A. f	*Aspergillus fumigatus*
B. s	*Bacillus subtilis*
C. a	*Candida albicans*
E. c	*Enterobacter cloacae*
M. c	*Microsporum canis*
P. i	*Penicillium italicum*
P. a	*Pseudomonas aeruginosa*
S. c	*Saccharomyces cerevisiae*
S. g	*Salmonella gallinarum*
S. a	*Staphylococcus aureus*
Gly	glycine
NTA	nitrilotriacetate
PHEN	1,10-phenanthroline
SHAM	salicylhydroxamic
SMZ	sulfamethoxazole (4-amino-N-(5-methyl-3-isoxazolyl)-benzenesulfonamide

Author details

Wafaa Mahmoud Hosny

Address all correspondence to: whosny@sci.cu.edu.eg

Department of Chemistry, Faculty of Science, Cairo University, Giza, Egypt

References

[1] Sunda WG. Iron uptake and growth limitation in oceanic and coastal phytoplankton.; Huntsman SA. Marine Chemistry. 1995;**50**:189-206

[2] Schneider W. Iron hydrolysis and the biochemistry of iron: The interplay of hydroxide and biogenic ligands. Chimia. 1988;**42**:9-20

[3] Ussher SJ, Achterberg EP, Worsfold PJ. Mrine biogeochemistry of iron.; Environmental Chemistry. 2004;**1**:67-80

[4] Crumbliss AL, Garrison JM. A comparison of some aspects of the aqueous coordination chemistry of Al(III) and Fe (III). Comments on Inorganic Chemistry. 1988;**8**:1-26

[5] Byrne RH, Luo YR, Young RW. Iron hydrolysis and solubility revisited: Observations and comments on iron hydrolysis characterizations. Marine Chemistry. 2000;**70**:23-35

[6] Millero FJ. Geochemical Transactions. 2001;**2**:57-64

[7] Guggenheim KY. Chlorosis: The rise and disappearance of a nutritional disease. Journal of Nutrition. 1995;**125**:1822-1825

[8] Yip R, Dallman PR. Iron. In: Ziegler EE, Filer LJ, editors. Present Knowledge in Nutrition. 7th ed. Washington, DC: ILSI Press; 1996. pp. 278-292

[9] Underwood EJ, Suttle NF. The Mineral Nutrition of Livestock. 3rd ed. Wallingford: CABI Publishing, CAB International; 1999. p. 614. ISBN: 0 85199 128 9

[10] Allen AP, Gillooly JF, Savage VM and Brown JH. Kinetic effects of temperature on rates of genetic divergence and speciation. PNAS. 2006;**103**:9130-9135

[11] Brabin BJ, Premji Z, Verhoeff F. An analysis of anemia and child mortality. Journal of Nutrition. 2001;**131**:636-45S

[12] Cornell RM, Schwertmann U. The Iron Oxides. Weinheim, Germany: Wiley-VCH; 1996. ISBN: 978-3-527-60644-3

[13] Lorenz WJ, Heusler KE. Corrosion Mechanisms. New York: Marcel Decker; 1987

[14] Sapieszko RS et al. J. Phys. Chem. 1977;**81**:1061

[15] Motekaitis RJ, Martell AE. J. Am. Chem. Soc. 1988;**110**:7715

[16] Legget DJ. Computational Methods for the Determination of Formation Constants. New York: Plenum Press; 1985. pp. 291-353

[17] Meloum M, Havel J, Hgfeldt E. Computation of Solution Equilibria. Chichester, UK: Ellis Horwood; 1994

[18] Gampp H, Maeder M, Meyer CJ, Zuberbühler AD. Calculation of equilibrium constants from multiwavelength spectroscopic data—II: SPECFIT: two user-friendly programs in basic and standard FORTRAN 77. Talanta. 1985;**32**(4):257-264

[19] Sabatini A, Vacca A, Gans P. Miniquad—A general computer program for the computation of formation constants from potentiometric data. Talanta. 1974;**21**:53-77

[20] Gans P, Sabatini A, Vacca A. An improved computer program for the computation of formation constants from potentiometric data. Inorganica Chimica Acta. 1976;**18**: 237-239

[21] Chandler JP, Thomson RE, Spivey HO, Li EL-F. An improved computer program for calculating formation constants of ligand complexes from pH data. Analytica Chimica Acta. 1984;**162**:399-402

[22] Gans P, Sabatini A, Vacca A. SUPERQUAD, an improved general program for computation of formation constants from potentiometric data. Journal of the Chemical Society Dalton Transactions. 1985;1195-1200

[23] Zekany L, Nagypal I, Leggett DJ. Computational Methods for the Determination of Formation Constants. New York: Plenum Press; 1985. p. 71

[24] Sabatini A, Vacca A, Gans P. Mathematical algorithms and computer programs for the determination of the equilibrium constants from potentiometric and spectrophotometric measurements. Coordination Chemistry Reviews. 1992;**120**:389-405

[25] Gordon WE. Data analysis for acid-base titration of an unknown solution. Analytical Chemistry. 1982;**54**(9):1595-1601

[26] Perrin DD, Stunzi H, Leggett DJ. Computational Methods for the Determination Of Formation Constants. New York: Plenum Press; 1985. p. 71

[27] Tauler R, Casassas E, Izquierdo-Ridorsa A. Self-modelling curve resolution in studies of spectrometric titrations of multi-equilibria systems by factor analysis. Analytica Chimica Acta. 1991;**248**:447-458

[28] Hernández F, Morell I, Beltrán J, López FJ. Multi-residue procedure for the analysis of pesticides in groundwater: Application to samples from the comunidad Valenciana. Chromatographia. 1993;**37**(5):303-312

[29] Beltrán JL, Codony R, Prat MD. Evaluation of stability constants from Multiwavelength absorbance data: Program STAR. Analytica Chimica Acta. 1993;**276**:441

[30] Frassineti C, Ghelli S, Gans P, Sabatini A, Moruzzi MS, Vacca A. Nuclear magnetic resonance as a tool for determining protonation constants of natural polyprotic bases in solution. Analytical Biochemistry. 1995;**231**:374-382

[31] Klaus F. Analytical Profiles of Drug Substances. Vol. 2, 2nd ed. TNC: Academic Press; 1984. p. 467

[32] Vree TB, Hekster YA, Karger S. Pharmacokinetics of sulfonamides revisited. Antimicrobial Chemotherapy. 1985;**34**:5-65

[33] Vree TB, Hekster YA, Karger S. Clinical Pharmacokinetics of Sulfonamides and their Metabolites. An Encyclopedia, Antibiotics and Chemotherapy. Basel, 1987; VIII: 216. ISBN: 3-8955-4511-8

[34] El-Sherif AA, Shoukry MM, Hosny WM, Abd-Elmoghny M. Complex formation equilibria of unusual seven-coordinate Fe(EDTA) complexes with DNA constituents and related bio-relevant ligands. Journal of Solution Chemistry. 2012;**41**:813-827

[35] Van Uitert GL, Hass CG. Studies on the coordination compounds. A method for determining thermodynamic equilibrium constants in mixed solvents. Journal of the American Chemical Society. 1971;**75**:451-455

[36] Motekaitis RJ, Martell AE, Nelson DA. Formation and stabilities of cobalt(II) chelates of N-benzyl triamine Schiff bases and their dioxygen complexes. Inorganic Chemistry. 1984;**23**:275-283

[37] Alousi AS, Shehata MR, Shoukry MM, Mohamed NM. Interaction of dimethyltin(IV) and trimethyltin(IV) with dehydroacetic acid. Journal of Chemical Speciation & Bioavailability. 2009;**21**:1-6

[38] Maskos K. The interaction of metal ions with nucleic acids. A nuclear magnetic Resonance relaxation time study of the copper(II)–inosine 5-monophosphate system in solution. Acta Biochimica Polonica. 1981;**28**:183-200

[39] Maskos K. Spectroscopic studies on the copper(II)–inosine system. Journal of Inorganic Biochemistry. 1985;**25**:1-14

[40] Martin RB, Mariam YH. Metal Ions in Biological Systems. Vol. 8. New York: Marcel Dekker; 1979. pp. 57-124

[41] Sigel H, Massoud SS, Corfu NA. Comparison of the extent of macrochelate formation in complexes of divalent metal ions with guanosine (GMP2-), inosine (IMP2-), and adenosine 5-monophosphate (AMP2-). The crucial role of N-7 basicity in metal ion–nucleic base recognition. Journal of the American Chemical Society. 1994;**116**:2958-2971

[42] Kramer-Schnabel U, Linder PW. Substituent effects in the protonation and complexation with copper(II) ions of organic monophosphate esters. A potentiometric and calorimetric study. Inorganic Chemistry. 1991;**30**:1248-1254

[43] Rees DC. Experimental evaluation of the effective dielectric constant of proteins. Journal of Molecular Biology. 1980;**141**(3):323-326

[44] Rogersa NK, Mooreb GR, Sternberga MJE. Electrostatic interactions in globular proteins: Calculation of the pH dependence of the redox potential of cytochrome C551. Journal of Molecular Biology. 1985;**182**:613-616

[45] Åkerlöf G, Short OA. The dielectric constant of dioxane–water mixtures between 0 and 80_-correction. Journal of the American Chemical Society. 1953;**75**:6357

[46] Dogan A, Köseoglu F, Kılıç E. The stability constants of copper(II) complexes with some α-amino acids in dioxane water mixtures. Analytical Biochemistry. 2001;**295**:237-239

[47] Fazaryl AE. Bulletin of the Chemical Society of Ethiopia Journal. 2014;**3**:28

Chapter 3

Controlled Thiolate-Protected Gold and Alloy Clusters

Bharat Kumar, Yoshiki Niihori,
Wataru Kurashige and Yuichi Negishi

Additional information is available at the end of the chapter

http://dx.doi.org/10.5772/67833

Abstract

Small metal clusters exhibit physical and chemical properties that differ substantially from those of corresponding bulk metals. Furthermore, the properties of clusters vary greatly depending on the number of constituent atoms. Metal clusters with these characteristics currently attract great attention in a wide range of fields as new nanoscale functional materials. In recent years, the techniques to precisely synthesize metal clusters protected with organic ligands and polymers with atomic precision have advanced dramatically. In addition, substantial knowledge of the size-specific physical/chemical properties exhibited by these metal clusters has been accumulated. In this chapter, we describe the precise synthesis methods of the most studied thiolate (SR)-protected gold clusters $Au_n(SR)_m$ and their heteroatom-substituted clusters (alloy clusters).

Keywords: gold clusters, alloy clusters, precise synthesis, fractionation, size focusing, metal exchange

1. Introduction

Substances in our surroundings are composed of assemblies of atoms. For example, a metal is a conglomerate of a nearly infinite number of metal atoms. By contrast, certain substances are made up of a countable number of metal atoms. These substances are called "metal clusters" because their shape resembles grape clusters. Although no clear definition of metal clusters has been established, the term generally refers to an aggregate of two to several hundred metal atoms (**Figure 1**); most such aggregates have a superfine size of 2 nm or less.

The proportion of surface atoms in metal clusters differs substantially from that in bulk metals. Taking a metal cluster with an icosahedral structure as an example, a metal cluster with

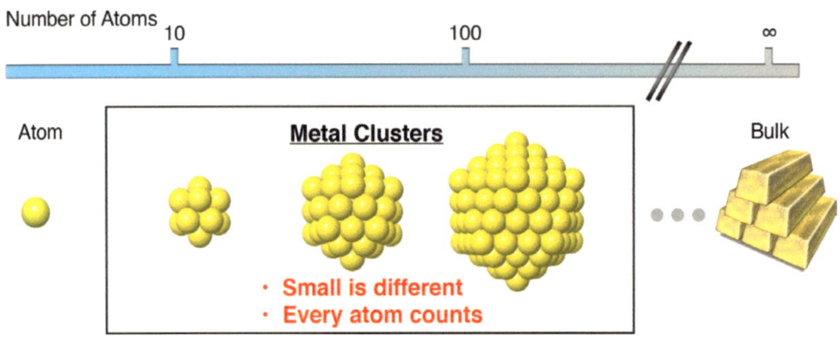

Figure 1. Relation of metal clusters discussed in this chapter to a single atom and the bulk metal.

55 atoms (**Figure 1**) has 42 surface atoms, corresponding to 76.3% of the total atom number. In the case of a 13-atom metal cluster (**Figure 1**), 12 atoms are on the surface, corresponding to 92.3% of the total atoms. In bulk metals (**Figure 1**), the proportion of surface atoms is only approximately 0.00001% in a cube of 1 cm^3. Thus, compared with bulk metals, metal clusters have a much higher proportion of surface atoms available to react with other substances. Moreover, in addition to these geometric features, metal clusters also exhibit particular characteristics related to their electronic structures. Bulk metals have an electronic structure in which the valence and conduction bands are connected. Conversely, discretization of the electronic structure occurs in metal clusters because of the small number of constituent atoms.

Because of these geometric and electronic features, metal clusters exhibit physical and chemical properties that differ from those of the corresponding bulk metals. For example, although bulk gold (Au) is an inactive metal, as its size decreases to the cluster level, Au exhibits high catalytic activity in various oxidation and reduction reactions [1, 2]. Furthermore, the size-specific properties of clusters greatly vary depending on the number of constituent atoms. **Figure 2** shows a photograph of aqueous solutions of thiolate (SR)-protected Au clusters (approximately 1 nm in size) with 10–39 gold atoms [3]. The color of the cluster solutions differs substantially depending on the number of constituent atoms in the clusters. This diversity of colors can be attributed to the aforementioned discrete electronic structure of clusters.

As illustrated above, metal clusters exhibit physical and chemical properties that differ substantially from those of bulk metals despite being composed of the same elements. Furthermore, the properties of clusters vary greatly depending on the number of constituent atoms. Because of their very small size, clusters contribute to the miniaturization of materials and conservation of resources. Thus, metal clusters currently attract great attention in a wide range of fields as new nanoscale functional materials.

Figure 2. Photograph of aqueous solutions of glutathionate-protected Au_n clusters [3].

http://dx.doi.org/10.5772/67833

In recent years, the atomically precise synthesis of metal clusters protected with organic ligands [4–19] and polymers [20, 21] has advanced dramatically. In addition, substantial knowledge about the size-specific physical/chemical properties exhibited by these metal clusters has been gathered. In this chapter, we describe the precise synthesis methods of the most-studied SR-protected Au clusters, $Au_n(SR)_m$, and their heteroatom-substituted clusters, which are called alloy clusters.

2. Gold clusters

As described in Section 1, the properties of metal clusters vary greatly depending on the number of constituent atoms (**Figure 2**). Therefore, it is important to synthesize clusters with atomic precision to produce clusters with controlled functions. Typically, $Au_n(SR)_m$ clusters with a defined number of constituent atoms are synthesized by one of the following four methods [22]:

(i) High-resolution separation of a mixture of clusters of various sizes according to the number of constituent atoms (**Figure 3(a)**).

(ii) Exposure of a mixture of clusters of various sizes to extreme conditions followed by the collection of only those clusters stable under such conditions (**Figure 3(b)**).

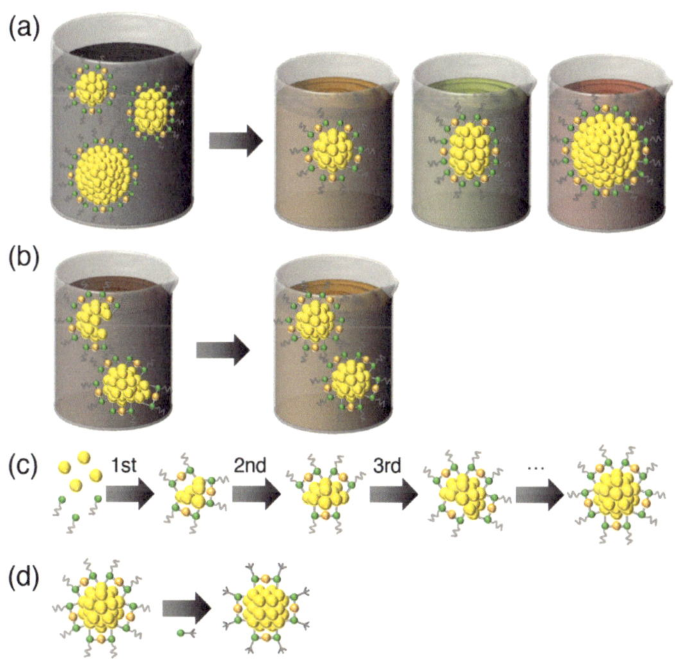

Figure 3. Typical methods for the precise synthesis of $Au_n(SR)_m$ clusters [22]: (a) fractionation, (b) size focusing, (c) slow reduction, and (d) transformation from one stable size to another.

(iii) Controlling the growth rate of the clusters to obtain a uniform chemical composition (**Figure 3(c)**).

(iv) Replacing the ligands of the cluster with ligands having different bulkiness to render clusters with a different chemical composition stable (**Figure 3(d)**).

Hereafter, each of these methods is explained in detail.

2.1. Fractionation

$Au_n(SR)_m$ clusters are generally prepared by adding a reducing agent to a solution containing a thiol and Au ions [23]. The aggregation of the resultant Au atoms leads to the formation of $Au_n(SR)_m$ clusters, and the products obtained by this method normally have a distribution in their number of constituent atoms [3, 4, 24–29]. The high-resolution separation of such a mixture according to the number of atoms is one of the most efficient methods of obtaining $Au_n(SR)_m$ clusters with well-defined numbers of constituent atoms (**Figure 3(a)**).

Polyacrylamide gel electrophoresis is a highly effective technique for separating hydrophilic SR-protected $Au_n(SR)_m$ clusters [3, 25, 26 30, 31]. Using this method, clusters, such as $Au_{10}(SG)_{10}$, $Au_{15}(SG)_{13}$, $Au_{18}(SG)_{14}$, $Au_{22}(SG)_{16}$, $Au_{22}(SG)_{17}$, $Au_{25}(SG)_{18}$, $Au_{29}(SG)_{20}$, $Au_{33}(SG)_{22}$, and $Au_{39}(SG)_{24}$ (SG = glutathionate), have been isolated with high purity (**Figure 4(a)**) [3]. Fractional precipitation [32] and fractional extraction [33–35] have been primarily employed to separate hydrophobic RS-protected $Au_n(SR)_m$ clusters. In addition to these frequently used conventional methods, reverse-phase high-performance liquid chromatography has also recently proved very effective for the high-resolution separation of both types of $Au_n(SR)_m$ clusters [27–29, 36–38]. $Au_n(SC_{12}H_{25})_m$ clusters with a wide range of sizes, from $Au_{38}(SC_{12}H_{25})_{24}$ to $Au_{\sim520}(SC_{12}H_{25})_{\sim130}$, have been systematically isolated by this method (**Figure 4(b)**) [28]. Thus, several methodologies have been established to date for the fractionation of $Au_n(SR)_m$ clusters, and precise systematic isolation can be now achieved for $Au_n(SR)_m$ clusters protected with either hydrophilic or hydrophobic SR using these techniques.

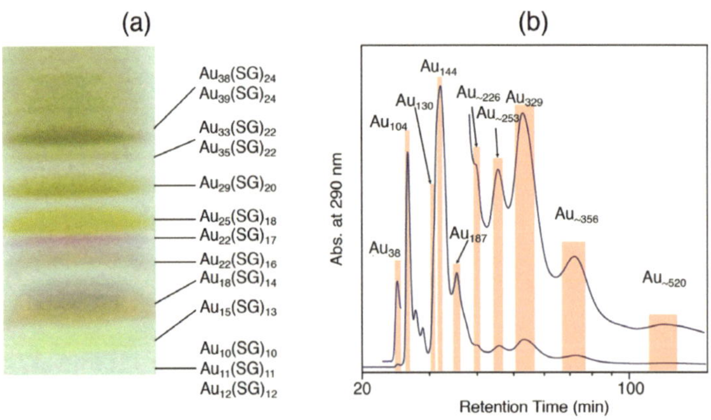

Figure 4. Examples of the precise synthesis of $Au_n(SR)_m$ clusters by fractionation: (a) separation of $Au_n(SG)_m$ clusters by polyacrylamide gel electrophoresis and (b) separation of $Au_n(SC_{12}H_{25})_m$ clusters by reverse-phase high-performance liquid chromatography. Figures were adapted from Refs. [3, 28], respectively.

2.2. Size focusing

The fractionation methods noted above are suitable for the systematic isolation of a series of $Au_n(SR)_m$ clusters. However, these methods are not suitable for mass production. To synthesize $Au_n(SR)_m$ clusters with a specific chemical composition on a large scale (~100 mg), it is necessary to conduct selective synthesis, and size focusing is an effective means of doing so (**Figure 3(b)**). Several $Au_n(SR)_m$ clusters show higher resistance to deterioration in solution [3] or thiol etching [39–42] than other clusters of the same type. The less stable clusters can be converted into stable clusters when exposed to extreme conditions [39–42]. As an example, when $Au_n(SR)_m$ clusters (SR = SC_2H_4Ph; $25 \leq n < 102$, SR = SG; $25 \leq n < 38$) are exposed to extreme conditions, unstable $Au_n(SR)_m$ clusters are transformed into stable $Au_{25}(SR)_{18}$ cluster. Stable clusters can be precisely synthesized on a large scale with this size-focusing method (**Figure 5(a)** and **(b)**).

2.3. Slow reduction

Typically, $NaBH_4$ is employed as the reducing agent to generate Au atoms. However, CO can also be used as the reducing agent. Au atoms are generated more slowly using CO than using $NaBH_4$ and so the $Au_n(SR)_m$ clusters are formed at a slower rate [43]. This slower synthesis rate tends to produce more uniform clusters (**Figure 3(c)**). The precise and size-selective synthesis of $Au_n(SR)_m$ clusters up to $Au_{25}(SG)_{18}$ has been realized by this method (**Figure 6(a)**). Even when using $NaBH_4$ as the reducing agent, slow reduction can be achieved by controlling the pH of the solution. Au_{76}(4-MEBA)$_{44}$ (4-MEBA = 4-(2-mercaptoethyl) benzoicacid) has been synthesized in this manner (**Figure 6(b)**) [44]. However, this method is only applicable to small hydrophilic $Au_n(SR)_m$ clusters. It is expected that size-selective synthetic methods based on this principle will be established for $Au_n(SR)_m$ clusters protected by hydrophobic SR as well as for large hydrophilic $Au_n(SR)_m$ clusters.

2.4. Transformation from one stable size to another

The chemical composition of stable clusters varies depending on the bulk of the SR functional group [45]. Therefore, when the ligand of a stable cluster is replaced with a bulkier

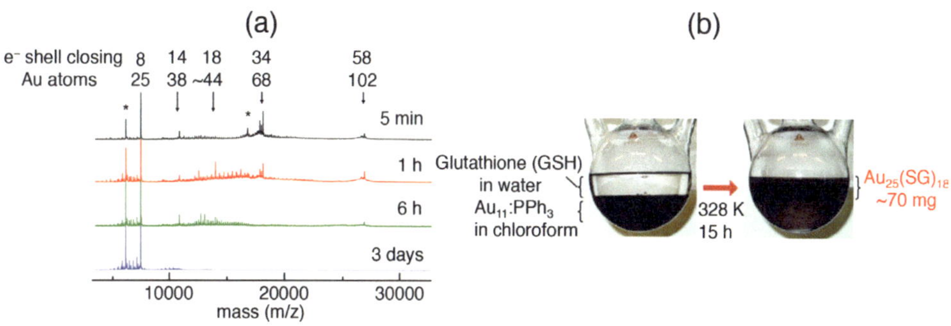

Figure 5. Examples of the precise synthesis of $Au_n(SR)_m$ clusters by size focusing: (a) synthesis of hydrophobic $Au_{25}(SC_2H_4Ph)_{18}$ and (b) synthesis of hydrophilic $Au_{25}(SG)_{18}$. Figures were adapted from Refs. [40–42], respectively.

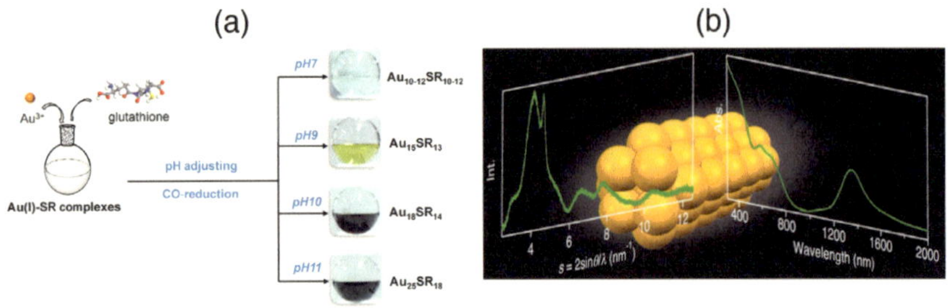

Figure 6. Examples of the precise synthesis of $Au_n(SR)_m$ clusters by slow reduction: (a) $Au_n(SG)_m$ clusters synthesized by CO reduction and (b) Au_{76}(4-(2-mercaptoethyl)benzoic acid)$_{44}$ synthesized by controlling the pH of the solution. Figures were adapted from Refs. [43, 44], respectively.

SR, distortion of the metal core is induced, resulting in the formation of clusters with a different composition (**Figure 3(d)**) [46]. An example is the reaction of phenylethanethiolate (SC_2H_4Ph)-protected $Au_{38}(SC_2H_4Ph)_{24}$ with 4-*tert*-butylbenzenthiol (HSPh-^{t}Bu) in solution, which generates Au_{36}(SPh-^{t}Bu)$_{24}$ as the primary product (yield of ~90%) (**Figure 7(a)**). This technique enables the synthesis of stable clusters with different chemical compositions from $Au_n(SC_2H_4Ph)_m$ or $Au_n(SC_xH_{2x+1})_m$ (SC_xH_{2x+1} = alkanethiolate) clusters. Clusters such as Au_{28}(SPh-^{t}Bu)$_{20}$ (**Figure 7(b)**), Au_{36}(SPh-^{t}Bu)$_{24}$, and Au_{133}(SPh-^{t}Bu)$_{52}$ have been synthesized by this method [46].

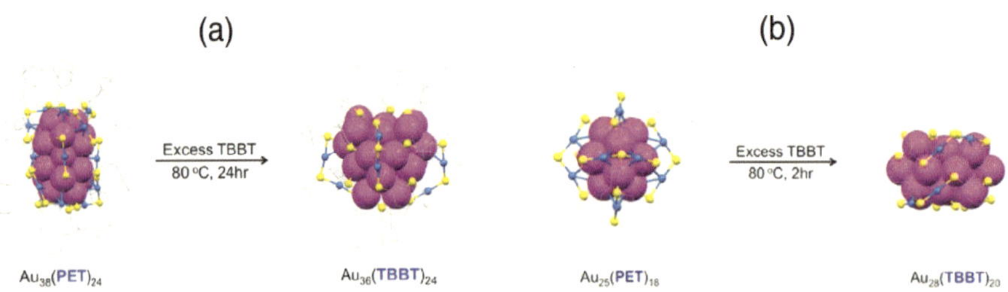

Figure 7. Examples of the precise synthesis of $Au_n(SR)_m$ clusters by transformation from one stable size to another: (a) from $Au_{38}(SC_2H_4Ph)_{24}$ to Au_{36}(SPh-^{t}Bu)$_{24}$ and (b) from $Au_{25}(SC_2H_4Ph)_{18}$ to Au_{28}(SPh-^{t}Bu)$_{20}$. Figures were adapted from Ref. [46].

3. Alloy clusters

The physical and chemical properties of metal clusters also strongly depend on the chemical composition as well as on the size of the metal core. For example, the catalytic activity of polymer-stabilized Pd_{147} clusters is remarkably improved when the Pd at the surface is partially substituted by Au [47]. In addition, alloy nanoclusters composed of Pd and Ru exhibit markedly different catalytic activities compared with those of their monometallic nanocluster counterparts. The catalytic activity obtained by mixing these two metals is higher than that of monometallic nanoclusters of Rh, which is located between these two elements in the periodic table [48]. As illustrated by these examples, synergistic effects caused by mixing different elements generate physical and chemical properties that differ from those of monometallic clusters. Thus, the composition control of metal clusters is very interesting from the viewpoint of modification of the physical and chemical properties of clusters, and results in new applications for clusters.

It is well known that SR forms strong bonds with Au (Section 2). Furthermore, stable $Au_n(SR)_m$ clusters can be produced, and a large number of methods have been established for their isolation (Section 2). Therefore, in the synthesis of alloy clusters protected by SR ligands, Au is often employed as one of the main metal elements, and the cluster size is controlled using a method similar to that used for $Au_n(SR)_m$ clusters. The important point in these syntheses is how to successfully mix other metallic elements with Au. In this section, we describe three typical procedures used to address this issue (**Figure 8**).

3.1. Co-reduction of multiple kinds of metal ions

The most common method for mixing other metallic elements with Au is the simultaneous reduction of the other metal ions with Au ions using a reducing agent (**Figure 8(a)**). This approach is called the co-reduction method. For example, to synthesize SR-protected alloy clusters, Au and other metal ions are mixed in solution, followed by the addition of thiol. A strong reducing agent ($NaBH_4$) is then added, resulting in the simultaneous reduction of all of the metal ions present. Examples of alloy clusters synthesized using this method include $Au_{25-x}Ag_x(SR)_{18}$ (R = $C_{12}H_{25}$ or C_2H_4Ph; x = 1–11; **Figure 9(a)**), $Au_{25-x}Cu_x(SR)_{18}$ (R = C_8H_{17} or C_2H_4Ph; x = 1–5; **Figure 9(b)**), $Au_{24}Pt(SC_2H_4Ph)_{18}$ (**Figure 9(c)**), $Au_{24}Pd(SR)_{18}$ (R = $C_{12}H_{25}$ or C_2H_4Ph; **Figure 9(d)**), $Au_{38-x}Ag_x(SC_2H_4Ph)_{24}$ (x = 1–12), $Au_{36}Pd_2(SC_2H_4Ph)_{24}$, $Au_{144-x}Ag_x(SC_2H_4Ph)_{60}$ (x ~ 30, 34, 52, 53), $Au_{143/144/145-x}Cu_x(SC_2H_4Ph)_{59/60/61}$ (x = 1–23), $Au_{144}Cu(SC_6H_{13})_{60}$, and $Au_{144-x}Pd_x(SC_2H_4Ph)_{60}$ (x = 1–7) [49–59]. Using the co-reduction method, it is also possible to synthesize $Au_{12}Ag_{32}(SR)_{30}$ (R = PhF, PhF_2, or $PhCF_3$) or $Au_{12+x}Cu_{32}(SPhCF_2)_{30+x}$ (x = 0, 2, 4, 6) alloy clusters, in which Ag or Cu is the base metal element [19, 60]. However, in this method, two or more types of metal atoms need to be generated simultaneously by reduction to successfully form alloy clusters.

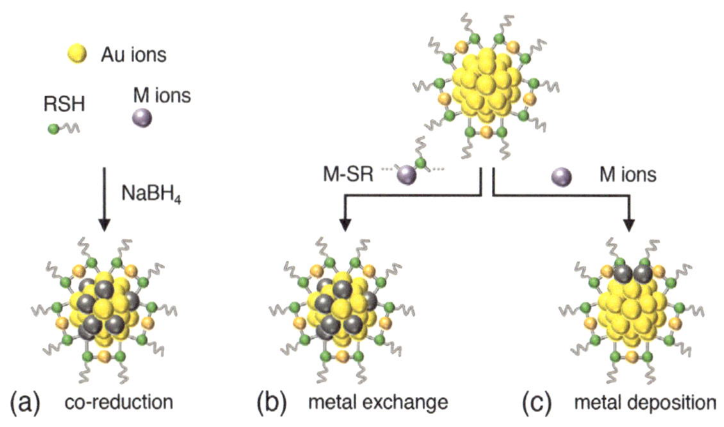

Figure 8. Representative synthesis methods of thiolate-protected alloy clusters: (a) co-reduction of multiple kinds of metal ions, (b) metal exchange with metal complexes, and (c) deposition of metal atoms onto metal clusters.

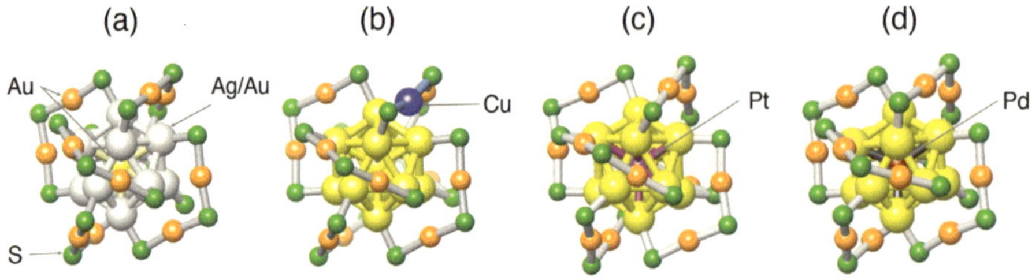

Figure 9. Thiolate-protected alloy clusters synthesized by the co-reduction method: (a) $Au_{25-x}Ag_x(SC_2H_4Ph)_{18}$, (b) $Au_{25-x}Cu_x(SC_2H_4Ph)_{18}$, (c) $Au_{24}Pt(SC_2H_4Ph)_{18}$, and (d) $Au_{24}Pd(SC_2H_4Ph)_{18}$. Ag/Au indicates Ag or Au. R groups are omitted for clarity. Figures were adapted from Refs. [50, 51, 54].

Therefore, it is difficult to form an alloy cluster using this method when there is a large difference in redox potential between the precursor metal ions. As a result, alloy clusters synthesized by this method are presently limited to those containing Au, Ag, Cu, Pt, and Pd.

3.2. Metal exchange with metal complexes

Metal clusters can exchange metal atoms with metal complexes (**Figure 8(b)**). This reaction enables heteroelements to be introduced into metal clusters to synthesize alloy clusters [61]. Although there are some exceptions [62], the number of constituent atoms of the metal core generally does not change during this exchange [63–71]. Therefore, this reaction enables some of the atoms in a cluster to be replaced with other elements while maintaining the original number of constituent atoms and geometry. In addition, this reaction allows heteroelements to be mixed more easily than the co-reduction method. The metal exchange reaction enables the synthesis of alloy clusters composed of metal elements with very different redox potentials, and a larger number of heteroatoms can be replaced than that achieved by the co-reduction method. Using this type of exchange reaction, alloy clusters such as $Au_{25-x}Ag_x(SR)_{18}$ (x = 1–8), $Au_{25-x}Cu_x(SR)_{18}$ (x = 1–9), $Au_{24}Cd(SR)_{18}$ (**Figure 10(a)**), $Au_{24}Hg(SR)_{18}$ (**Figure 10(b)**), $Au_{24-x}Ag_xCd(SR)_{18}$ (x = 2–6), $Au_{24-x}Ag_xHg(SR)_{18}$ (x = 1–8; **Figure 10(c)**), $Au_{24-x-y}Ag_xCu_yPd(SR)_{18}$ (x = 1–3, y = 1, 2; **Figure 10(d)**), $Ag_{25-x}Au_x(SR)_{18}$ (x = 1, 2), $Ag_{24-x}Au_xPt(SR)_{18}$ (x = 1, 2, 4–9), and $Au_{38-x}Ag_x(SR)_{24}$ (x = 1–11) have been synthesized to date [63–71].

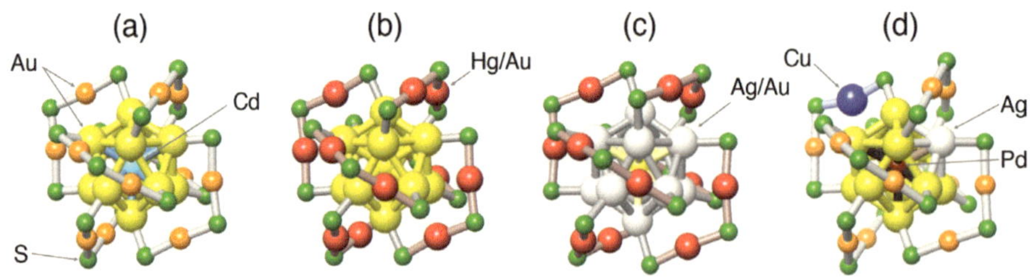

Figure 10. Thiolate-protected alloy clusters synthesized by the metal exchange method: (a) $Au_{24}Cd(SC_2H_4Ph)_{18}$, (b) $Au_{24}Hg(SC_2H_4Ph)_{18}$, (c) $Au_{24-x}Ag_xHg(SC_2H_4Ph)_{18}$, and (d) $Au_{22}AgCuPd(SC_{12}H_{25})_{18}$. Hg/Au indicates Hg or Au. R groups are omitted for clarity. Figures were adapted from Refs. [64, 66, 68, 71].

3.3. Deposition of metal atoms onto metal clusters

When $Au_n(SR)_m$ clusters with fine metal cores (<2 nm) react with Ag ions, the Ag ions are reduced by Au, and Ag is deposited on the cluster surface (**Figure 8(c)**). This type of reaction has been used to synthesize $Au_{25}Ag_2(SC_2H_4Ph)_{18}$ by depositing two Ag atoms on the surface of an $Au_{25}(SC_2H_4Ph)_{18}$ (**Figure 11**) [72, 73]. A particular feature of this synthesis is that the substrate clusters maintain their chemical composition while the other metal atoms are deposited on the cluster surface. However, the metal clusters and metal ions used for this reaction were the same as those used in the aforementioned metal exchange method (Section 3.2). In the future, it is expected that the reason why the reaction changes from metal exchange to metal deposition under slightly different experimental conditions will be elucidated.

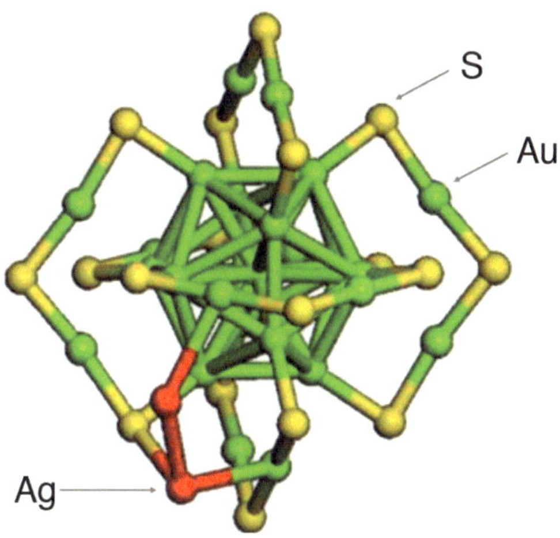

Figure 11. $Au_{25}Ag_2(SC_2H_4Ph)_{18}$ synthesized by the metal deposition method. R groups are omitted for clarity. This figure was adapted from Ref. [72].

4. Conclusions and prospects

In this chapter, we focused on $Au_n(SR)_m$ and related alloy clusters as examples of metal nanoclusters and described the latest techniques and knowledge regarding their precise synthesis. The study of $Au_n(SR)_m$ clusters has progressed with spectacular speed in recent years. Consequently, the associated synthetic techniques have also advanced dramatically, and a greater understanding of their characteristics has been obtained [74–76]. These clusters are now expected to be applied in various fields such as sensing, imaging, cancer radiation therapy, catalysis, photocatalysis, solar cells, fuel cells, photosensitizers, and single-electron devices. If these $Au_n(SR)_m$ clusters can be regularly assembled [77], further new functions could be induced and their fields of application might be further expanded. In the future, it is expected that intensive investigations will be conducted regarding the formation of various nanoarchitectures using $Au_n(SR)_m$ clusters as structural units in addition to research on the $Au_n(SR)_m$ clusters themselves.

Acknowledgements

The authors thank all the coauthors whose names appear in the cited references. This work was supported by JSPS KAKENHI (grant numbers JP15H00763, JP15H00883, JP16H04099, 16K17480, and 16K21402). Funding from the Nippon Sheet Foundation for Materials Science and Engineering, the Sumitomo Foundation, the Takahashi Industrial and Economic Research Foundation, the Tanaka Kikinzoku Memorial Foundation, and the Futaba Electronics Memorial Foundation is also gratefully acknowledged.

Author details

Bharat Kumar[1], Yoshiki Niihori[1], Wataru Kurashige[1] and Yuichi Negishi[1,2]*

*Address all correspondence to: negishi@rs.kagu.tus.ac.jp

1 Department of Applied Chemistry, Faculty of Science, Tokyo University of Science, Shinjuku-ku, Tokyo, Japan

2 Photocatalysis International Research Center, Tokyo University of Science, Noda, Chiba, Japan

References

[1] Taketoshi A, Haruta M. Size- and structure-specificity in catalysis by gold clusters. Chemistry Letters. 2014;**43**:380-387. DOI: 10.1246/cl.131232

[2] Tsukuda T, Tsunoyama H, Sakurai H. Aerobic oxidations catalyzed by colloidal nanogold. Chemistry – An Asian Journal. 2011;**6**:736-748. DOI: 10.1002/asia.201000611

[3] Negishi Y, Nobusada K, Tsukuda T. Glutathione-protected gold clusters revisited: Bridging the gap between gold(I)–thiolate complexes and thiolate-protected gold nanocrystals. Journal of the American Chemical Society. 2005;**127**:5261-5270. DOI: 10.1021/ja042218h

[4] Negishi Y. Toward the creation of functionalized metal nanoclusters and highly active photocatalytic materials using thiolate-protected magic gold clusters. Bulletin of the Chemical Society of Japan. 2014;**87**:375-389. DOI: 10.1246/bcsj.20130288

[5] Whetten RL, Shafigullin MN, Khoury JT, Schaaff TG, Vezmar I, Alvarez MM, Wilkinson A. Crystal structures of molecular gold nanocrystal arrays. Accounts of Chemical Research. 1999;**32**:397-406. DOI: 10.1021/ar970239t

[6] Parker JF, Fields-Zinna CA, Murray RW. The story of a monodisperse gold nanoparticle: $Au_{25}L_{18}$. Accounts of Chemical Research. 2010;**43**:1289-1296. DOI: 10.1021/ar100048c

[7] Luo Z, Nachammai V, Zhang B, Yan N, Leong DT, Jiang D, Xie J. Toward understanding the growth mechanism: Tracing all stable intermediate species from reduction of Au(I)–thiolate complexes to evolution of Au_{25} nanoclusters. Journal of the American Chemical Society. 2014;**136**:10577-10580. DOI: 10.1021/ja505429f

[8] Qian H, Zhu M, Wu Z, Jin R. Quantum sized gold nanoclusters with atomic precision. Accounts of Chemical Research. 2012;**45**:1470-1479. DOI: 10.1021/ar200331z

[9] Dass A. Nano-scaling law: Geometric foundation of thiolated gold nanomolecules. Nanoscale. 2012;**4**:2260-2263. DOI: 10.1039/c2nr11749e

[10] Zhang P. X-ray spectroscopy of gold–thiolate nanoclusters. The Journal of Physical Chemistry C. 2014;**118**:25291-25299. DOI: 10.1021/jp507739u

[11] Udayabhaskararao T, Pradeep T. New protocols for the synthesis of stable Ag and Au nanocluster molecules. The Journal of Physical Chemistry Letters. 2013;**4**:1553-1564. DOI: 10.1021/jz400332g

[12] Knoppe S, Wong OA, Malola S, Häkkinen H, Bürgi T, Verbiest T, Ackerson CJ. Chiral phase transfer and enantioenrichment of thiolate-protected Au_{102} clusters. Journal of the American Chemical Society. 2014;**136**:4129-4132. DOI: 10.1021/ja500809p

[13] Dolamic I, Knoppe S, Dass A, Bürgi T. First enantioseparation and circular dichroism spectra of Au_{38} clusters protected by achiral ligands. Nature Communications. 2012;**3**:798. DOI: 10.1038/ncomms1802

[14] Jadzinsky PD, Calero G, Ackerson CJ, Bushnell DA, Kornberg RD. Structure of a thiol monolayer-protected gold nanoparticle at 1.1 Å resolution. Science. 2007;**318**:430-433. DOI: 10.1126/science.1148624

[15] Kwak K, Kumar SS, Pyo K, Lee D. Ionic liquid of a gold nanocluster: A versatile matrix for electrochemical biosensors. ACS Nano. 2014;**8**:671-679. DOI: 10.1021/nn4053217

[16] Dainese T, Antonello S, Gascón JA, Pan F, Perera NV, Ruzzi M, Venzo A, Zoleo A, Rissanen K, Maran F. $Au_{25}(SEt)_{18}$, a nearly naked thiolate-protected Au_{25} cluster: Structural analysis by single crystal X-ray crystallography and electron nuclear double resonance. ACS Nano. 2014;**8**:3904-3912. DOI: 10.1021/nn500805n

[17] Konishi K, Iwasaki M, Sugiuchi M, Shichibu Y. Ligand-based toolboxes for tuning of the optical properties of subnanometer gold clusters. The Journal of Physical Chemistry Letters. 2016;**7**:4267-4274. DOI: 10.1021/acs.jpclett.6b01999

[18] Desireddy A, Conn BE, Guo J, Yoon B, Barnett RN, Monahan BM, Kirschbaum K, Griffith WP, Whetten RL, Landman U, Bigioni TP. Ultrastable silver nanoparticles. Nature. 2013;**501**:399-402. DOI: 10.1038/nature12523

[19] Yang H, Wang Y, Huang H, Gell L, Lehtovaara L, Malola S, Häkkinen H, Zheng N. All-thiol-stabilized Ag_{44} and $Au_{12}Ag_{32}$ nanoparticles with single-crystal structures. Nature Communications. 2013;**4**:2422. DOI: 10.1038/ncomms3422

[20] Zheng J, Petty JT, Dickson RM. High quantum yield blue emission from water-soluble Au_8 nanodots. Journal of the American Chemical Society. 2003;**125**:7780-7781. DOI: 10.1021/ja035473v

[21] Yamamoto K, Imaoka T, Chun WJ, Enoki O, Katoh H, Takenaga M, Sonoi A. Size-specific catalytic activity of platinum clusters enhances oxygen reduction reactions. Nature Chemistry. 2009;**1**:397-402. DOI: 10.1038/nchem.288

[22] Kurashige W, Niihori Y, Sharma S, Negishi Y. Precise synthesis, functionalization and application of thiolate–protected gold clusters. Coordination Chemistry Reviews. 2016;**320-321**:238-250. DOI: 10.1016/j.ccr.2016.02.013

[23] Brust M, Walker M, Bethell D, Schiffrin DJ, Whyman R. Synthesis of thiol-derivatised gold nanoparticles in a two-phase liquid–liquid system. Journal of the Chemical Society, Chemical Communications. 1994:801-802. DOI: 10.1039/c39940000801

[24] Tsukuda T. Toward an atomic-level understanding of size-specific properties of protected and stabilized gold clusters. Bulletin of the Chemical Society of Japan. 2012;**85**:151-168. DOI: 10.1246/bcsj.20110227

[25] Negishi Y, Takasugi Y, Sato S, Yao H, Kimura K, Tsukuda T. Magic-numbered Au_n clusters protected by glutathione monolayers (n = 18, 21, 25, 28, 32, 39): Isolation and spectroscopic characterization. Journal of the American Chemical Society. 2004;**126**:6518-6519. DOI: 10.1021/ja0483589

[26] Negishi Y, Takasugi Y, Sato S, Yao H, Kimura K, Tsukuda T. Kinetic stabilization of growing gold clusters by passivation with thiolates. The Journal of Physical Chemistry B. 2006;**110**:12218-12221. DOI: 10.1021/jp062140m

[27] Negishi Y, Sakamoto C, Ohyama T, Tsukuda T. Synthesis and the origin of the stability of thiolate-protected Au_{130} and Au_{187} clusters. The Journal of Physical Chemistry Letters. 2012;**3**:1624-1628. DOI: 10.1021/jz300547d

[28] Negishi Y, Nakazaki T, Malola S, Takano S, Niihori Y, Kurashige W, Yamazoe S, Tsukuda T, Häkkinen H. A critical size for emergence of nonbulk electronic and geometric structures in dodecanethiolate-protected Au clusters. Journal of the American Chemical Society. 2015;**137**:1206-1212. DOI: 10.1021/ja5109968

[29] Niihori Y, Uchida C, Kurashige W, Negishi Y. High-resolution separation of thiolate-protected gold clusters by reversed-phase high-performance liquid chromatography. Physical Chemistry Chemical Physics. 2016;**18**:4251-4265. DOI: 10.1039/c5cp04660b

[30] Schaaff TG, Knight G, Shafigullin MN, Borkman RF, Whetten RL. Isolation and selected properties of a 10.4 kDa gold:Glutathione cluster compound. The Journal of Physical Chemistry B. 1998;**102**:10643-10646. DOI: 10.1021/jp9830528

[31] Schaaff TG, Whetten RL. Giant gold–glutathione cluster compounds: Intense optical activity in metal-based transitions. The Journal of Physical Chemistry B. 2000;**104**:2630-2641. DOI: 10.1021/jp993691y

[32] Whetten RL, Khoury JT, Alvarez MM, Murthy S, Vezmar I, Wang ZL, Stephens PW, Cleveland CL, Luedtke WD, Landman U. Nanocrystal gold molecules. Advanced Materials. 1996;**8**:428-433. DOI: 10.1002/adma.19960080513

[33] Donkers RL, Lee D, Murray RW. Synthesis and isolation of the molecule-like cluster $Au_{38}(PhCH_2CH_2S)_{24}$. Langmuir. 2004;**20**:1945-1952. DOI: 10.1021/la035706w

[34] Negishi Y, Chaki NK, Shichibu Y, Whetten RL, Tsukuda T. Origin of magic stability of thiolated gold clusters: A case study on $Au_{25}(SC_6H_{13})_{18}$. Journal of the American Chemical Society. 2007;**129**:11322-11323. DOI: 10.1021/ja073580+

[35] Chaki NK, Negishi Y, Tsunoyama H, Shichibu Y, Tsukuda T. Ubiquitous 8 and 29 kDa gold:Alkanethiolate cluster compounds: Mass-spectrometric determination of molecular formulas and structural implications. Journal of the American Chemical Society. 2008;**130**:8608-8610. DOI: 10.1021/ja8005379

[36] Wolfe RL, Murray RW. Analytical evidence for the monolayer-protected cluster $Au_{225}[(S(CH_2)_5CH_3)]_{75}$. Analytical Chemistry. 2006;**78**:1167-1173. DOI: 10.1021/ac051533z

[37] Choi MMF, Douglas AD, Murray RW. Ion-pair chromatographic separation of water-soluble gold monolayer-protected clusters. Analytical Chemistry. 2006;**78**:2779-2785. DOI: 10.1021/ac052167m

[38] Black DM, Bhattarai N, Bach SBH, Whetten RL. Selection and identification of molecular gold clusters at the nano(gram) scale: Reversed phase HPLC–ESI–MS of a mixture of Au-peth MPCs. The Journal of Physical Chemistry Letters. 2016;**7**:3199-3205. DOI: 10.1021/acs.jpclett.6b01403

[39] Shichibu Y, Negishi Y, Tsunoyama H, Kanehara M, Teranishi T, Tsukuda T. Extremely high stability of glutathionate-protected Au_{25} clusters against core etching. Small. 2007;**3**:835-839. DOI: 10.1002/smll.200600611

[40] Dharmaratne AC, Krick T, Dass A. Nanocluster size evolution studied by mass spectrometry in room temperature $Au_{25}(SR)_{18}$ synthesis. Journal of the American Chemical Society. 2009;**131**:13604-13605. DOI: 10.1021/ja906087a

[41] Jin R, Qian H, Wu Z, Zhu Y, Zhu M, Mohanty A, Garg N. Size focusing: A methodology for synthesizing atomically precise gold nanoclusters. The Journal of Physical Chemistry Letters. 2010;**1**:2903-2910. DOI: 10.1021/jz100944k

[42] Shichibu Y, Negishi Y, Tsukuda T, Teranishi T. Large-scale synthesis of thiolated Au_{25} clusters via ligand exchange reactions of phosphine-stabilized Au_{11} clusters. Journal of the American Chemical Society. 2005;**127**:13464-13465. DOI: 10.1021/ja053915s

[43] Yu Y, Chen X, Yao Q, Yu Y, Yan N, Xie J. Scalable and precise synthesis of thiolated Au_{10-12}, Au_{15}, Au_{18}, and Au_{25} nanoclusters via pH controlled CO reduction. Chemistry of Materials. 2013;**25**:946-952. DOI: 10.1021/cm304098x

[44] Takano S, Yamazoe S, Koyasu K, Tsukuda T. Slow-reduction synthesis of a thiolate-protected one-dimensional gold cluster showing an intense near-infrared absorption. Journal of the American Chemical Society. 2015;**137**:7027-7030. DOI: 10.1021/jacs.5b03251

[45] Nishigaki J, Tsunoyama R, Tsunoyama H, Ichikuni N, Yamazoe S, Negishi Y, Ito M, Matsuo T, Tamao K, Tsukuda T. A new binding motif of sterically demanding thiolates on a gold cluster. Journal of the American Chemical Society. 2012;**134**:14295-14297. DOI: 10.1021/ja305477a

[46] Zeng C, Chen Y, Das A, Jin R. Transformation chemistry of gold nanoclusters: From one stable size to another. The Journal of Physical Chemistry Letters. 2015;**6**:2976-2986. DOI: 10.1021/acs.jpclett.5b01150

[47] Zhang H, Watanabe T, Okumura M, Haruta M, Toshima N. Catalytically highly active top gold atom on palladium nanocluster. Nature Materials. 2012;**11**:49-52. DOI: 10.1038/nmat3143

[48] Kusada K, Kobayashi H, Ikeda R, Kubota Y, Takata M, Toh S, Yamamoto T, Matsumura S, Sumi N, Sato K, Nagaoka K, Kitagawa H. Solid solution alloy nanoparticles of immiscible Pd and Ru elements neighboring on Rh: Changeover of the thermodynamic behavior for hydrogen storage and enhanced CO-oxidizing ability. Journal of the American Chemical Society. 2014;**136**:1864-1871. DOI: 10.1021/ja409464g

[49] Negishi Y, Iwai T, Ide M. Continuous modulation of electronic structure of stable thiolate-protected Au_{25} cluster by Ag doping. Chemical Communications. 2010;**46**:4713-4715. DOI: 10.1039/c0cc01021a

[50] Kumara C, Aikens CM, Dass A. X-ray crystal structure and theoretical analysis of $Au_{25-x}Ag_x(SCH_2CH_2Ph)_{18}^-$ alloy. The Journal of Physical Chemistry Letters. 2014;**5**:461-466. DOI: 10.1021/jz402441d

[51] Negishi Y, Munakata K, Ohgake W. Nobusada K. Effect of copper doping on electronic structure, geometric structure, and stability of thiolate-protected Au_{25} nanoclusters. The Journal of Physical Chemistry Letters. 2012;**3**:2209-2214. DOI: 10.1021/jz300892w

[52] Negishi Y, Kurashige W, Niihori Y, Nobusada K. Toward the creation of stable, functionalized metal clusters. Physical Chemistry Chemical Physics. 2013;**15**:18736-18751. DOI: 10.1039/C3CP52837E

[53] Qian H, Jiang D, Li G, Gayathri C, Das A, Gil RR, Jin R. Monoplatinum doping of gold nanoclusters and catalytic application. Journal of the American Chemical Society. 2012;**134**:16159-16162. DOI: 10.1021/ja307657a

[54] Tian S, Liao L, Yuan J, Yao C, Chen J, Yang J, Wu Z. Structures and magnetism of mono-palladium and mono-platinum doped $Au_{25}(PET)_{18}$ nanoclusters. Chemical Communications. 2016;**52**:9873-9876. DOI: 10.1039/c6cc02698b

[55] Negishi Y, Kurashige W, Niihori Y, Iwasa T, Nobusada K. Isolation, structure, and stability of a dodecanethiolate-protected Pd_1Au_{24} cluster. Physical Chemistry Chemical Physics. 2010;**12**:6219-6225. DOI: 10.1039/b927175a

[56] Niihori Y, Matsuzaki M, Pradeep T, Negishi Y. Separation of precise compositions of noble metal clusters protected with mixed ligands. Journal of the American Chemical Society. 2013;**135**:4946-4949. DOI: 10.1021/ja4009369

[57] Niihori Y, Matsuzaki M, Uchida C, Negishi Y. Advanced use of high-performance liquid chromatography for synthesis of controlled metal clusters. Nanoscale. 2014;**6**:7889-7896. DOI: 10.1039/c4nr01144a

[58] Kumara C, Dass A. AuAg alloy nanomolecules with 38 metal atoms. Nanoscale. 2012;**4**:4084-4086. DOI: 10.1039/c2nr11781a

[59] Negishi Y, Igarashi K, Munakata K, Ohgake W, Nobusada K. Palladium doping of magic gold cluster $Au_{38}(SC_2H_4Ph)_{24}$: Formation of $Pd_2Au_{36}(SC_2H_4Ph)_{24}$ with higher stability than $Au_{38}(SC_2H_4Ph)_{24}$. Chemical Communications. 2012;**48**:660-662. DOI: 10.1039/c1cc15765e

[60] Yang H, Wang Y, Yan J, Chen X, Zhang X, Häkkinen H, Zheng N. Structural evolution of atomically precise thiolated bimetallic $[Au_{12+n}Cu_{32}(SR)_{30+n}]^{4-}$ ($n = 0, 2, 4, 6$) nanoclusters. Journal of the American Chemical Society. 2014;**136**:7197-7200. DOI: 10.1021/ja501811j

[61] Shon Y-S, Dawson GB, Porter M, Murray RW. Monolayer-protected bimetal cluster synthesis by core metal galvanic exchange reaction. Langmuir. 2002;**18**:3880-3885. DOI: 10.1021/la025586c

[62] Li Q, Wang S, Kirschbaum K, Lambright KJ, Das A, Jin R. Heavily doped $Au_{25-x}Ag_x(SC_6H_{11})_{18}^-$ nanoclusters: Silver goes from the core to the surface. Chemical Communications. 2016;**52**:5194-5197. DOI: 10.1039/c6cc01243d

[63] Wang S, Song Y, Jin S, Liu X, Zhang J, Pei Y, Meng X, Chen M, Li P, Zhu M. Metal exchange method using Au_{25} nanoclusters as templates for alloy nanoclusters with atomic precision. Journal of the American Chemical Society. 2015;**137**:4018-4021. DOI: 10.1021/ja511635g

[64] Yao C, Lin Y, Yuan J, Liao L, Zhu M, Weng L, Yang J, Wu Z. Mono-cadmium vs Mono-mercury doping of Au_{25} nanoclusters. Journal of the American Chemical Society. 2015;**137**:15350-15353. DOI: 10.1021/jacs.5b09627

[65] Liao L, Zhou S, Dai Y, Liu L, Yao C, Fu C, Yang J, Wu Z. Mono-mercury doping of Au_{25} and the HOMO/LUMO energies evaluation employing differential pulse voltammetry. Journal of the American Chemical Society. 2015;**137**:9511-9514. DOI: 10.1021/jacs.5b03483

[66] Yang S, Wang S, Jin S, Chen S, Sheng H, Zhu M. A metal exchange method for thiolate-protected tri-metal $M_1Ag_xAu_{24-x}(SR)_{18}^0$ (M = Cd/Hg) nanoclusters. Nanoscale. 2015;**7**:10005-10007. DOI: 10.1039/c5nr01965f

[67] Yan N, Liao L, Yuan J, Lin Y, Weng L, Yang J, Wu Z. Bimetal doping in nanoclusters: Synergistic or counteractive?. Chemistry of Materials. 2016;**28**:8240-8247. DOI: 10.1021/acs.chemmater.6b03132

[68] Bootharaju MS, Joshi CP, Parida MR, Mohammed OF, Bakr OM. Templated atom-precise galvanic synthesis and structure elucidation of a $[Ag_{24}Au(SR)_{18}]^-$ nanocluster. Angewandte Chemie International Edition. 2016;**55**:922-926. DOI: 10.1002/anie.2015 09381

[69] Bootharaju MS, Sinatra L, Bakr OM. Distinct metal-exchange pathways of doped Ag_{25} nanoclusters. Nanoscale. 2016;**8**:17333-17339. DOI: 10.1039/C6NR06353E

[70] Kang X, Xiong L, Wang S, Yu H, Jin S, Song Y, Chen T, Zheng L, Pan C, Pei Y, Zhu M. Shape-controlled synthesis of trimetallic nanoclusters: Structure elucidation and properties investigation. Chemistry – A European Journal. 2016;**22**:17145-17150. DOI: 10.1002/chem.201603893

[71] Sharma S, Yamazoe S, Ono T, Kurashige W, Niihori Y, Nobusada K, Tsukuda T, Negishi Y. Tuning the electronic structure of thiolate-protected 25-atom clusters by co-substitution with metals having different preferential sites. Dalton Transactions. 2016;**45**:18064-18068. DOI: 10.1039/c6dt03214a

[72] Yao C, Chen J, Li M-B, Liu L, Yang J, Wu Z. Adding two active silver atoms on Au_{25} nanoparticle. Nano Letters. 2015;**15**:1281-1287. DOI: 10.1021/nl504477t

[73] Tian S, Yao C, Liao L, Xia N, Wu Z. Ion-precursor and ion-dose dependent anti-galvanic reduction. Chemical Communications. 2015;**51**:11773-11776. DOI: 10.1039/c5cc03267a

[74] Chen LY, Wang CW, Yuan Z, Chang HT. Fluorescent gold nanoclusters: Recent advances in sensing and imaging. Analytical Chemistry. 2015;**87**:216-229. DOI: 10.1021/ac503636j

[75] Yang X, Yang M, Pang B, Vara M, Xia Y. Gold nanomaterials at work in biomedicine. Chemical Reviews. 2015;**115**:10410-10488. DOI: 10.1021/acs.chemrev.5b00193

[76] Mathew A, Pradeep T. Noble metal clusters: Applications in energy, environment, and biology. Particle & Particle Systems Characterization. 2014;**31**:1017-1053. DOI: 10.1002/ppsc.201400033

[77] Ariga K, Vinu A, Yamauchi Y, Ji Q, Hill JP. Nanoarchitectonics for mesoporous materials. Bulletin of the Chemical Society of Japan. 2012;**85**:1-32. DOI: 10.1246/bcsj.20110162

Chapter 4

Transition Metal Complexes with Antipyrine-Derived Schiff Bases: Synthesis and Antibacterial Activity

Elena Mihaela Pahontu

Additional information is available at the end of the chapter

http://dx.doi.org/10.5772/67584

Abstract

The increase of death rate, associated with infectious diseases, is directly linked to the bacteria that have multiple resistance to antibiotics. The lack of efficient medical treatment is the main cause of this problem. The synthesis of new antibacterial agents, through various methods, is, for sure, an emergency medical issue. Recent research focuses more and more on the synthesis of complexes of the transitional metals with ligands of Schiff-base type, as a result of the biological properties which they have. This article presents the synthesis of several complexes with base Schiff ligands, derived from 4-aminoantipyrine and in vitro research of their antibacterial activities. The new compounds were tested for their in vitro antibacterial activity against *Staphylococcus aureus var. Oxford 6538, Klebsiella pneumoniae ATCC* 100131, *Escherichia coli ATCC 10536,* and *Pseudomonas aeruginosa ATCC 9027* strains. Based on the "*in vitro*" studies, we can say that ten of the complexes synthetized can be successfully used instead of streptomycin, where there is resistance to this antibiotic.

Keywords: 4-aminoantipyrine, Schiff bases, metal complexes, antibacterial agents, streptomycin

1. Introduction

The synthesis of new antibacterial agents, through various methods, is, for sure, an emergency medical issue [1–3]. Schiff bases are important precursors for the synthesis of some bioactive compounds [4, 5]. Schiff bases have received considerable attention since the discovery of their antibacterial [6, 7], antifungal [8], anti-HIV [9, 10], anti-inflammatory [11], anticonvulsant [12, 13], antiviral [14], and anticancer properties [15–17]. The presence of the inimical grouping in these organic ligands plays an important part in manifesting these biological characteristics

[18–20]. Schiff bases can be regarded as promising antimicrobial agents. For example, N-(salicylidene)-2-hydroxyaniline proved efficiency against *Mycobacterium tuberculosis* H37Rv, exhibiting an MIC value of 8 μg/mL [21]. The 5-chloro-salicylaldehyde-Shiff base derivatives are efficient against *Pseudomonas fluorescence* (MIC=2.5–5.2 μg/mL), *Escherichia coli* (MIC=1.6–5.7 μg/mL), *Bacillus subtilis* (MIC=1.8 μg/mL), and *Staphylococcus aureus* (MIC=1.6 and 3.1 μg/mL), respectively, while the MIC values for the reference drug kanamycin against the same bacterial strains were 3.9 μg/mL [22]. Some of the isatin-derived Schiff bases have shown antibacterial activity against *Escherichia coli* NCTC 10418 (MIC=2.4 μg/mL), *Vibrio cholerae* non-01(MIC=0.3 μg/mL), *Enterococcus faecalis* (MIC=1.2 μg/mL), and *Proteus shigelloides* (MIC=4.9 μg/mL). The MIC values for the reference drug sulfamethoxazole against the same bacterial strain were in the range of 312–5000 μg/mL. Therefore, these compounds were proven to be 1040-, 1040-, 4160-, and 1020-fold more potent than sulfamethoxazole [23, 24]. The studies run on the Schiff bases, derived from the isoniazid have allowed to identify a compound which has turned out to have a therapeutical effectiveness and safety, that is, 4000 times higher than that of isoniazid [25].

The morpholine-derived Schiff bases was effective against *Staphylococcus aureus* (MIC=20 μg/mL), *Micrococcus luteus* (MIC=32 μg/mL), *Streptococcus epidermidis* (MIC=17 μg/mL), *Bacillus cereus* (MIC=21 μg/mL), and *Escherichia coli* (MIC=16 μg/mL).

Schiff bases with a 2,4-dichloro-5-fluorophenyl moiety completely inhibited the growth of *Staphylococcus aureus, Escherichia coli, Pseudomonas aeruginosa,* and *Klebsiella pneumoniae.* MIC values for these compounds varied from 6.3 to 12.5 μg /mL, which are comparable to those obtained for the reference drug ciprofloxacin [26]. Lately, within the last couple of years, a special attention has been paid to the chemistry of the metal complexes of the Schiff bases. This is due to the chemical stability of the complexes as well as to the possibility of using them in the most varied fields. To a great extent, remarkable successes, in this field, have been obtained due to the various synthesis methods of the complexes. Recent research focuses more and more on the synthesis of complexes of the transitional metals with ligands of Schiff-base type, as a result of the biological properties which they have. In many cases, the conclusion has been that, through the coordination of the Schiff bases, to the metal ions, which are present in the biological systems, the biological activity of the respective Schiff base increases. A large number of Schiff bases and the corresponding metal complexes have proven antibacterian, antifungal, antitumor, and antileukemia activity [27–29].

Ever since it was synthesized [30], antipyrine (1-fenil-2,3-dimetil-5-pirazolona) has enjoyed a lot of attention due to its analgesic and antipyretic properties. The discovery of these properties has allowed for deeper research on antipyrine and its derivatives. Thus, 4-amino-2,3-dimethyl-1-phenyl-3-pyrazolin-5-one (4-aminoantipyrine) was discovered, a derivative with analgesic action, antipyretic, anti-inflammatory, antibacterian, and antineoplastic [31, 32]. The derivatives of 4-aminoantipyrine are used in the synthesis of azo-colorant, in analytical chemistry for spectrophotometric determination of metal ions [33], in pharmacology, as an effective antitumor [34], analgesic [35], antiviral [36], anti-inflammatory [37], anticancer [38], and antimicrobial drugs [39–42].

Lately, the research has been conducted in order to get metal complexes with a wide range of biological activities and with the lowest level of toxicity. In this work, the synthesis of some complexes with base Schiff ligands is presented, derived from 4-aminoantipyrine and in vitro research of their antibacterial activities.

2. Experimental

2.1. Metal complexes with aminoantipyrine Schiff bases: structure and methods of synthesis

Complexes of Cu(II), Co(II), Ni(II), Zn(II), Mn(II), VO(II), and Fe(III) were prepared by direct reaction between Schiff base ligand and the corresponding metal salts.

The next Schiff bases were synthesized:

- 1-phenyl-2,3-dimethyl-4-(N- acetoacetanilide)-3-pyrazolin-5-one (**HL^1**)
- 1,5-dimethyl-2-phenyl-4-(1-(pyridin-2-yl)ethylideneamino)-1H-pyrazol-3(2H)-one (**HL^2**)
- 5-nitro-salicylidene-4-aminoantipyrine (**HL^3**)
- 4-((E)-4-((E)-(4-chlorophenyl)diazenyl)-2-hydroxybenzylideneamino)-1,5-dimethyl-2-phenyl-1H-pyrazol-3(2H)-one (**HL^4**)
- 1-phenyl-2,3-dimethyl-4-(N-imidazole-2-carboxaldehyd)-3-pyrazolin-5-one (**HL^5**)
- 4-(2-pyrrolylmethylideneamino)antipyrine (**HL^6**)
- 4[(benzylidene)amino]antipyrine (**HL^7**)
- 4[(cinnamalidene)amino]antipyrine (**HL^8**)
- 4[(2-chlorobenzylidene)amino]antipyrine (**HL^9**)
- 4-[(furan-2-ylmethylene)amino]-1,2-dihydro-1,5-dimethyl-2-phenyl-3Hpyrazol-3-one(**HL^{10}**)
- 1-phenyl-2,3–dimethyl-4-(2-hydroxy-5-nitro-benzylideneamino)-3-pyrazolin-5-one (**HL^{11}**)
- salicylidene-4-aminoantipyrine (**HL^{12}**)
- salicylidene-4-aminoantipyrinyl-2-amino-3-hydroxypyridine (**HL^{13}**)
- napthylidene-4-aminoantipyrine (**HL^{14}**)
- 5-[(E)-[(3Z)-3-[2-[(Z)-[4-[(E)-(3-hydroxy-4-nitro-phenyl)methyleneamino]-1,5-dimethyl-2-phenyl-pyrazol-3-ylidene]amino]phenyl]imino-1,5-dimethyl-2-phenyl-pyrazol-4-yl] iminomethyl]-2-nitro-phenol (**HL^{15}**)
- 4-[(E)-[(3Z)-3-[2-[(Z)-[4-[(E)-(4-hydroxy-3-methoxy-phenyl)methyleneamino]-1,5-dimethyl-2-phenyl-pyrazol-3-ylidene]amino]phenyl]imino-1,5-dimethyl-2-phenyl-pyrazol-4-yl]iminomethyl]-2-methoxy-phenol (**HL^{16}**)
- 3-salicylideneacetylacetone-2,4-di(imino-4′-antipyrinyl)pentane (**HL^{17}**)
- 2,6-diformyl-4-methylphenol bis(4-amino-3-antipyrine) (**HL^{18}**)
- 4,4′-(1E,1′E)-(1,1′-(pyridine-2,6-diyl)bis(ethan-1-yl-1-ylidene))bis(azan-1-yl-1-ylidene) bis (1,5-dimethyl-2-phenyl-1H-pyrazol-3(2H)-one) (**HL^{19}**)
- 3(3′-hydroxy-4′-nitrobenzalidene)-2,4-di(imino-4′-antipyrinyl)pentane (**HL^{20}**)

2.1.1. Synthesis of the complexes with **HL**$^{1–5}$ *ligands*

The metal complexes with these Schiff bases are obtained by adding a methanolic or ethanolic ligand solution to a solution of metal salt, in a molar ratio L:M=2:1 [43, 44] or 1:1 [45–47]. The mixture of reaction is refluxed for 2–5h or stirring for 12h. The precipitate is obtained that is filtered, washed with ether, methanol or ethanol, and dried *in vacuo* (**Figures 1–4**).

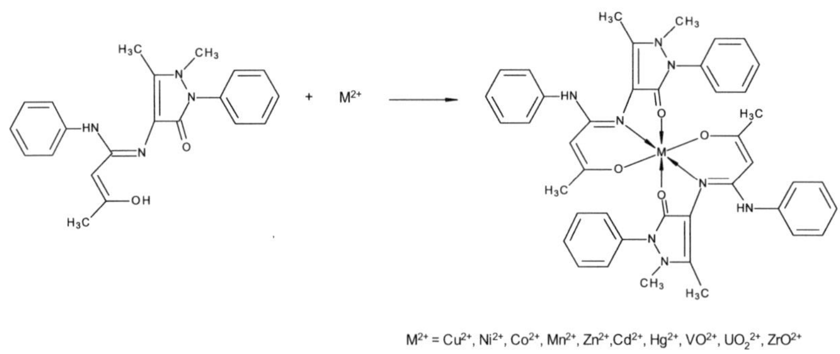

Figure 1. Scheme of synthesis of complexes with ligand **HL**^{1}.

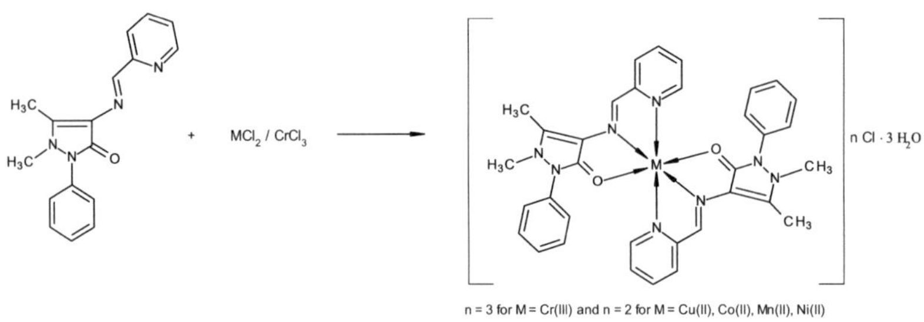

Figure 2. Scheme of synthesis of complexes with ligand **HL**^{2}.

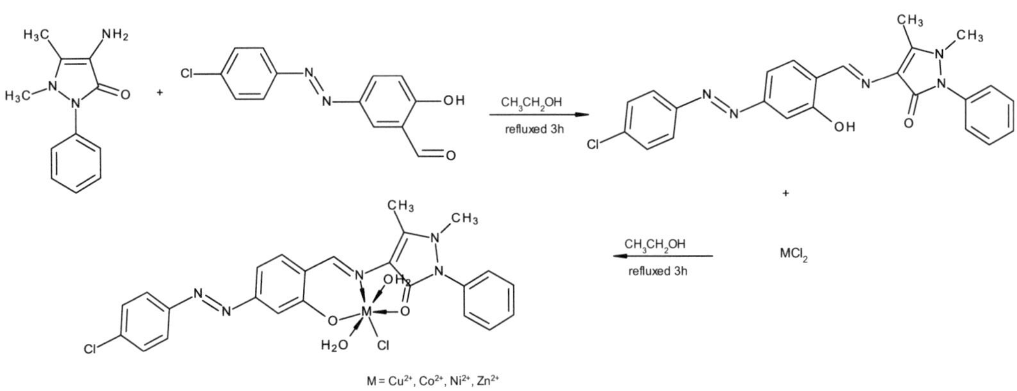

Figure 3. Scheme of synthesis of complexes with ligand **HL**^{4}.

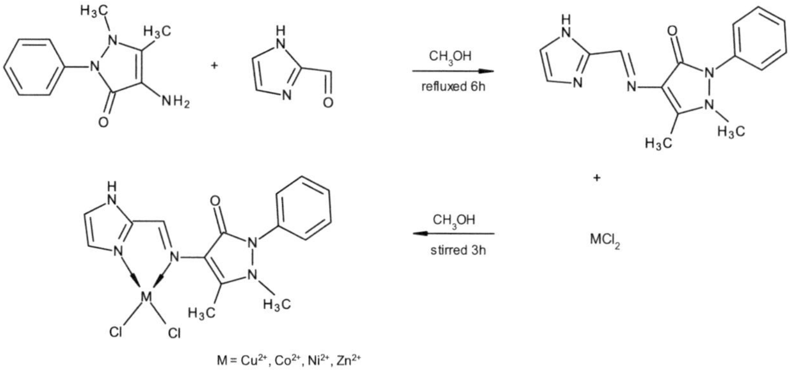

Figure 4. Scheme of synthesis of complexes with ligand **HL**^{5}.

2.1.2. *Synthesis of the complexes with ligands* $\boldsymbol{HL^{6-11}}$ *and various co-ligands*

The metal complexes with ligands base Schiff $\boldsymbol{HL^{6-11}}$ are obtained through three methods:

Method 1. Previously, the complex combination with the Schiff base is obtained to which the co-ligand is added(α-picoline, β-picoline, γ-picoline, n-propylamine). After the complete precipitation, the solid compound is obtained that is filtered, washed with ether, and dried in the exicator (**Figure 5**) [48].

Method 2. The mixture of reaction which contains the alcoholic ligand solutions (the Schiff base and the co-ligand) and the alcoholic solution of metal salt is refluxed for 6–8h. After the concentration of the solution to a third of its volume, on the water bath, a precipitate is obtained which is filtered, washed with alcohol, and dried in vacuo (**Figure 6**) [49]; (**Figure 7**) [50, 51].

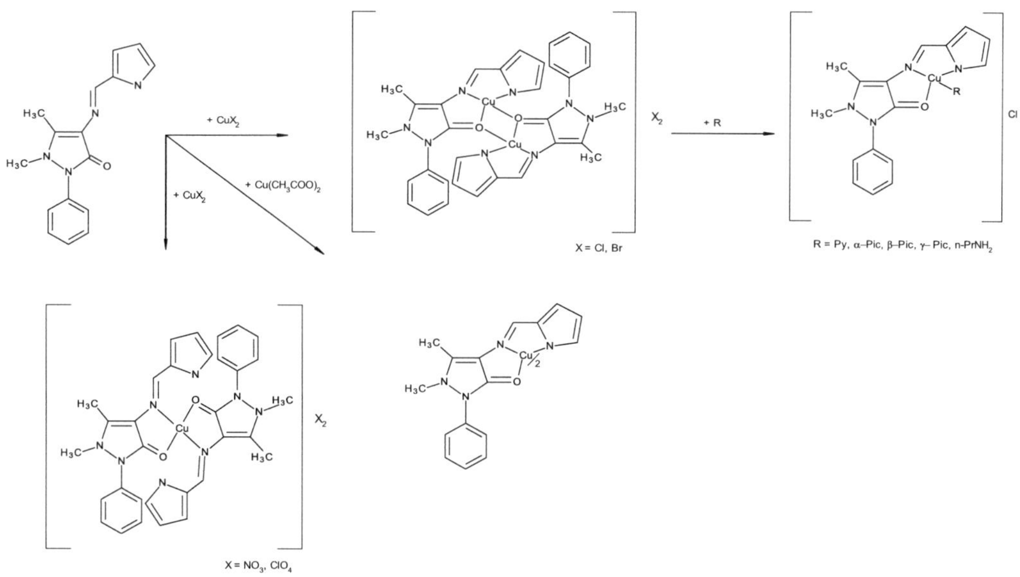

Figure 5. Scheme of synthesis of complexes with ligand **HL**^{6}.

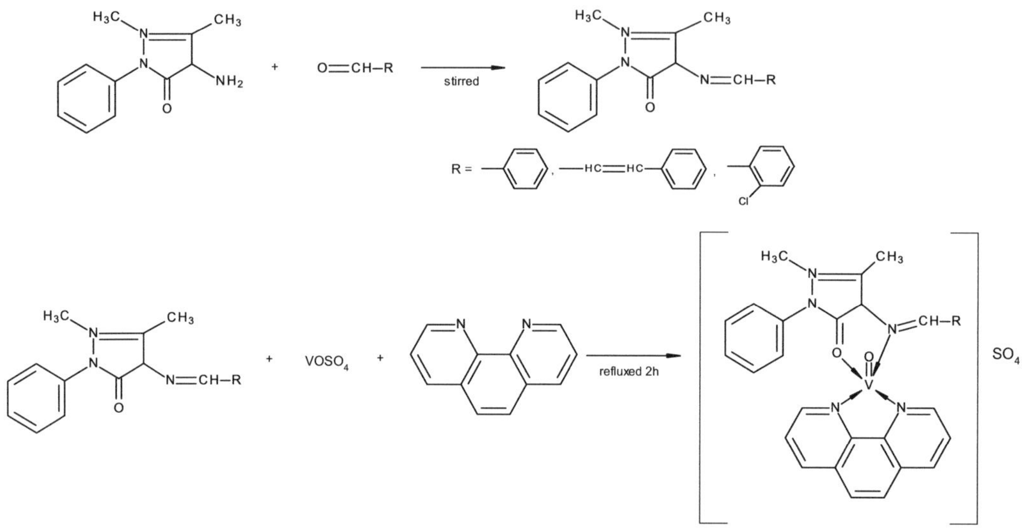

Figure 6. Scheme of synthesis of complexes with ligands **HL**[7–9].

Method 3. By adding an ethanolic solution of Schiff base to a metal salt solution, in a molar ratio 1:1, a mixture is obtained which is refluxed for 3–8h. An ethanol solution of co-ligand (amino acids; 1,10-phenanthroline; 2,2′- bipyridine; etc.) is added in the reaction environment, and the reflux is kept going on for another 1–3h. The precipitate is obtained which is filtered, washed in ether, and dried in vacuo (**Figure 7**) [52, 53]; (**Figure 8**) [54].

*2.1.3. Synthesis of the complexes with ligands **HL**[10, 12–14] and various aromatic amine*

The metal complexes with these Schiff bases are obtained through refluxing, lasting for 3–4h of a mixture that contains the metal salt dissolved in ethanol and the ligand dissolved in the same solvent. The ligand can be previously obtained through two different methods (**Figure 9**) [55], respectively (**Figure 10**) [53, 56–58].

*2.1.4. Synthesis of the complexes with ligands **HL**[15] and **HL**[16]*

The metal complexes with ligands Schiff bases **HL**[15] and **HL**[16] are obtained by treating a ligand solution with a solution of metal salt, in a molar ratio L:M=1:1. The mixture is refluxed for 5–6h. After the concentration of the solution to a third of its volume, on the water bath, a precipitate is obtained which is filtered, washed with ether, and dried in vacuo (**Figure 11**) [59, 60].

*2.1.5. Synthesis of the complexes with ligands **HL**[17–20]*

The metal complexes with Schiff base ligands **HL**[17–20] are obtained through treating a solution that contains the ligand dissolved in ethanol or acetonitrile with the solution of metal salt, in a molar ratio of L:M=1:1. The mixture is refluxed for 5–6h (**Figures 12, 13**) [61–64] or, in other cases, even 12h (**Figure 14**) [45]. The precipitation begins immediately or after the concentration of the solution to a third of its volume, on a water bath. The precipitate is obtained which is filtered, washed with ether, and dried in vacuo.

http://dx.doi.org/10.5772/67584

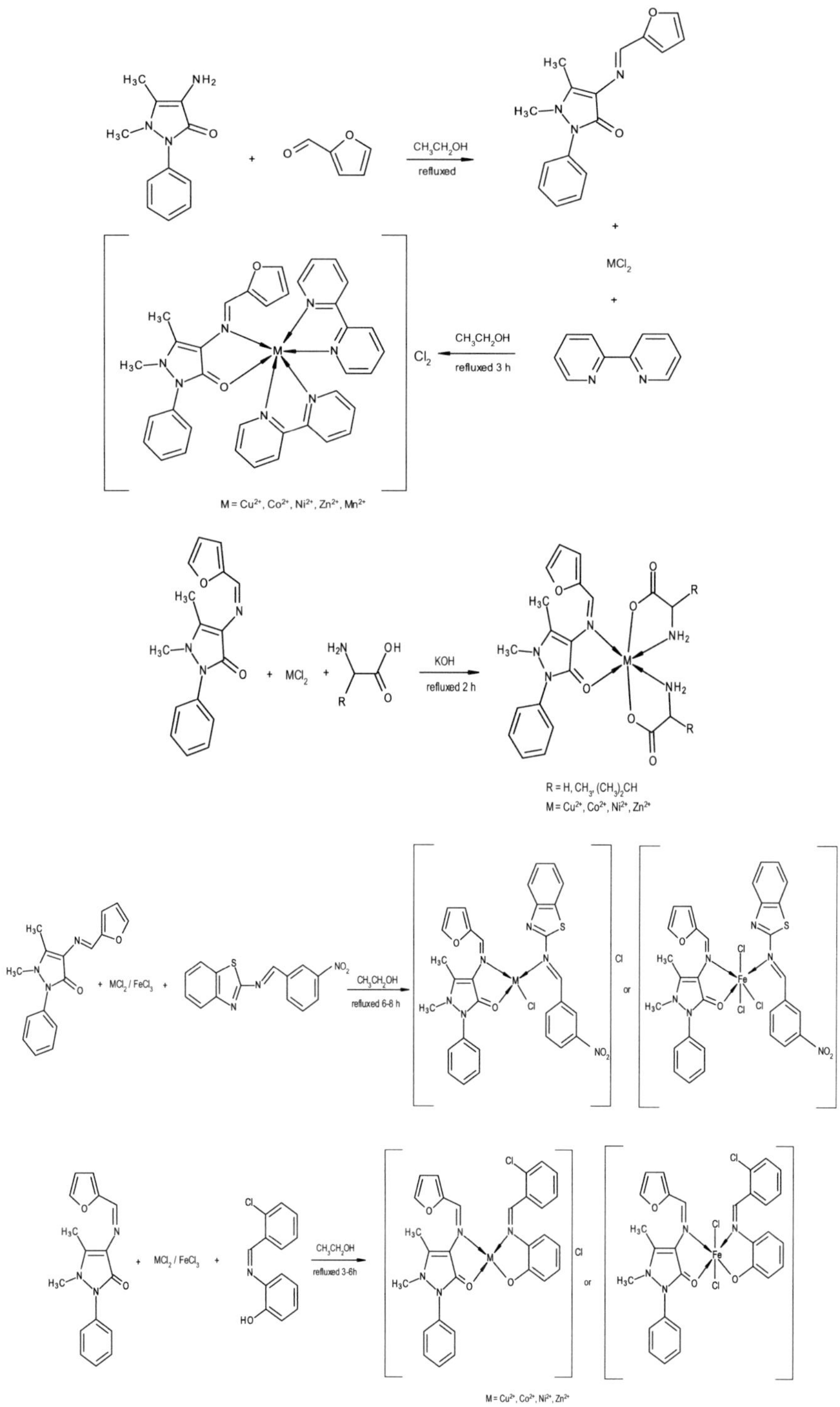

Figure 7. Scheme of synthesis of complexes with ligand **HL[10]**.

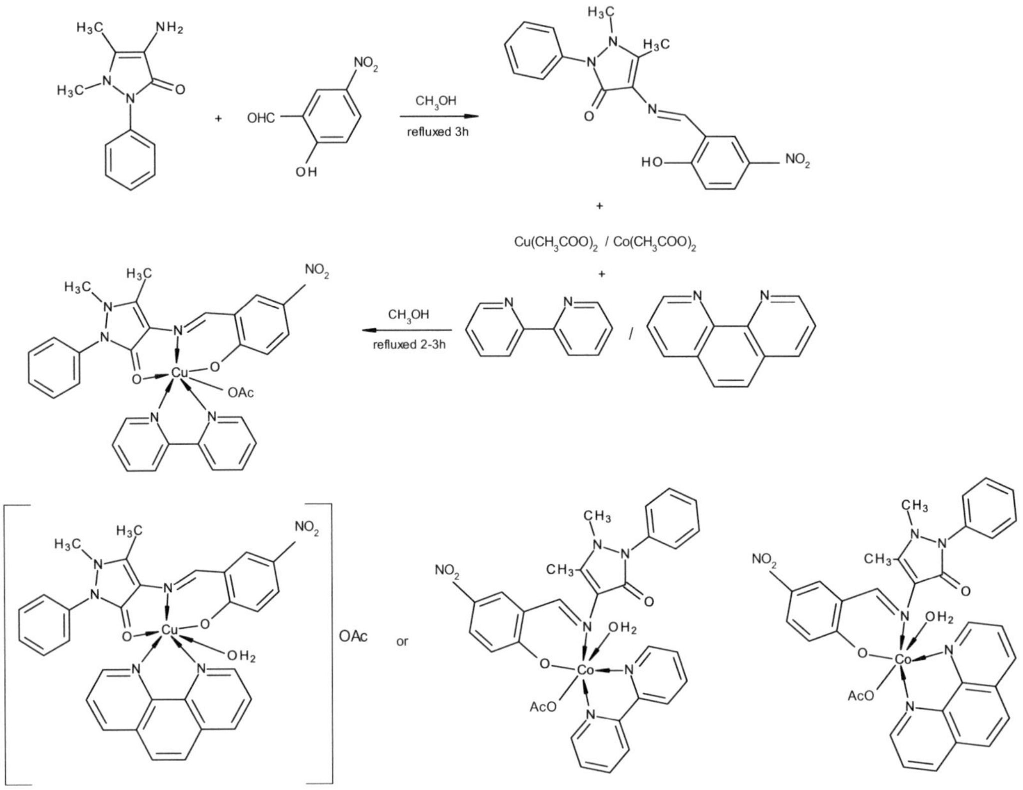

Figure 8. Scheme of synthesis of complexes with ligand **HL**[11].

The study methods used to describe the complexes were as follows: the basic chemical analysis, spectrometry IR, UV-VIS, EPR, the thermogravimetric analysis, the magnetic susceptibility, and the molar electric conductibility. The complexes synthetized were tested from the point of view of the antibacterian activity; the obtained results were presented in the respective papers.

2.2. New compounds: structure and antibacterial activity

2.2.1. Structure and synthesis of the compounds 1–28

With a view to obtaining new compounds with significant antibacterian activity, we have synthesized and characterized a series of complexes of Cu(II), Ni(II), Mn(II), and V(IV) with ligands Schiff bases, derived from 4-aminoantipyrine.

In this regard, we have synthesized four ligands, with chromophore groups ONO, respectively, ON, and different volumes of the aldehyde which is a part of Schiff base: 1-phenyl-2,3-dimethyl-4-(N-3-formyl-6-methyl-chromone)-3-pyrazolin-5-one (**HL**[21]); 1-phenyl-2,3-dimethyl-4-(N-2-hydroxy-4-methoxy-benzaldehyde)-3-pyrazolin-5-one (**HL**[22]); 1-phenyl-2,3-dimethyl-4-(1H-indole-3-carboxaldehyde)-3-pyrazolin-5-one (**HL**[23]); 1-phenyl-2,3-dimethyl-4-(N-pyridoxal hydrochloride)-3-pyrazolin-5-one (**HL**[24]) as well as their complex combinations with transitional metals: Cu^{2+}, Ni^{2+}, Mn^{2+}, VO^{2+} (28 metal complexes).

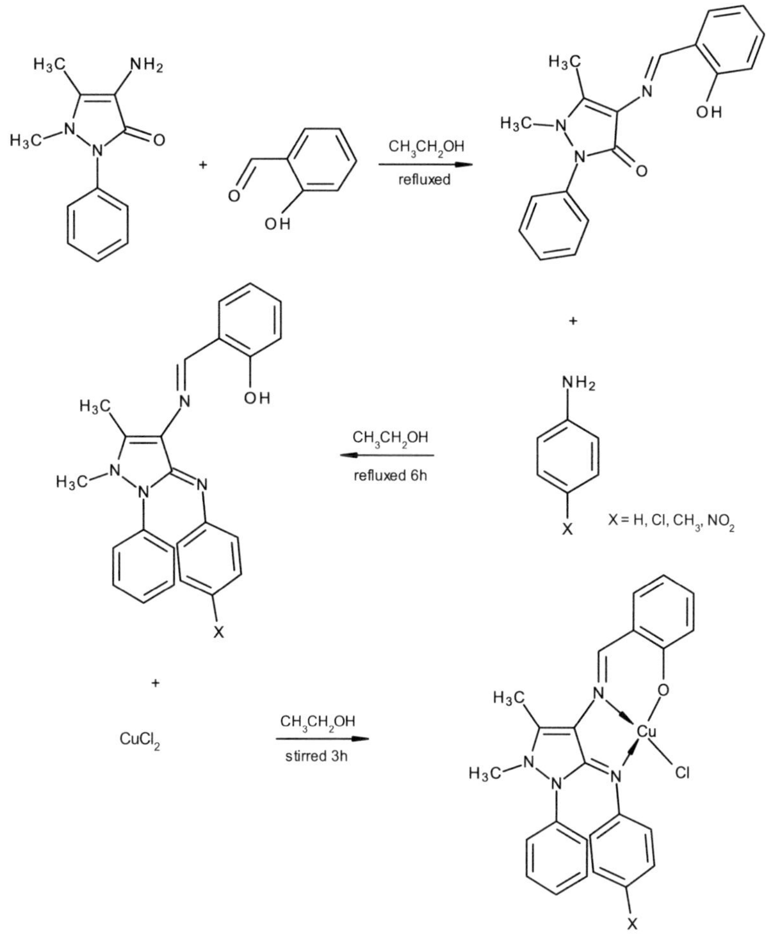

Figure 9. Scheme of synthesis of complexes with ligand **HL[12]**.

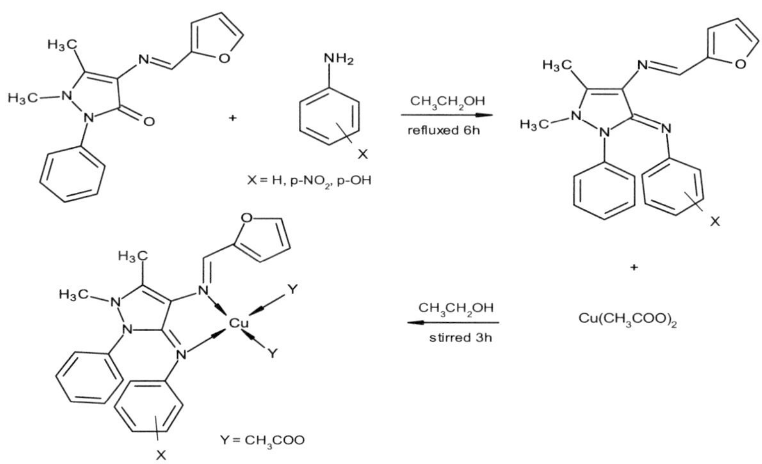

Figure 10. Scheme of synthesis of complexes with ligand **HL[10]**.

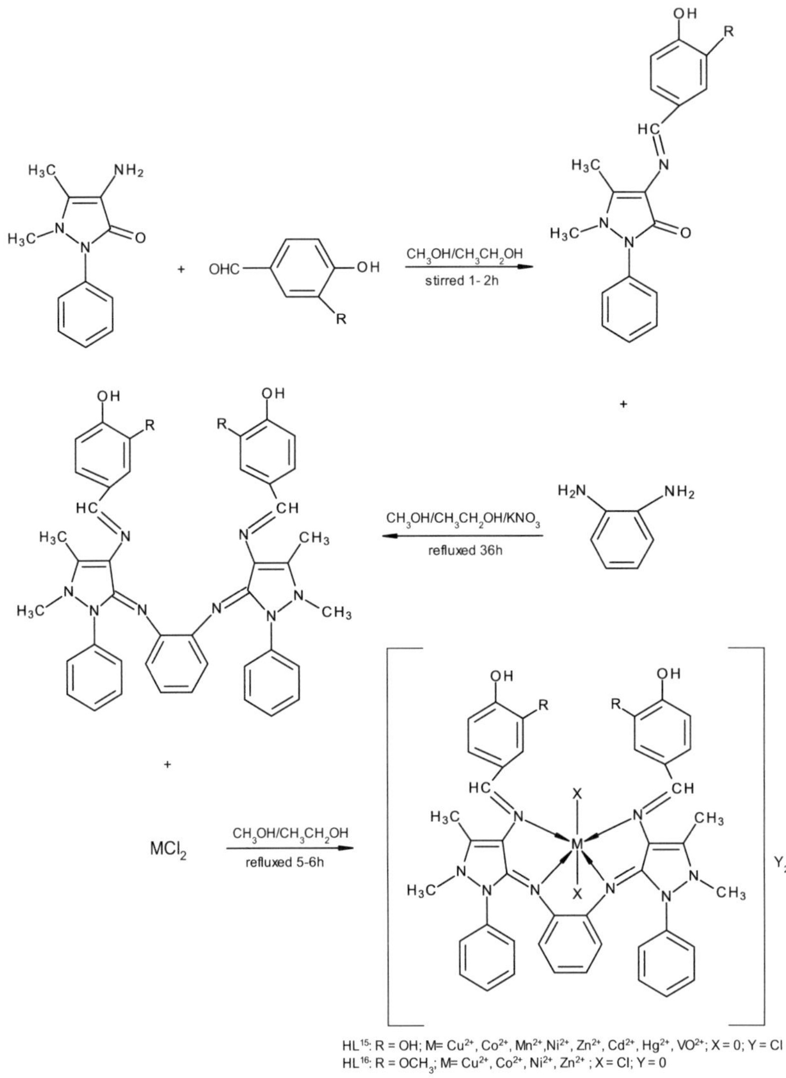

Figure 11. Scheme of synthesis of complexes with ligands **HL**[15] and **HL**[16].

The study methods used to characterize the metal complexes were as follows: elemental analysis, the thermogravimetric analysis, IR, UV-Vis, EPR spectroscopy, the molar electric conductibility, the magnetic susceptibility, and the X-ray diffraction.

*2.2.1.1. Synthesis of the complexes with ligand **HL**[21]*

Ligand $C_{24}H_{25}N_3O_4$, (HL[21])

Ethanolic solution of 3-formyl-6-methyl-chromone (1 mmol) and 4-amino-2,3-dimethyl-1-phenyl-3-pyrazolin-5-one (1 mmol) was stirred at room temperature, then refluxed for 2h, and kept at

http://dx.doi.org/10.5772/67584

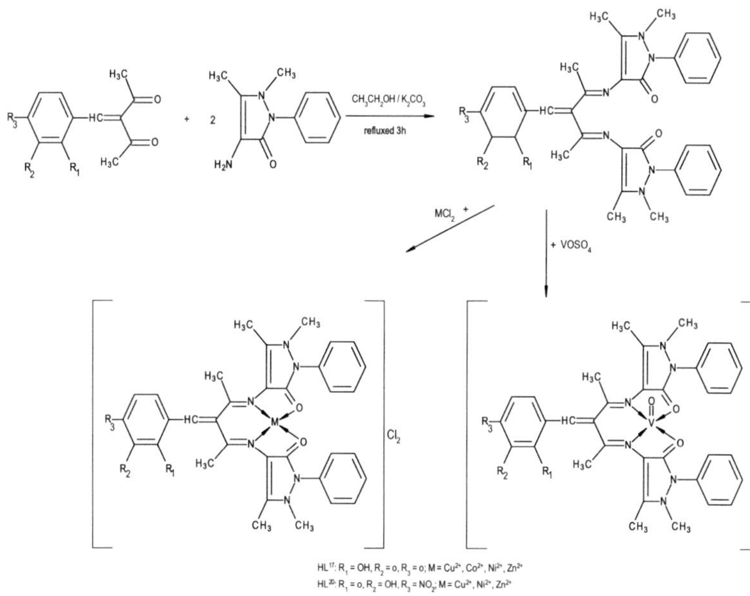

Figure 12. Scheme of synthesis of complexes with ligands **HL**[17] and **HL**[20].

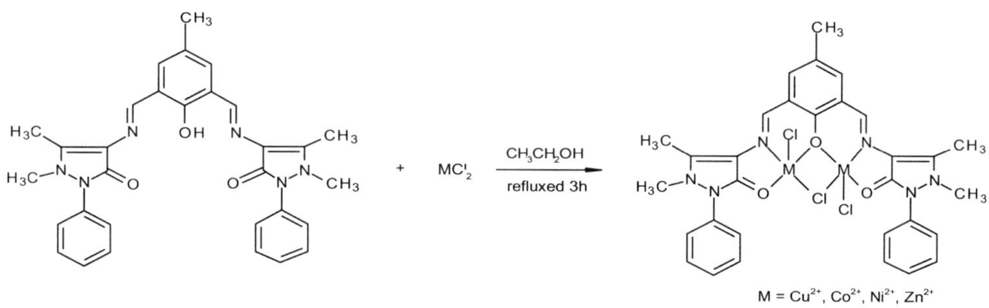

Figure 13. Scheme of synthesis of complexes with ligand **HL**[18].

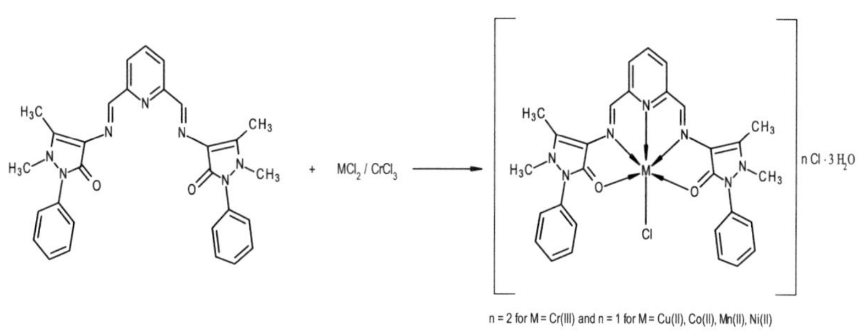

Figure 14. Scheme of synthesis of complexes with ligand **HL**[19].

4°C for 2 days. The resulting precipitate of intense yellow color was filtered, washed with methanol, and dried. Yellow single crystals suitable for structure determination were obtained from methanolic solution upon slow evaporation at room temperature [65].

Complexes **1–3** and **5–9** were prepared by direct reaction between the ligand and the corresponding metal salts, while complex **4** was prepared by the metathetical displacement of the acetate ion, in $Cu(OAc)_2{\cdot}H_2O$, by the thiocyanate ion [65] (**Figures 15, 16**).

$$[\mathbf{Cu_2L^{21}Cl_2}] \tag{1}$$

To $CuCl_2{\cdot}2H_2O$ (2 mmol) dissolved in aqueous/ethanol solution (1:2 v/v) was added ligand $\mathbf{HL^{21}}$ (2 mmol) dissolved in hot ethanol and refluxing for 2 h. The green-brown precipitate, which separated on cooling, was filtered, washed with hot water, ethanol followed by ether, and dried in vacuo.

$$[\mathbf{CuL^{21}(NO_3)}] \tag{2}$$

Complex **2** was prepared similarly, using $Cu(NO_3)_2{\cdot}3H_2O$ (2 mmol). Green solid.

$$[\mathbf{CuL^{21}(OAc)]CH_3OH} \tag{3}$$

Complex **3** was prepared similarly, using $Cu(OAc)_2{\cdot}H_2O$ (2 mmol). Brown solid, X-ray quality single crystals were obtained.

$$[\mathbf{CuL^{21}(SCN)}] \tag{4}$$

For the synthesis of complex **4**, the acetate complex was first prepared and the acetate ion was then displaced by thiocyanate ion by using KSCN (2 mmol). Dark-green solid.

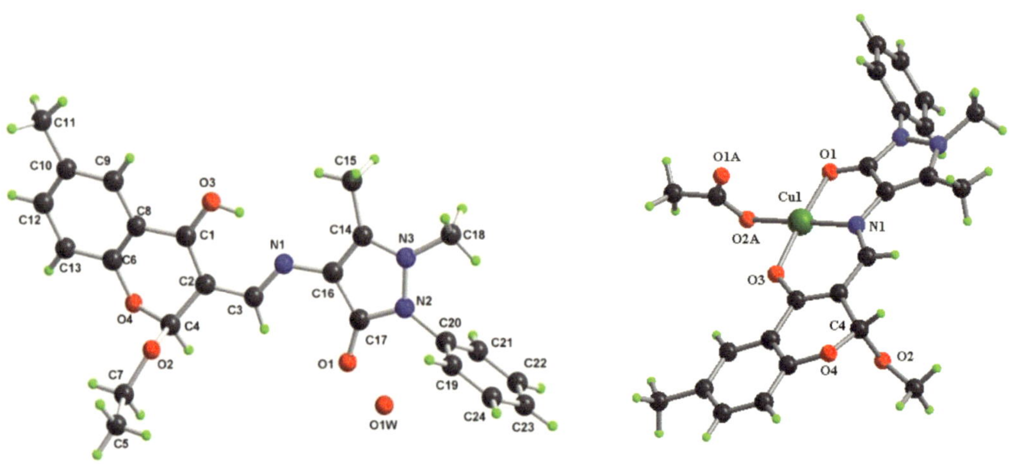

Figure 15. X-ray molecular structure of ligand $\mathbf{HL^{21}}$ and complex **3**.

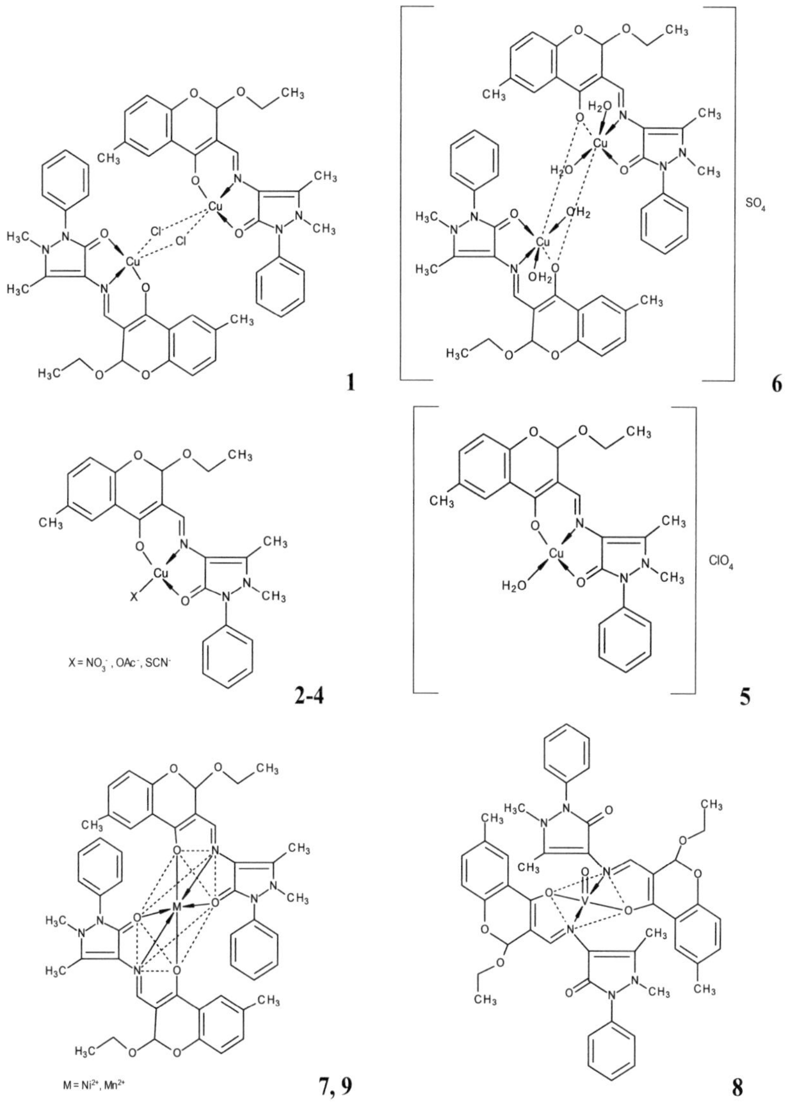

Figure 16. Proposed structures of the metal complexes **1–9**.

$$[\mathbf{CuL^{21}(H_2O)(ClO_4)}] \quad (5)$$

Complex **5** was prepared similarly, using $Cu(ClO_4)_2{\cdot}6H_2O$ (2 mmol). The mixture was stirred at room temperature for 1 h, when a dark-green precipitate appeared immediately.

$$[\mathbf{Cu_2(L^{21})_2(H_2O)_4]SO_4} \quad (6)$$

Complex **6** was prepared similarly, using $CuSO_4{\cdot}5H_2O$ (2 mmol). The mixture was stirred at reflux temperature for 4 h, when appeared a dark-green precipitate.

$$[\mathbf{Ni(L^{21})_2}] \quad (7)$$

Complex **7** was prepared similarly, using $NiCl_2{\cdot}6H_2O$ (2 mmol). Green to yellow solid.

$$[\mathbf{VO(L^{21})_2}] \quad (8)$$

Complex **8** was prepared similarly, using $VOSO_4{\cdot}2H_2O$ (2 mmol). Brown solid.

$$[\mathbf{Mn(L^{21})_2}] \quad (9)$$

Complex **9** was prepared similarly, using $Mn(ClO_4)_2{\cdot}6H_2O$ (2 mmol). Orange solid.

2.2.1.2. Synthesis of the complexes with ligand ***HL***^{22}

Ligand $C_{19}H_{19}N_3O_3$, (HL^{22})

The ligand **HL**^{22} was synthesized by refluxing equimolar amounts of 4-amino-2,3-dimethyl-1-phenyl-3-pyrazolin-5-one and 2-hydroxy-4-methoxy-benzaldehyde in ethanol according to the experimental protocol described in Ref. [66].

Complexes **10–12**, **14,** and **15** were prepared by the direct reaction between the ligand and the corresponding metal salts. Complex **13** was obtained by refluxing a mixture of $CuCl_2{\cdot}2H_2O$ and 1-phenyl-2,3-dimethyl-4-(N-2-hydroxy-4-methoxy-benzaldehyde)-3-pyrazolin-5-one with the addition of KSCN (**Figures 17, 18**) [66].

$$[\mathbf{CuL^{22}Cl(H_2O)}] \quad (10)$$

An ethanol solution of $CuCl_2{\cdot}2H_2O$ (2 mmol, 15 mL aqueous/ethanol 1:2 v/v) was added dropwise to a stirred ethanol solution of the Schiff base ligand **HL**^{22} (2 mmol, 15 mL). The resulting solution was stirring for 3 h at room temperature. The green-brown colored solid was filtered, washed with hot water, ethanol followed by ether, and dried in vacuo.

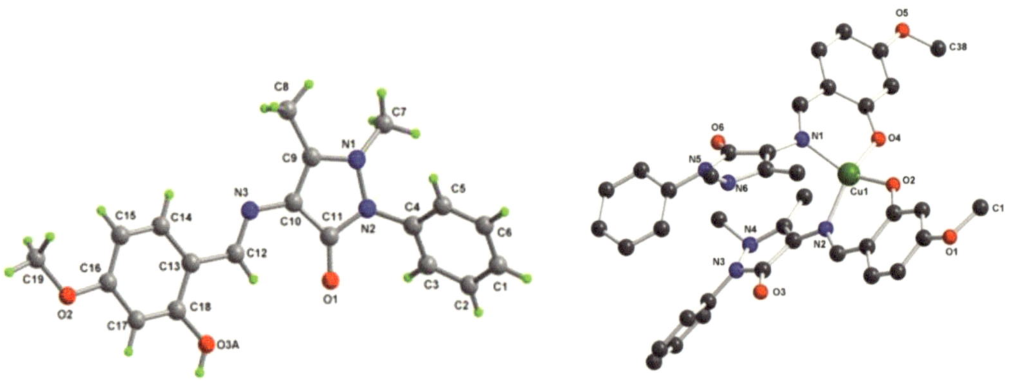

Figure 17. X-ray molecular structure of ligand **HL**^{22} and complex **12**.

http://dx.doi.org/10.5772/67584

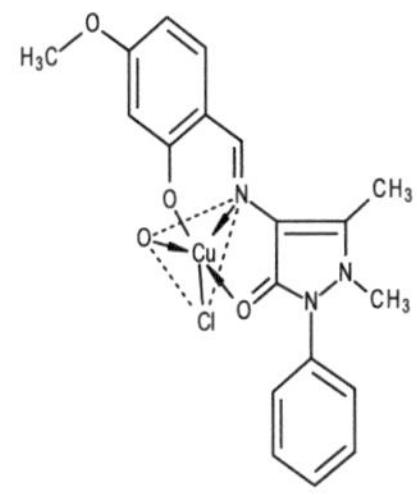

10

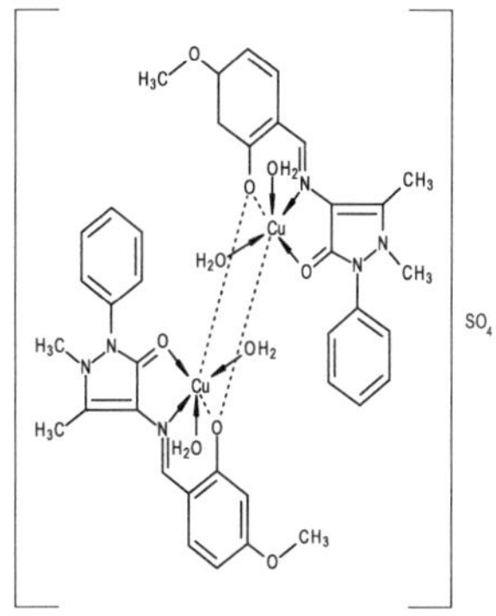

15

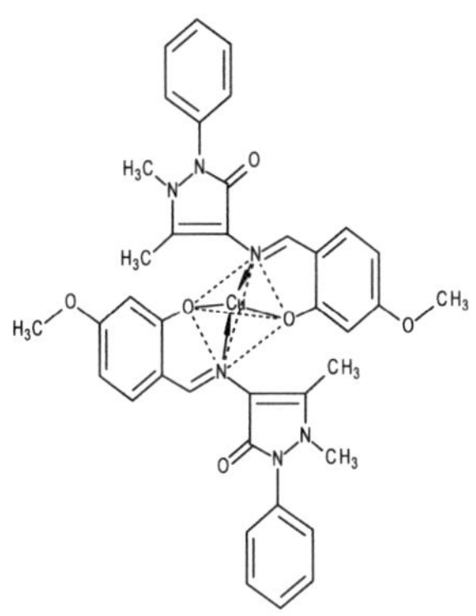

12

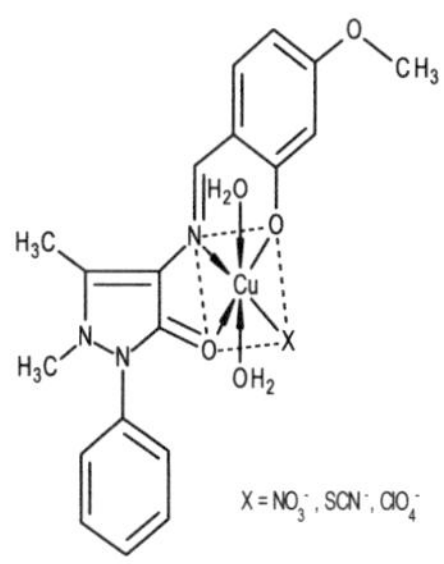

11, 13, 14

Figure 18. Proposed structures of the metal complexes **10–15**.

$$[\mathbf{CuL^{22}(NO_3)(H_2O)_2}] \tag{11}$$

Complex **11** was prepared similarly, using $Cu(NO_3)_2{\cdot}3H_2O$ (2 mmol). Dark-green solid.

$$[\mathbf{Cu(L^{22})_2}] \tag{12}$$

Complex **12** was prepared similarly, using $Cu(OAc)_2{\cdot}H_2O$ (2 mmol). Brown solid, X-ray quality single crystals were obtained.

$$[\mathbf{CuL^{22}(SCN)(H_2O)_2}] \tag{13}$$

For the synthesis of complex **13**, the chloride complex was first prepared and chloride ion was then displaced by thiocyanate ion by using KSCN (2 mmol). The green colored solid, which separated on cooling, were filtered, washed with hot water, ethanol followed by ether and dried in vacuo.

$$[\mathbf{CuL^{22}(ClO_4)(H_2O)_2}] \tag{14}$$

Complex **14** was prepared similarly, using $Cu(ClO_4)_2{\cdot}6H_2O$ (2 mmol). Green solid.

$$[Cu_2(L^{22})_2(H_2O)_4]\ SO_4 \quad (15)$$

Complex **15** was prepared similarly, using $CuSO_4 \cdot 5H_2O$ (2 mmol). Dark-green solid.

2.2.1.3. Synthesis of the complexes with ligand ***HL***^{23}

Ligand $C_{20}H_{18}N_4O$, (HL^{23})

The ligand **HL**^{23} was synthesized by refluxing equimolar amounts of 4-amino-2,3-dimethyl-1-phenyl-3-pyrazolin-5-one and indole-3-carboxaldehyde in methanol according to the experimental protocol described in Ref. [67].

Complexes **16–21** were prepared by direct reaction between the ligand and the corresponding metal salts (**Figures 19, 20**) [67].

$$[Cu(L^{23})_2Cl_2] \quad (16)$$

To a stirred solution of $CuCl_2 \cdot 2H_2O$ (1 mmol) in ethanol (15 mL) was added a solution of ligand **HL**^{23} (1 mmol) in ethanol (15 mL). The mixture was stirred at reflux temperature for 2h. The resulting precipitate of green-brown color was filtered, washed with ethanol, and dried.

$$[Cu(L^{23})_2]\ (NO_3)_2 \quad (17)$$

Complex **17** was prepared similarly, using $Cu(NO_3)_2 \cdot 3H_2O$ (1 mmol). Brown solid.

$$[Cu(L^{23})_2(OAc)_2] \quad (18)$$

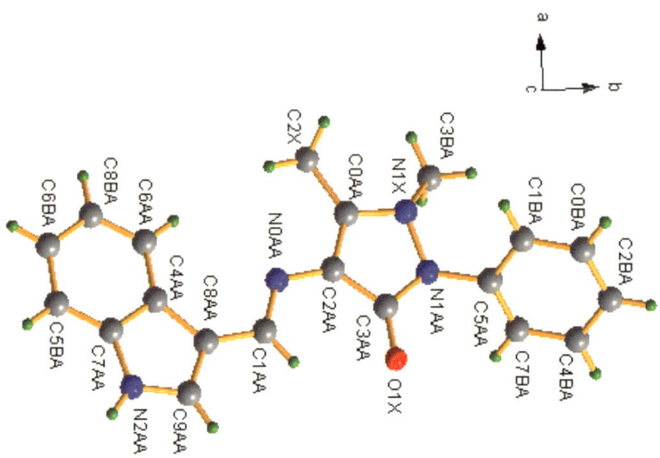

Figure 19. X-ray molecular structure of ligand **HL**^{23}.

http://dx.doi.org/10.5772/67584

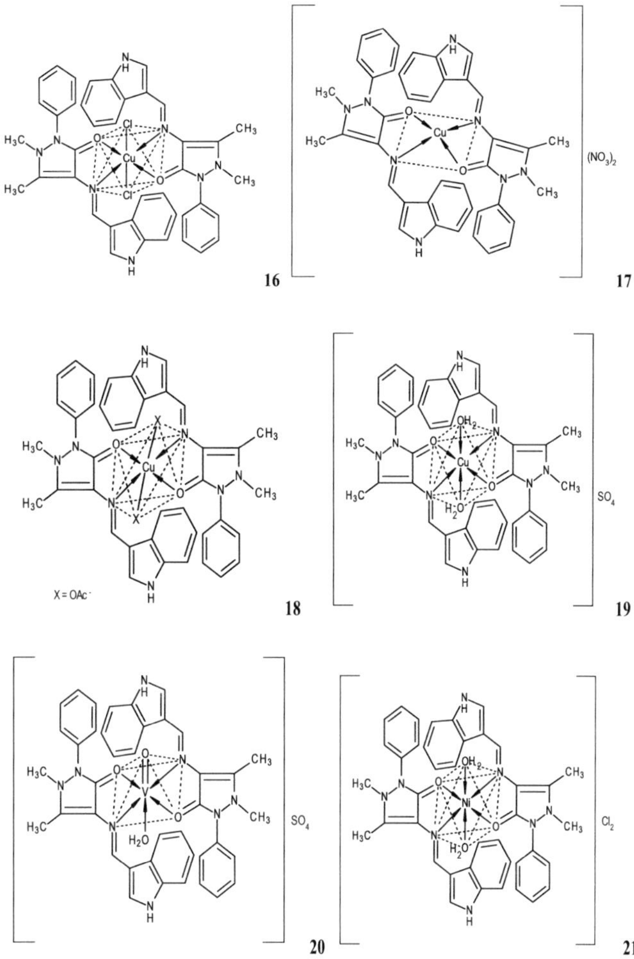

Figure 20. Proposed structures of the metal complexes **16–21**.

Complex **18** was prepared similarly, using $Cu(OAc)_2{\cdot}H_2O$ (1 mmol). Dark-green solid.

$$[\mathbf{Cu}(\mathbf{L}^{23})_2(\mathbf{H_2O})_2]\ \mathbf{SO_4} \tag{19}$$

Complex **15** was prepared similarly, using $CuSO_4{\cdot}5H_2O$ (1 mmol). Green solid.

$$[\mathbf{VO}(\mathbf{L}^{23})_2(\mathbf{H_2O})]\ \mathbf{SO_4} \tag{20}$$

Complex **20** was prepared in a similarly, using $VOSO_4{\cdot}2H_2O$. Green solid.

$$[\mathbf{Ni}(\mathbf{L}^{23})_2(\mathbf{H_2O})_2]\mathbf{Cl_2} \tag{21}$$

Complex **21** was prepared in a similarly, using $NiCl_2{\cdot}6H_2O$. Dark-green solid.

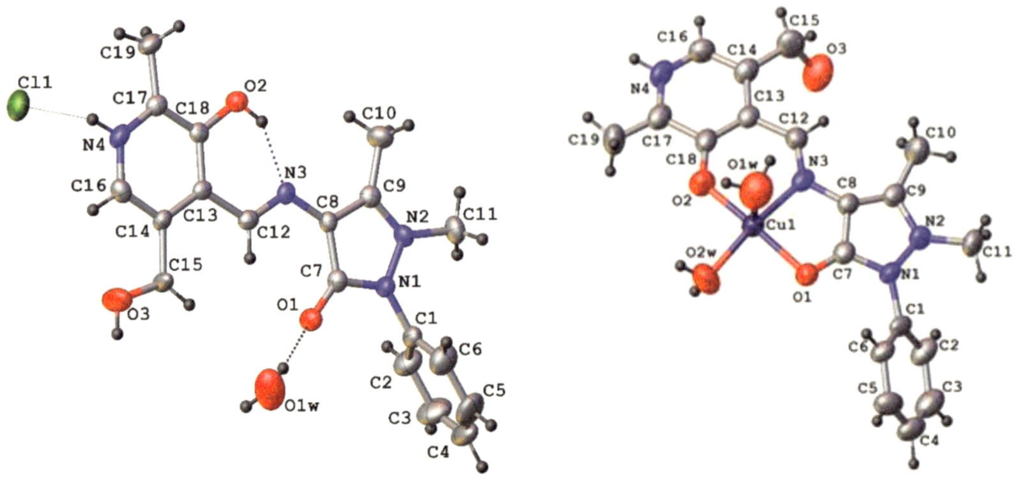

Figure 21. X-ray molecular structure of ligand **HL**[24] and complex **23**.

2.2.1.4. Synthesis of the complexes with ligand ***HL***[24]

Ligand $C_{19}H_{20}N_4O_3$, (HL^{24})

The ligand **HL**[24] was synthesized by refluxing equimolar amounts of 4-amino-2,3-dimethyl-1-phenyl-3-pyrazolin-5-one and pyridoxal hydrochloride in methanol according to the experimental protocol described in Ref. [68].

Complexes **22–28** were prepared by direct reaction between the ligand and the corresponding metal salts (**Figures 21**, **22**) [68].

$$[CuL^{24}Cl]_2 \tag{22}$$

To a hot solution of **HL**[24] (1 mmol) in methanol was added a hot solution of $CuCl_2 \cdot 2H_2O$ (2 mmol) in aqueous/methanol (1:2 v/v), and the mixture was stirred at reflux temperature for 4 h. Brown solid.

$$[CuL^{24}(H_2O)_2]\ NO_3 \cdot 2.25H_2O \tag{23}$$

Complex **23** was prepared similarly, using $Cu(NO_3)_2 \cdot 3H_2O$ (1 mmol). The mixture was stirred at room temperature for 7 h. Brown solid, X-ray quality single crystals were obtained.

$$[Cu(L^{24})(OAc)(H_2O)] \tag{24}$$

Complex **24** was prepared similarly, using $Cu(OAc)_2 \cdot H_2O$ (2 mmol). The mixture was stirred at reflux temperature for 4 h. Green-brown solid.

http://dx.doi.org/10.5772/67584

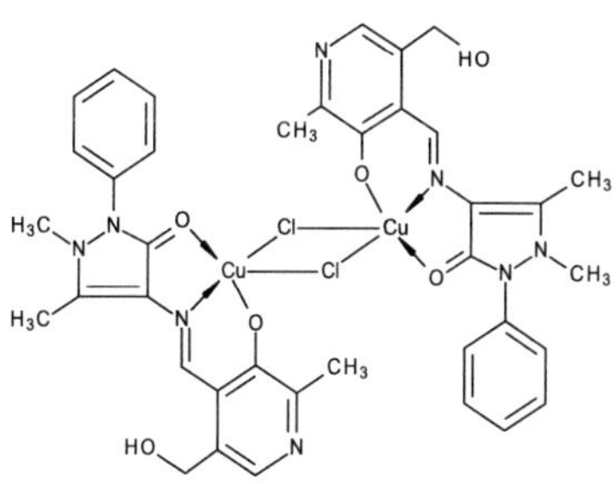

22

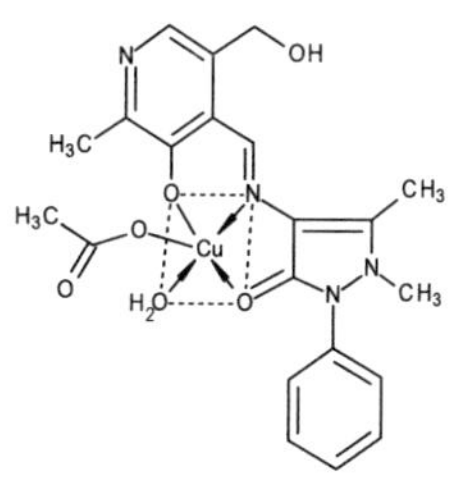

24

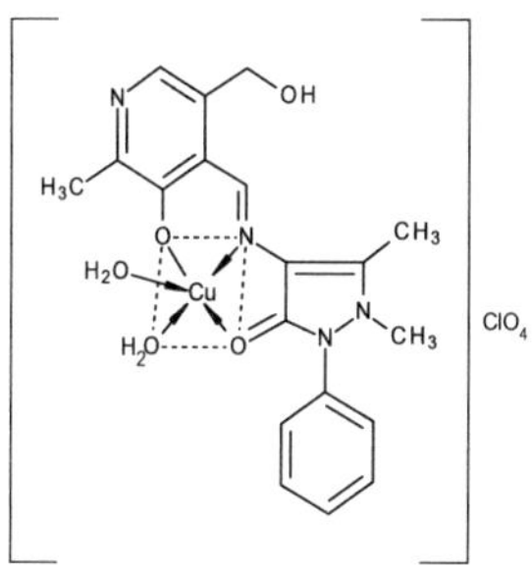

25

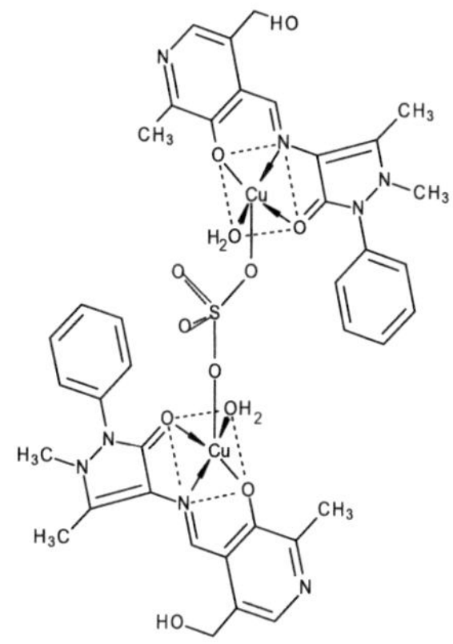

26

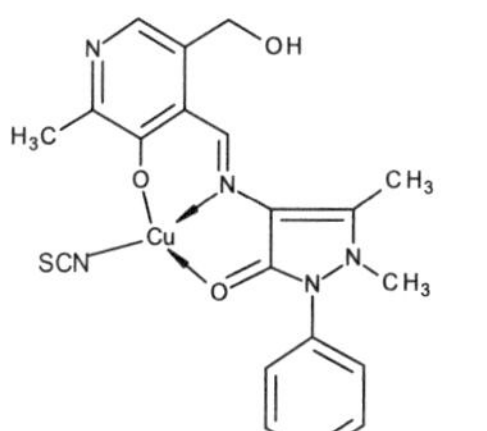

27

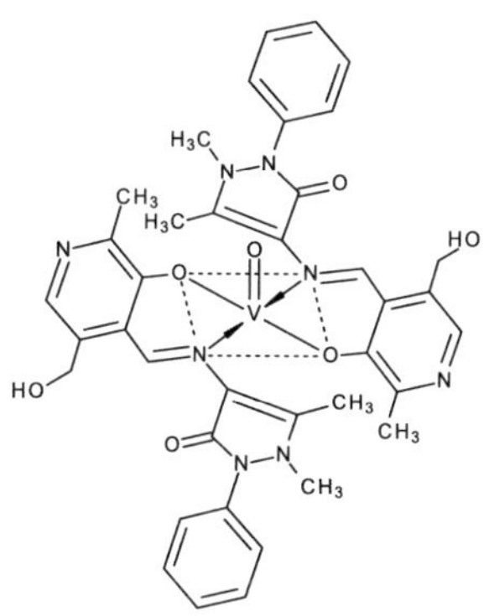

28

Figure 22. Proposed structures of the metal complexes **22**, **24–28**.

$$[\mathbf{CuL^{24}(H_2O)_2}]\ \mathbf{ClO_4} \tag{25}$$

Complex **25** was prepared similarly, using $Cu(ClO_4)_2{\cdot}6H_2O$ (2mmol). The mixture was stirred at reflux temperature for 5h. Brown solid.

$$[\mathbf{Cu_2(L^{24})_2(SO_4)(H_2O)_2}] \tag{26}$$

Complex **26** was prepared in a similar fashion to complex **24**, using $CuSO_4{\cdot}5H_2O$. The mixture was stirred at reflux temperature for 4h, giving a dark-red precipitate.

$$[\mathbf{CuL^{24}(NCS)}] \cdot \mathbf{2H_2O} \tag{27}$$

For the synthesis of complex **27**, the acetate complex was first prepared and the acetate ion was then displaced by thiocyanate ion by using KSCN (2mmol). Green solid.

$$[\mathbf{VO(L^{24})_2}] \tag{28}$$

Complex **28** was prepared in a similar fashion to complex **24**, using $VOSO_4{\cdot}2H_2O$. Brown solid.

2.3. Antibacterial activity

The complexes and ligands $\mathbf{HL^{21-24}}$ were tested for their in vitro antibacterial activity against de *Staphylococcus aureus var. Oxford 6538, Klebsiella pneumoniae ATCC* 100131, *Escherichia coli ATCC 10536,* and *Pseudomonas aeruginosa ATCC 9027* strains using the paper disc diffusion method (for the qualitative determination) and the serial dilutions in liquid broth method (for determination of MIC) [66]. Streptomycin was used as internal standard.

The results of the antibacterial activity point out the fact that the activity of the Schiff bases $\mathbf{HL^{21-24}}$ is more pronounced when it coordinates at the metal ion (**Table 1**). In case of complexes **1, 6, 10, 12, 15, 18, 19, 20, 23, 25, 26,** and **28,** there can be seen a visible increase in the antibacterial action.

Missing a clear action mechanism, in vitro, of the respective ligand of the complexes obtained on a microbian stem, there can be made the following stipulations:

-the structure of the tested complexes seems to be the main element that influences the antibacterial activity. Thus, for complexes **1, 6, 10, 12, 15, 19, 20, 23, 25, 26,** and **28,** there has been determined an increased activity against all bacterial species, probably due to the presence of the monomeric form in DMSO solution and also due to the tetracoordination of the metal center.

-the presence of the anions with a large volume, in the outer coordination sphere of the complexes, can be deemed as another main element that can influence the antibacterial activity. The complexes **5, 6, 15, 19, 20,** and **25** contain the groups ClO_4^- and SO_4^{2-}, respectively, and prevent a visible increase in their activity against all species of bacteria used.

-if there is drawn a comparison between the coefficients of the molecular orbitals, computed on the basis of the transitions noticed in spectra UV-Vis and of the parameters g and A in spectra EPR [69] and the antibacterial activity, the conclusion is the fact that, for the complexes with the most pronounced activity, the values of δ^2 parameter to a weak covalent character of the link π out of the plan.

http://dx.doi.org/10.5772/67584

	Minimum inhibitory concentration (μg/ml)			
Compusul	*K. pneumoniae G(+)*	*S. aureus* *G(+)*	*P. aeruginosa* *G(−)*	*E. coli* *G(−)*
HL^{21}	256	256	512	128
$[Cu_2(L^{21})_2(Cl)_2]$ **(1)**	32	8	32	32
$[CuL^{21}(NO_3)]$ **(2)**	64	32	256	512
$[CuL^{21}(OAc)]$ **(3)**	64	32	256	512
$[CuL^{21}(SCN)]$ **(4)**	64	32	256	512
$[CuL^{21}(H_2O)]ClO_4$ **(5)**	32	32	128	256
$[Cu_2(L^{21})_2(H_2O)_4]SO_4$ **(6)**	16	8	32	32
$[Ni(L^{21})_2]$ **(7)**	512	512	256	512
$[VO(L^{21})_2]$ **(8)**	265	256	64	256
$[Mn(L^{21})_2]$ **(9)**	256	512	256	128
HL^{22}	128	256	256	512
$[Cu(L^{22})Cl(H_2O)]$ **(10)**	16	8	128	16
$[Cu(L^{22})(NO_3)(H_2O)_2]$ **(11)**	128	128	256	256
$[Cu(L^{22})_2]$ **(12)**	8	4	16	64
$[CuL^{22}(SCN)(H_2O)_2]$ **(13)**	64	64	256	512
$[CuL^{22}(ClO_4)(H_2O)_2]$ **(14)**	256	128	256	512
$[Cu_2(L^{22})_2(H_2O)_4]SO_4$ **(15)**	8	16	128	16
HL^{23}	64	128	64	32
$[Cu(L^{23})_2Cl_2]$ **(16)**	64	128	64	64
$[Cu(L^{23})_2](NO_3)_2$ **(17)**	128	64	32	64
$[Cu(L^{23})_2(OAc)_2]$ **(18)**	8	4	16	16
$[Cu(L^{23})_2(H_2O)_2]SO_4$ **(19)**	8	8	64	16
$[VO(L^{23})_2(H_2O)]SO_4$ **(20)**	64	128	16	256
$[Ni(L^{23})_2(H_2O)_2]Cl_2$ **(21)**	256	256	256	128
HL^{24}	128	256	256	512
$[CuL^{24}Cl]_2$ **(22)**	64	64	256	512
$[CuL^{24}(H_2O)_2]NO_3{\cdot}2.25H_2O$ **(23)**	16	8	128	16
$[Cu(L^{24})(OAc)(H_2O)]$ **(24)**	128	128	256	256
$[CuL^{24}(H_2O)_2]\ ClO_4$ **(25)**	8	4	16	64
$[Cu_2(L^{24})_2(SO_4)(H_2O)_2]$ **(26)**	8	64	16	64
$[CuL^{24}(NCS)]{\cdot}2H_2O$ **(27)**	256	128	256	512
$[VO(L^{24})_2]$ **(28)**	8	16	128	16
Streptomycin	8	4	16	8

K. pneumoniae (*Klebsiella pneumoniae* ATCC 31488); *S. aureus* (*Staphylococus aureus* var. Oxford ATCC 6538); *P. aeruginosa* (*Pseudomonas aeruginosa* ATCC 9027); *E. coli* (*Escherichia coli* ATCC 10536). G(−): Gram-negative bacteria; G(+): Gram-positive bacteria.

Table 1. "*In vitro*" antibacterial activity of the ligands and corresponding complexes.

3. Conclusion

The investigations of the antibacterial screening, carried out for these new classes of compounds, reveal the fact that they present activity, especially toward the gram-positive bacteria, in comparison with the standard streptomycin. The increased antibacterial activity of the metal complexes can be accounted for by a cluster of reasons that refer to the chelation theory, nature of the ligand and of the metal ion, the geometry of the metal complexes, liposolubility, the presence of the co-ligands, and a series of sterical and pharmacokinetic factors. We can say that, ten of the complexes synthetized can be successfully used instead of streptomycin, where there is resistance to this antibiotic.

Author details

Elena Mihaela Pahontu

Address all correspondence to: elenaandmihaela@yahoo.com

Inorganic Chemistry Department, Faculty of Pharmacy, University of Medicine and Pharmacy "Carol Davila", Bucharest, Romania

References

[1] Baquero F., Gram-positive resistance: challenge for the development of new antibiotics. Journal of Antimicrobial Chemotherapy. 1997; **39**: 1–6.

[2] Alekshun M.N., Levy S.B., Molecular mechanisms of antibacterial Multidrug resistance. Cell. 2007; **128**: 1037–1050.

[3] Rice L.B., Unmet medical needs in antibacterial therapy. Biochemical Pharmacology. 2006; **71**: 991–995.

[4] Dhar D.N., Taploo C.L., Schiff bases and their applications. Journal of Scientific and Industrial Research. 1982; **41**: 501–506.

[5] Przybylski P., Huczynski A., Pyta K., Brzezinski B., Bartl F., Biological properties of Schiff bases and azo derivatives of phenols. Current Organic Chemistry. 2009; **13**: 124–148.

[6] Wang P.H., Keck J.G., Lien E.J., Lai M.M.C., Design, synthesis, testing, and quantitative structure-activity relationship analysis of substituted salicylaldehyde Schiff bases of 1-amino-3-hydroxyguanidine tosylate as new antiviral agents against coronavirus. Journal of Medicinal Chemistry. 1990; **33**: 608–614.

[7] Jarrahpour A., Motamedifar M., Pakshir K., Hadi N., Zarei M., Synthesis of novel azo Schiff bases and their antibacterial and antifungal activities. Molecules. 2004; **9**: 815–824.

[8] Ceyhana G., Urus S., Demirtas I., Elmastas M., Tumera M., Antioxidant, electrochemical, thermal, antimicrobial and alkane oxidation properties of tridentate Schiff base ligands and their metal complexes. Spectrochimica Acta Part A: Molecular and Biomolecular Spectroscopy. 2011; **81**:184–198.

[9] Xiong Y.Z., Chen F.E., Balzarini J., Clercq E.D., Pannecouque C., Non-nucleoside HIV-1 reverse transcriptase inhibitors. Part 11: Structural modulations of diaryltriazines with potent anti-HIV activity. European Journal of Medicinal Chemistry. 2008; **43**: 1230–1236.

[10] Sriram D., Yogeeswari P., Sirisha N., Saraswat V., Abacavir prodrugs: Microwave-assisted synthesis and their evaluation of anti-HIV activities. Bioorganic & Medicinal Chemistry Letters. 2006; **16**: 2127–2129.

[11] Bhandari S.V., Bothara K.G., Raut M.K., Patil A.A., Sarkate A.P., Mokale V.J., Design, synthesis and evaluation of antiinflammatory, analgesic and ulcerogenicity studies of novel s-substituted phenacyl-1,3,4-oxadiazole-2-thiol and Schiff bases of diclofenac acid as nonulcerogenic derivatives. Bioorganic & Medicinal Chemistry. 2008; **16**: 1822–1831.

[12] Sridhar K., Pandeya N., Stables P., Ramesh A., Anticonvulsant activity of hydrazones, Schiff and Mannich bases of isatin derivatives. European Journal of Pharmaceutical Sciences. 2002; **16**:129–132.

[13] Kaplan J.P., Raizon B.M., Desarmenien M., Feltz P., Headley P.M., Worms P., Lloyd K.G., Bartholini G., New anticonvulsants: Schiff bases of.gamma.-aminobutyric acid and. gamma.-aminobutyramide. Journal of Medicinal Chemistry. 1980; **23**: 702–704.

[14] Das A., Trousdale M.D., Ren S., Lien E.J., Inhibition of herpes simplex virus type 1 and adenovirus type 5 by heterocyclic Schiff bases of aminohydroxyguanidine tosylate. Antiviral Research. 1999; **44**: 201–208.

[15] Mladenova R., Ignatova M., Manolova N., Petrova T., Rashkov I., Preparation, characterization and biological activity of Schiff base compounds derived from 8-hydroxyquinoline-2-carboxaldehyde and Jeffamines ED. European Polymer Journal. 2002; **38**: 989–1000.

[16] Walsh O.M., Meegan M.J., Prendergast R.M., Nakib T.A., Synthesis of 3-acetoxyazetidin-2-ones and 3-hydroxyazetidin-2-ones with antifungal and antibacterial activity. European Journal of Medicinal Chemistry. 1996; **31**: 989–1000.

[17] Liu Y., Yang Z., Synthesis, crystal structure, antioxidation and DNA binding properties of binuclear Ho(III) complexes of Schiff-base ligands derived from 8-hydroxyquinoline-2 carboxy-aldehyde and four aroyl-hydrazines. Journal of Organometallic Chemistry. 2009; **694**: 3091–3101.

[18] Bringmann G., Dreyer M., Faber J.H., Dalsgaard P.W., Staerk D., Jaroszewski J.W., Ndangalasi F., Mbago R., Brun S., Brøgger C., Ancistrotanzanine C and related 5,1′- and 7,3′-coupled naphthyliso-quinoline alkaloids from *Ancistrocladus tanzaniensis*. Journal of Natural Products. 2004; **67**:743–748.

[19] Rehman W., Baloch M.K., Muhammad B., Badshah A., Khan K.M., Characteristic spectral studies and in vitro antifungal activity of some Schiff bases and their organotin(IV) complexes. Chinese Science Bulletin. 2004; **49**:119–22.

[20] Guo Z., Xing R., Liu S., Zhong Z., Ji X., Wang L., Pengcheng L., Antifungal properties of Schiff bases of chitosan, N-substituted chitosan and quaternized chitosan. Carbohydrate Research. 2007; **342**: 1329–1332.

[21] Souza A.O., Galetti F.C.S., Silva C.L., Bicalho B., Parma M.M., Fonseca S.F., Marsaioli A.J., Trindade A.C.L.B., Gil R.P.F., Bezerra F.S., Andrade-Neto M., Oliveira M.C.F., Antimycobacterial and cytotoxicity activity of synthetic and natural compounds. Química Nova. 2007; **30**: 1563–1566.

[22] Shi L., Ge H.M., Tan S.H., Li H.Q., Song Y.C., Zhu H.L., Tan R.X., Synthesis and antimicrobial activities of Schiff bases derived from 5-chloro-salicylaldehyde. European Journal of Medicinal Chemistry. 2007; **42**: 558–564.

[23] Pandeya S.N., Sriram D., Nath G., De Clercq E., Synthesis and antimicrobial activity of Schiff and Mannich bases of isatin and its derivatives with pyrimidine. IL Farmaco. 1999; **54**: 624–628.

[24] Jarrahpour A., Khalili D., De Clercq E., Salmi C., Brunel J.M., Synthesis, antibacterial, antifungal and antiviral activity evaluation of some new bis-Schiff bases of isatin and their derivatives. Molecules. 2007; **12**: 1720–1730.

[25] Hearn M.J., Cynamon M.H., Design and synthesis of antituberculars: preparation and evaluation against *Mycobacterium tuberculosis* of an isoniazid Schiff base. Journal of Antimicrobial Chemotherapy. 2004; **53**: 185–191.

[26] Karthikeyan M.S., Prasad D.J., Poojary B., Bhat K.S., Holla B.S., Kumari N.S., Synthesis and biological activity of Schiff and Mannich bases bearing 2,4-dichloro-5-fluorophenyl moiety. Bioorganic & Medicinal Chemistry. 2006; **14**: 7482–7489.

[27] Chandra S., Synthesis of new fungicides. 2-(4′-Arylthiazolyl-2'-imino)-3-aryl-4-thiazolidones. Bulletin of the Chemical Society of Japan. 1967; **40**: 2422–2424.

[28] Modi D., Sabnis S.S., Deliwala C.V., Potential anticancer agents. III. Schiff bases from benzaldehyde nitrogen mustards and aminophenylthiazoles. Journal of Medicinal Chemistry.1970; **13**: 935–941.

[29] Johnson D.K., Murphy T.B., Rose T.B., Goodwin W.H., Pickart L., Cytotoxic chelators and chelates. Inhibition of DNA synthesis in cultured rodent and human cells by aroylhydrazones and by a copper(II) complex of salicylaldehyde benzoyl hydrazone. Inorganica Chimica Acta. 1982; **67**: 159–165.

[30] Knorr L.On the constitution of quinizine derivatives. Reports of the German Chemical Society (A and B Series). 1883; **17**: 2032–2049.

[31] Girges M.M, Abou El-Zahab M.M, Hanna M.A, Synthesis of novel 4-substituted phenazone derivatives as potential antibacterial and antineoplastic agents. Archives of Pharmacal Research. 1988; **11**: 169–174.

[32] Rubtsov A.E., Makhmudov R.R., Kovylyaeva N.V., Prosyanik N.I., Bobrov A.V., Zalesov V.V., Synthesis, antiinflammatory and analgesic activity of 4-antipyrine derivatives. Pharmaceutical Chemistry Journal. 2002; **36**: 608–612.

[33] Ito T., Goto C., Noguchi K. Lanthanoid ion-selective solvent polymeric membrane electrode based on 1-phenyl-3-methyl-4-octadecanoyl-5-pyrazolone. Analytica Chimica Acta. 2001; **443**: 41–51.

[34] Radzikowska E., Onish K., Chojak E., Prospective assessment of cancer incidence and antipyrine metabolism. European Journal of Cancer. 1995; **31**: S225.

[35] Ismail M.M.F., Ammar Y.A., El-Zahaby H.S.A., Eisa S.I., Barakat S.E.S., Synthesis of novel 1-pyrazolylpyridin-2-ones as potential anti-inflammatory and analgesic agents. Archiv der Pharmazie – Chemistry in Life Sciences. 2007; **340**: 476–482.

[36] Mahmoud M., Abdel-Kader R., Hassanein M., Saleh S., Botros S., Antipyrine clearance in comparison to conventional liver function tests in hepatitis C virus patients. European Journal of Pharmacology. 2007; **569**: 222–227.

[37] Sondhi S.M., Sharma V.K., Verma R.P., Singhal N., Shukla R., Raghubir R., Dubey M.P., Synthesis, anti-inflammatory and analgesic activity evaluation of some mercapto pyrimidine and pyrimidobenzimidazole derivatives. Synthesis-Stuttgart. 1999; **5**: 878–884.

[38] Sondhi S.M., Singhal N., Verma R.P., Arora S.K., Dastidar S.G., Synthesis of hemin and porphyrin derivatives and their evaluation for anticancer activity. Indian Journal of Chemistry Section B Organic and Medicinal Chemistry. 2001; **40**: 113–119.

[39] Bondock S., Rabie R., Etman H.A., Fadda A.A., Synthesis and antimicrobial activity of some new heterocycles incorporating antipyrine moiety. European Journal of Medicinal Chemistry. 2008; **43**: 2122–2129.

[40] Cunha S., Oliveira S.M., Rodrigues M.T., Bastos R.M., Ferrari J., Oliveira C.M.A., Kato L., Napolitano H.B., Vencato I., Lariucci C., Structural studies of 4-aminoantipyrine derivatives. Journal of Molecular Structure, 2005; **752**: 32–39.

[41] Mishra A.P., Physicochemical and antimicrobial studies on nickel(II) and copper(II) Schiff base complexes derived from 2-furfuraldehyde. Journal of the Indian Chemical Society. 1999; **76**: 35–37.

[42] Raman N., Kulandaisamy A., Jeyasubramanian K., Synthesis, spectral, redox, and antimicrobial activity of Schiff base transition metal(II) complexes derived from 4-aminoantipyrine and benzil. Synthesis and Reactivity in Inorganic and Metal-Organic Chemistry. 2002; **32**: 1583–1610.

[43] Raman N., Kulandaisamy A., Shunmugasundaram A., Jeyasubramanian K., Synthesis, spectral, redox and antimicrobial activities of Schiff base complexes derived from 1-phenyl-2,3-dimethyl-4-aminopyrazol-5-one and acetoacetanilide. Transition Metal Chemistry. 2001; **26**: 131–135.

[44] Anupama B., Kumari G.C., Synthesis, characterization, DNA binding and antimicrobial activity of 4-amino antipyrine Schiff base metal complexes. Research Journal of Pharmaceutical, Biological and Chemical Sciences, 2011; **2**: 140–159.

[45] Ispir E., Toroğlu S., Kayraldiz A., Syntheses, characterization, antimicrobial and genotoxic activities of new Schiff bases and their complexes. Transition Metal Chemistry. 2008; **33**: 953–960.

[46] Anitha C., Sheela C.D., Tharmaraj P., Sumathi S., Spectroscopic studies and biological evaluation of some transition metal complexes of azo Schiff-base ligand derived from (1-phenyl-2,3-dimethyl-4-aminopyrazol-5-one) and 5-((4-chlorophenyl)diazenyl)-2 hydroxybenzaldehyde. Spectrochimica Acta Part A: Molecular and Biomolecular Spectroscopy. 2012; **96**: 493–500.

[47] Selwin Joseyphus R., Shiju C., Joseph J., Justin Dhanaraj C., Arish D., Synthesis and characterization of metal complexes of Schiff base ligand derived from imidazole-2-carboxaldehyde and 4-aminoantipyrine. Spectrochimica Acta Part A: Molecular and Biomolecular Spectroscopy. 2014; **133**: 149–155.

[48] Ismail Kamal Z., Synthesis, spectroscopic, magnetic and biological activity studies of copper(II) complexes of an antipyrine Schiff base. Transition Metal Chemistry. 2000; **25**: 522–528.

[49] Raman N., Johnson Raja S., Joseph J., Dhaveethu Raja J., Molecular designing, structural elucidation, and comparison of the cleavage ability of oxovanadium(IV) Schiff Base Complexes. Russian Journal of Coordination Chemistry. 2007; **33**: 7–11.

[50] Joseph J., Ayisha Bibin Rani G., Metal based SOD mimetic therapeutic agents: synthesis, characterization and biochemical studies of metal complexes. Arabian Journal of Chemistry. 2013; **6**: 1–9.

[51] Joseph J., Ayisha Bibin Rani G., Antioxidant and biochemical activities of mixed ligand complexes. Applied Biochemistry and Biotechnology. 2014; **172**: 867–890.

[52] Sakthivel A., Raman N., Mitu L., DNA interaction studies of pyrazolone- and diimine-incorporated Mn(II), Co(II), Ni(II), Cu(II), and Zn(II) complexes: synthesis, spectroscopic characterization, and antimicrobial study. Monatshefte für Chemie. 2013; **144**: 605–620.

[53] Raman N., Sakthivel A., Selvaganapathy M., Mitu L., Effect of DNA interaction involving antioxidative 4-aminoantipyrine incorporating mixed ligand complexes having alpha-amino acid as co-ligand. Journal of Molecular Structure. 2013; **1060**: 63–74.

[54] Shiva Leela D., Ushaiah B., Anupama G., Sunitha M., Gyana Kumari C., Synthesis, characterization, antimicrobial, DNA binding and cleavage studies of mixed ligand Cu (II), Co(II) complexes. Journal of Fluorescence. 2015; **25**: 185–197.

[55] Raman N., Kulandaisamy A., Thangaraja C., Manisankar P., Viswanathan S., Vedhi C., Synthesis, structural characterisation and electrochemical and antibacterial studies of Schiff base copper complexes. Transition Metal Chemistry. 2004; **29**: 129–135.

[56] Raman N., Johnson Raja S., Joseph J., Dhaveethu Raja J., Synthesis, spectral characterization and DNA cleavage study of heterocyclic Schiff base metal complexes. Journal of the Chilean Chemical Society. 2007b; **52**: 1138–1141.

[57] Joseph J., Nagashri K., Ayisha Bibin Rani G., Synthesis, characterization and antimicrobial activities of copper complexes derived from 4-amino-antipyrine derivatives. Journal of Saudi Chemical Society. 2013; **17**: 285–294.

[58] Kavitha T., Kulandaisamy A., Thillaiarasu P., Synthesis, spectroscopic characterization, electrochemical and antimicrobial studies of copper(II), nickel(II), cobalt(II), zinc(II) and oxovanadium(II) complexes derived from napthylidene-4-aminoantipyrine and tryptophan. International Journal of Innovative Research in Science, Engineering and Technology. 2015; **4**: 12221–12231.

[59] Raman N., Dhaveethu Raja J., Sakthivel A., Synthesis, spectral characterization of Schiff base transition metal complexes: DNA cleavage and antimicrobial activity studies. Journal of Chemical Sciences. 2007c; **119**: 303–310.

[60] Sivasankaran Nair M., Arish D., Synthesis, characterization and biological studies of Co (II), Ni(II), Cu(II) and Zn(II) complexes involving a potentially tetradentate Schiff base ligand. Transactions of the Indian Institute of Metals. 2011; **64**: 287–292.

[61] Raman N., Thangaraja C., Johnsonraja S., Synthesis, spectral characterization, redox and antimicrobial activity of Schiff base transition metal(II) complexes derived from 4-aminoantipyrine and 3-salicylidene-acetylacetone. Central European Journal of Chemistry. 2005; **3**: 537–555.

[62] Annigeri S.M., Sathisha M.P., Revankar V.K., Spectroscopic studies of bridged binuclear complexes of Co(II), Ni(II), Cu(II) and Zn(II). Transition Metal Chemistry. 2007; **32**: 81–87.

[63] Raman N., Mitu L., Sakthivel A., Pandi M.S.S., Studies on DNA cleavage and antimicrobial screening of transition metal complexes of 4-amino-antipyrine derivatives of N_2O_2 type. Journal of the Iranian Chemical Society. 2009; **6**: 738–748.

[64] Rosu T., Negoiu M., Pasculescu S., Pahontu E., Poirier D., Gulea A., Metal-based biologically active agents: synthesis, characterization, antibacterial and antileukemia activity evaluation of Cu(II), V(IV) and Ni(II) complexes with antipyrine-derived compounds. European Journal of Medicinal Chemistry. 2010; **45**: 774–781.

[65] Rosu T., Pahontu E., Maxim C., Georgescu R., Stanica N., Almajan G.L., Gulea A., Synthesis, characterization and antibacterial activity of some new complexes of Cu(II), Ni(II), VO(II), Mn(II) with Schiff base derived from 4-amino-2,3-dimethyl-1-phenyl-3-pyrazolin-5-one. Polyhedron. 2010; **29**: 757–766.

[66] Rosu T., Pahontu E., Maxim C., Georgescu R., Stanica N., Gulea A., Some new Cu(II) complexes containing an ON donor Schiff base: Synthesis, characterization and antibacterial activity. Polyhedron. 2011; **30**: 154–162.

[67] Rosu T., Pahontu E., Ilies D.-C., Georgescu R., Mocanu M., Leabu M., Shova S., Gulea A., Synthesis and characterization of some new complexes of Cu(II), Ni(II) and V(IV) with Schiff base derived from indole-3-carboxaldehyde. Biological activity on prokaryotes and eukaryotes. European Journal of Medicinal Chemistry. 2012; **53**: 380–389.

[68] Rosu T., Pahontu E., Mezey R.-S., Ilies D.-C., Georgescu R., Shova S., Gulea A., Synthesis, structural and spectral studies of Cu(II) and V(IV) complexes of a novel Schiff base derived from pyridoxal. Antimicrobial activity. Polyhedron. 2012; **31**: 352–360.

[69] Maki A.H., McGarvey B.R., Electron spin resonance in transition metal chelates. I. Copper(II) bis-acetylacetonate. Journal of Physical Chemistry. 1958; **29**: 31–34.

Chapter 5

Solubility Products and Solubility Concepts

Anna Maria Michałowska-Kaczmarczyk,

Aneta Spórna-Kucab and Tadeusz Michałowski

Additional information is available at the end of the chapter

http://dx.doi.org/10.5772/67840

Abstract

The chapter refers to a general concept of solubility product K_{sp} of sparingly soluble hydroxides and different salts and calculation of solubility of some hydroxides, oxides, and different salts in aqueous media. A (criticized) conventional approach, based on stoichiometry of a reaction notation and the solubility product of a precipitate, is compared with the unconventional/correct approach based on charge and concentration balances and a detailed physicochemical knowledge on the system considered, and calculations realized according to generalized approach to electrolytic systems (GATES) principles. An indisputable advantage of the latter approach is proved in simulation of static or dynamic, two-phase nonredox or redox systems.

Keywords: electrolytic two-phase systems, solubility, dissolution, static systems, dynamic systems, computer simulation, GATES, GEB

1. Introduction

The problem of solubility of various chemical compounds occupies a prominent place in the scientific literature. This stems from the fact that among various properties determining the use of these compounds, the solubility is of the paramount importance. Among others, this issue has been the subject of intense activities initiated in 1979 by the Solubility Data Commission V.8 of the IUPAC Analytical Chemistry Division established and headed by S. Kertes [1], who conceived the IUPAC-NIST Solubility Data Series (SDS) project [2, 3]. Within 1979–2009, the series of 87 volumes, concerning the solubility of gases, liquids, and solids in liquids or solids, were issued [3]; one of the volumes concerns the solubility of various oxides and hydroxides [4]. An extensive compilation of aqueous solubility data provides the Handbook of Aqueous Solubility Data [5].

A remark. Precipitates are marked in **bold** letters; soluble species/complexes are marked in normal letters.

The distinguishing feature of a chemical compound sparingly soluble in a particular medium is the solubility product K_{sp} value. In practice, the known K_{sp} values are referred only to aqueous media. One should note, however, that the expression for the solubility product and then the K_{sp} value of a precipitate depend on the notation of a reaction in which this precipitate is involved. From this it follows the apparent multiplicity of K_{sp}'s values referred to a particular precipitate. Moreover, as will be stated below, the expression for K_{sp} must not necessarily contain ionic species. On the other hand, factual or seeming lack of K_{sp}'s value for some precipitates is perceived; the latter issue be addressed here to $\mathbf{MnO_2}$, taken as an example.

Solubility products refer to a large group of sparingly soluble salts and hydroxides and some oxides, e.g., $\mathbf{Ag_2O}$, considered overall as hydroxides. Incidentally, other oxides, such as $\mathbf{MnO_2}$, $\mathbf{ZrO_2}$, do not belong to this group, in principle. For $\mathbf{ZrO_2}$, the solubility measurements showed quite low values even under a strongly acidic condition [6]. The solubility depends on the prior history of these oxides, e.g., prior roasting virtually eliminates the solubility of some oxides. Moderately soluble iodine ($\mathbf{I_2}$) dissolves due to reduction or oxidation, or disproportionation in alkaline media [7–12]; for $\mathbf{I_2}$, minimal solubility in water is a reference state. For 8-hydroxyquinoline, the solubility of the neutral molecule HL is a reference state; a growth in solubility is caused here by the formation of ionic species: H_2L^{+1} in acidic and L^{-1} in alkaline media.

The K_{sp} is the main but not the only parameter used for calculation of solubility s of a precipitate. The simplifications [13] practiced in this respect are unacceptable and lead to incorrect/false results, as stated in [14–18]; more equilibrium constants are also involved with two-phase systems. These objections, formulated in the light of the generalized approach to electrolytic systems (GATES) [8], where s is the "weighed" sum of concentrations of all soluble species formed by the precipitate, are presented also in this chapter, related to nonredox and redox systems.

Calculation of s gives an information of great importance, e.g., from the viewpoint of gravimetry, where the primary step of the analysis is the quantitative transformation of a proper analyte into a sparingly soluble precipitate (salt, hydroxide). Although the precipitation and further analytical operations are usually carried out at temperatures far greater than the room temperature, at which the equilibrium constants were determined, the values of s obtained from the calculations made on the basis of equilibrium data related to room temperature are helpful in the choice of optimal *a priori* conditions of the analysis, ensuring the minimal, summary concentration of all soluble forms of the analyte, remaining in the solution, in equilibrium with the precipitate obtained after addition of an excess of the precipitating agent; this excess is referred to as relative to the stoichiometric composition of the precipitate. The ability to perform appropriate calculations, based on all available physicochemical knowledge, in accordance with the basic laws of matter conservation, deepens our knowledge of the relevant systems. At the same time, it produces the ability to acquire relevant knowledge in

http://dx.doi.org/10.5772/67840

an organized manner—not just imitative, but focused on heuristics. This viewpoint is in accordance with constructivist teaching, based on the belief that learning occurs, as learners are actively involved in a process of meaning and knowledge construction, as opposed to passively receiving information [19].

2. Definitions and formulation of solubility products

The K_{sp} value refers to a two-phase system where the equilibrium solid phase is a sparingly soluble precipitate, whose K_{sp} value is measured/calculated according to defined expression for the solubility product. This assumption means that the solution with defined species is saturated against this precipitate, at given temperature and composition of the solution. However, often a precipitate, when introduced into aqueous media, is not the equilibrium solid phase, and then this fundamental requirement is not complied, as indicated in examples of the physicochemical analyses of the systems with struvite $\mathbf{MgNH_4PO_4}$ [20, 21], dolomite $\mathbf{CaMg(CO_3)_2}$ [22, 23], and $\mathbf{Ag_2Cr_2O_7}$.

The values of solubility products K_{sp} (usually represented by solubility constant pK_{sp} = $-\log K_{sp}$ value) are known for stoichiometric precipitates of $\mathbf{A_aB_b}$ or $\mathbf{A_aB_bC_c}$ type, related to dissociation reactions:

$$K_{sp} = [A]^a[B]^b \text{for } \mathbf{A_aB_b} = aA + bB, \text{or} \tag{1}$$

$$K_{sp} = [A]^a[B]^b[C]^c \text{for } \mathbf{A_aB_bC_c} = aA + bB + cC \tag{2}$$

where A and B or A, B, and C are the species forming the related precipitate; charges are omitted here, for simplicity of notation. The solubility products for more complex precipitates are unknown in the literature. The precipitates $\mathbf{A_aB_bC_c}$ are known as ternary salts [24], e.g., struvite, dolomite, and hydroxyapatite $\mathbf{Ca_5(PO_4)_3OH}$.

The solubility products for precipitates of $\mathbf{A_aB_b}$ type are most frequently met in the literature. In these cases, for A are usually put simple cations of metals, or oxycations [25]; e.g., BiO^{+1} and UO_2^{+2} form the precipitates: **BiOCl** and $\mathbf{(UO_2)_2(OH)_2}$. As B, simple or more complex anions are considered, e.g., Cl^{-1}, S^{-2}, PO_4^{-3}, $Fe(CN)_6^{-4}$, in **AgCl, HgS,** $\mathbf{Zn_3(PO_4)_2}$, and $\mathbf{Zn_2Fe(CN)_6}$.

In different textbooks, the solubility products are usually formulated for dissociation reactions, with ions as products, also for **HgS**

$$\mathbf{HgS} = Hg^{+2} + S^{-2}(K_{sp} = [Hg^{+2}][S^{-2}]) \tag{3}$$

although polar covalent bond exists between its constituent atoms [26]. Very low solubility product value (pK_{sp} = 52.4) for **HgS** makes the dissociation according to the scheme presented by Eq. (3) impossible, and even verbal formulation of the solubility product is unreasonable. Namely, the ionic product $x = [Hg^{+2}][S^{-2}]$ calculated at $[Hg^{+2}] = [S^{-2}] = 1/N_A$ exceeds K_{sp}, $1/N_A^2 > K_{sp}$ (N_A – Avogadro's number); the concentration $1/N_A = 1.66 \cdot 10^{-23}$ mol/L corresponds to 1 ion in 1 L of the solution. The scheme of dissociation into elemental species [14]

$$\mathbf{HgS} = \mathbf{Hg} + \mathbf{S}\ (K_{sp1} = [\mathbf{Hg}][\mathbf{S}]) \tag{4}$$

is far more favored from thermodynamic viewpoint; nonetheless, the solubility product (K_{sp}) for **HgS** is commonly formulated on the basis of reaction (3). We obtain $pK_{sp1} = pK_{sp} - 2A(E_{01}-E_{02})$, where $E_{01} = 0.850$ V for $Hg^{+2} + 2e^{-1} = \mathbf{Hg}$, $E_{02} = -0.48$ V for $\mathbf{S} + 2e^{-1} = S^{-2}$, $1/A = RT/F{\cdot}\ln10$, $A = 16.92$ for 298 K; then $pK_{sp1} = 7.4$.

Equilibrium constants are usually formulated for the simplest reaction notations. However, in this respect, Eq. (4) is simpler than Eq. (3). Moreover, we are "accustomed" to apply solubility products with ions (cations and anions) involved, but this custom can easily be overthrown. A similar remark may concern the notation referred to elementary dissociation of mercuric iodide precipitate

$$\mathbf{HgI_2} = \mathbf{Hg} + I_2(K_{sp1} = [\mathbf{Hg}][I_2]) \tag{5}$$

where I_2 denotes a soluble form of iodine in a system. From

$$\mathbf{HgI_2} = Hg^{+2} + 2I^{-1}(K_{sp} = [Hg^{+2}][I^{-1}]^2, pK_{sp} = 28.55) \tag{6}$$

we obtain $pK_{sp1} = pK_{sp} - 2A(E_{01}-E_{03})$, where

$$E_{01} = 0.850 \text{ V for } Hg^{+2} + 2e^{-1} = \mathbf{Hg}, E_{03} = 0.621 \text{ V for } I_2 + 2e^{-1} = 2I^{-1}; \text{ then } pK_{sp1} = 20.80.$$

The species in the expression for solubility products do not predominate in real chemical systems, as a rule. However, the precipitation of **HgS** from acidified (HCl) solution of mercury salt with H_2S solution can be presented in terms of predominating species; we have

$$HgCl_4^{-2} + H_2S = \mathbf{HgS} + 4Cl^{-1} + 2H^{+1} \tag{7}$$

Eq. (7) can be applied to formulate the related solubility product, K_{sp2}, for **HgS**. To be *online* with customary requirements put on the solubility product formulation, Eq. (7) should be rewritten into the form

$$\mathbf{HgS} + 4Cl^{-1} + 2H^{+1} = HgCl_4^{-2} + H_2S \tag{7a}$$

Applying the law of mass action to Eq. (7a), we have

$$K_{sp2} = \frac{[HgCl_4^{-2}][H_2S]}{[Cl^{-1}]^4[H^{+1}]^2}, \quad (pK_{sp2} = 17.33) \tag{8}$$

where $[HgCl_4^{-2}] = 10^{15.07}[Hg^{+2}][Cl^{-1}]^4$, $[H_2S] = 10^{20.0}[H^{+1}]^2[S^{-2}]$, K_{sp} (Eq. (3)).

The solubility product for **MgNH₄PO₄** can be formulated on the basis of reactions:

$$\mathbf{MgNH_4PO_4} = Mg^{+2} + NH_4^{+1} + PO_4^{-3}(K_{sp} = [Mg^{+2}][NH_4^{+1}][PO_4^{-3}]) \tag{9}$$

$$\mathbf{MgNH_4PO_4} = Mg^{+2} + NH_3 + HPO_4^{-2}(K_{sp1} = [Mg^{+2}][NH_3][HPO_4^{-2}] = K_{sp}K_{1N}/K_{3P}) \tag{10}$$

http://dx.doi.org/10.5772/67840

$$\mathbf{MgNH_4PO_4} + H_2O = MgOH^{+1} + NH_3 + H_2PO_4^{-1}(K_{sp2} = [MgOH^{+1}][NH_3][H_2PO_4^{-1}] = K_{sp}K_1^{OH}K_{1N}K_W/(K_{2P}K_{3P}) \quad (11)$$

where $K_{1N} = [H^{+1}][NH_3]/[NH_4^{+1}]$, $K_{2P} = [H^{+1}][HPO_4^{-2}]/[H_2PO_4^{-1}]$, $K_{3P} = [H^{+1}][PO_4^{-3}]/[HPO_4^{-2}]$, $[MgOH^{+1}] = K_1^{OH}[Mg^{+2}][OH^{-1}]$, $K_W = [H^{+1}][OH^{-1}]$.

Note that only uncharged (elemental) species are involved in Eqs. (4) and (5); H_2S enters Eq. (8), and NH_3 enters Eqs. (10) and (11). This is an extension of the definition/formulation commonly met in the literature, where only charged species were involved in expression for the solubility product. Note also that small/dispersed mercury drops are neutralized with powdered sulfur, according to thermodynamically favored reaction [27]

$$\mathbf{Hg} + \mathbf{S} = \mathbf{HgS}$$

reverse to Eq. (4). Some precipitates can be optionally considered as the species of $\mathbf{A_aB_b}$ or $\mathbf{A_aB_bC_c}$ type. For example, the solubility product for $\mathbf{MgHPO_4}$ can be written as $K_{sp} = [Mg^{+2}][HPO_4^{-2}]$ or $K_{sp1} = [Mg^{+2}][H^{+1}][PO_4^{-3}] = K_{sp}K_{3P}$.

The ferrocyanide ion $Fe(CN)_6^{-4}$ (with evaluated stability constant K_6 ca. 10^{37}) can be considered as practically undissociated, i.e., $Fe(CN)_6^{-4}$ is kinetically inert [28], and then it does not give Fe^{+2} and CN^{-1} ions. The solubility product of $\mathbf{Zn_2Fe(CN)_6}$ is $K_{sp} = [Zn^{+2}]^2[Fe(CN)_6^{-4}]$. Therefore, consideration of $\mathbf{Zn_2Fe(CN)_6}$ as a ternary salt with $K_{sp1} = [Zn^{+2}]^2[Fe^{2+}][CN^{-1}]^6 = K_{sp}/K_6$ is not acceptable.

In principle, the solubility product values are formulated for stoichiometric compounds, and specified as such in the related tables. However, some precipitates obtained in laboratory have nonstoichiometric composition, e.g., dolomite $\mathbf{Ca_{1+x}Mg_{1-x}(CO_3)_2}$ [22, 23], $\mathbf{Fe_xS}$ [29]. In particular, $\mathbf{Fe_xS}$ can be rewritten as $\mathbf{Fe^{+2}_pFe^{+3}_qS}$; from the relations: $2p + 3q - 2 = 0$ and $p + q = x$, we get $q/p = 2(1 - x)/(3x - 2)$.

In this context, some remark needs a formulation of K_{sp} for some hydroxyoxides (e.g., **FeOOH**) and oxides (e.g., $\mathbf{Ag_2O}$). The related solubility products are formulated after completion of the corresponding reactions with water, e.g., $\mathbf{FeOOH} + H_2O = \mathbf{Fe(OH)_3}$, $\mathbf{Fe_2O_3{\cdot}xH_2O} + (3 - x)H_2O = \mathbf{2Fe(OH)_3} \Rightarrow \mathbf{Fe(OH)_3} = Fe^{+3} + 3OH^{-1} \Rightarrow K_{sp} = [Fe^{+3}][OH^{-1}]^3$; $\mathbf{Ag_2O} + H_2O = \mathbf{2AgOH} \Rightarrow \mathbf{AgOH} = Ag^{+1} + OH^{-1} \Rightarrow K_{sp} = [Ag^{+1}][OH^{-1}]$, see it in the context with $gcd(a,b) = 1$.

The solubility product can be involved not only with dissociation reaction. For example, the dissolution reaction $\mathbf{Ca(OH)_2} + 2H^{+1} = Ca^{+2} + 2H_2O$ [30], characterized by $K_{sp1} = [Ca^{+2}]/[H^{+1}]^2$, is involved with $K_{sp} = [Ca^{+2}][OH^{-1}]^2$ in the relation $K_{sp1} = K_{sp}/K_W^2$. In Ref. [31], the solubility product is associated with formation (not dissociation) of a precipitate.

3. Solubility product(s) for MnO_2

The scheme presented above cannot be extended to all oxides. For example, one cannot recommend the formulation of this sequence for $\mathbf{MnO_2}$, i.e., $\mathbf{MnO_2} + 2H_2O = \mathbf{Mn(OH)_4} \Rightarrow$

Mn(OH)₄ = $Mn^{+4} + 4OH^{-1} \Rightarrow K_{sp0} = [Mn^{+4}][OH^{-1}]^4$; Mn^{+4} ions do not exist in aqueous media, and $\mathbf{MnO_2}$ is the sole Mn(+4) species present in such systems. In effect, K_{sp0} for $\mathbf{MnO_2}$ is not known in the literature, compare with Ref. [32]. However, the K_{sp} for $\mathbf{MnO_2}$ can be formally calculated according to an unconventional approach, based on the disproportionation reaction

$$5\mathbf{MnO_2} + 4H^{+1} = 2MnO_4^{-1} + 3Mn^{+2} + H_2O \tag{12}$$

reverse to the symproportionation reaction $2MnO_4^{-1} + 3Mn^{+2} + H_2O = 5\mathbf{MnO_2} + 4H^{+1}$. The $K_{sp} = K_{sp1}$ value can be found there on the basis of E_{01} and E_{02} values [33], specified for reactions:

$$MnO_4^{-1} + 4H^{+1} + 3e^{-1} = \mathbf{MnO_2} + 2H_2O(E_{01} = 1.692\text{ V}) \tag{13}$$

$$\mathbf{MnO_2} + 4H^{+1} + 2e^{-1} = Mn^{+2} + 2H_2O(E_{02} = 1.228\text{ V}) \tag{14}$$

Eqs. (13) and (14) are characterized by the equilibrium constants:

$$K_{e1} = \frac{[MnO_2][H_2O]^2}{[MnO_4^{-1}][H^{+1}]^4[e^{-1}]^3}, \qquad K_{e2} = \frac{[Mn^{+2}][H_2O]^2}{[\mathbf{MnO_2}][H^{+1}]^4[e^{-1}]^2} \tag{15}$$

defined on the basis of mass action law (MAL) [14], where $\log K_{e1} = 3{\cdot}A{\cdot}E_{01}$, $\log K_{e2} = 2{\cdot}A{\cdot}E_{02}$, $A = 16.92$. From Eqs. (13) and (14), we get

$$2\mathbf{MnO_2} + 4H_2O + 3\mathbf{MnO_2} + 12H^{+1} + 6e^{-1} = 2MnO_4^{-1} + 8H^{+1} + 6e^{-1} + 3Mn^{+2} + 6H_2O \tag{16}$$

Assuming $[\mathbf{MnO_2}] = 1$ and $[H_2O] = 1$ on the stage of the K_{sp1} formulation for reaction (16), equivalent to reaction (12), we have

$$K_{sp1} = \frac{[MnO_4^{-1}]^2[Mn^{+2}]^3}{[H^{+1}]^4} \tag{17}$$

and then

$$K_{sp1} = (K_{e2})^3 \cdot (K_{e1})^{-2} \tag{18}$$

$$pK_{sp1} = 3\log K_{e2} - 2\log K_{e1} = 6A(E_{01} - E_{02}) = 6 \cdot 16.92 \cdot (1.692 - 1.228) = 47.11 \tag{19}$$

The solubility products with $\mathbf{MnO_2}$ involved can be formulated on the basis of other reactions. For example, addition of

$$Mn^{+2} = Mn^{+3} + e^{-1} \tag{20}$$

to Eq. (14) gives

http://dx.doi.org/10.5772/67840

$$\mathbf{MnO_2} + 4H^{+1} + 2e^{-1} + Mn^{+2} = Mn^{+2} + 2H_2O + Mn^{+3} + e^{-1} \tag{21}$$

Multiplication of Eq. (21) by 3, and then addition to Eq. (13a)

$$\mathbf{MnO_2} + 2H_2O = MnO_4^{-1} + 4H^{+1} + 3e^{-1}$$

(reverse to Eq. (13)) gives the equation

$$\begin{aligned} &3\mathbf{MnO_2} + 12H^{+1} + 6e^{-1} + 3Mn^{+2} + \mathbf{MnO_2} + 2H_2O \\ &= 3Mn^{+2} + 6H_2O + 3Mn^{+3} + 3e^{-1} + MnO_4^{-1} + 4H^{+1} + 3e^{-1} \end{aligned} \tag{22}$$

and its equivalent form, obtained after simplifications,

$$4\mathbf{MnO_2} + 8H^{+1} = 3Mn^{+3} + MnO_4^{-1} + 4H_2O \tag{22a}$$

Eq. (22) and then Eq. (22a) is characterized by the solubility product

$$K_{sp2} = \frac{[MnO_4^{-1}][Mn^{+3}]^3}{[H^{+1}]^8} = (K_{e2})^3 \cdot (K_{e3})^{-3} \cdot (K_{e1})^{-1} \tag{23}$$

where

$$K_{e3} = \frac{[Mn^{+2}]}{[Mn^{+3}][e^{-1}]} \tag{24}$$

for $Mn^{+3} + e^{-1} = Mn^{+2}$ (E_{03} = 1.509 V) (reverse to Eq. (20)), $\log K_{e3} = A \cdot E_{03}$. Then

$$pK_{sp2} = 3A \cdot (E_{01} - 2E_{02} + E_{03}) + 37.82 \tag{25}$$

Formulation of K_{spi} for other combinations of redox and/or nonredox reactions is also possible. This way, some derivative solubility products are obtained. The choice between the "output" and derivative solubility product values is a matter of choice. Nevertheless, one can choose the K_{sp3} value related to the simplest expression for the solubility product $K_{sp3} = [Mn^{+2}][MnO_4^{-2}]$ involved with reaction $2\mathbf{MnO_2} = Mn^{+2} + MnO_4^{-2}$.

As results from calculations, the low K_{spi} (i = 1,2,3) values obtained from the calculations should be crossed, even in acidified solution with the related manganese species presented in **Figure 1**. In the real conditions of analysis, at C_a = 1.0 mol/L, the system is homogeneous during the titration, also after crossing the equivalence point, at $\Phi = \Phi_{eq} > 0.2$; this indicates that the corresponding manganese species form a metastable system [34], unable for the symproportionation reactions.

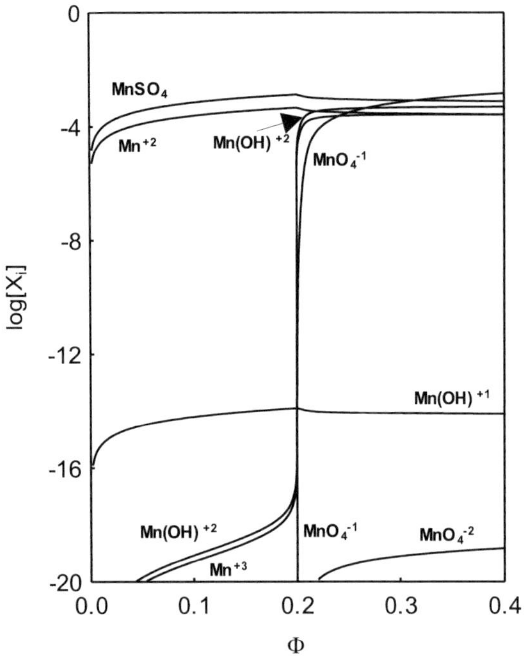

Figure 1. The $\log[X_i]$ versus Φ relationships for different manganese species X_i, plotted for titration of V_0 = 100 mL solution of $FeSO_4$ (C_0 = 0.01 mol/L) + H_2SO_4 (C_a = 1.0 mol/L) with V mL of C = 0.02 mol/L $KMnO_4$; $\Phi = C \cdot V/(C_0 \cdot V_0)$. The species X_i are indicated at the corresponding lines.

4. Calculation of solubility

In this section, we compare two options applied to the subject in question. The first/criticized option, met commonly in different textbooks, is based on the stoichiometric considerations, resulting from dissociation of a precipitate, characterized by the solubility product K_{sp} value, and considered *a priori* as an equilibrium solid phase in the system in question; the solubility value obtained this way will be denoted by s^* [mol/L]. The second option, considered as a correct resolution of the problem, is based on full physicochemical knowledge of the system, not limited only to K_{sp} value (as in the option 1); the solubility value thus obtained is denoted as s [mol/L]. The second option fulfills all requirements expressed in GATES and involved with basic laws of conservation in the systems considered. Within this option, we check, among others, whether the precipitate is really the equilibrium solid phase. The results (s^*, s) obtained according to both options (1 and 2) are compared for the systems of different degree of complexity. The unquestionable advantages of GATES will be stressed this way.

4.1. Formulation of the solubility s^*

The solubility s^* will be calculated for a pure precipitate of: (1°) $\mathbf{A_aB_b}$ or (2°) $\mathbf{A_aB_bC_c}$ type, when introduced into pure water. Assuming $[A] = a \cdot s^*$ and $[B] = b \cdot s^*$, from Eq. (1), we have

$$s^* = \left(\frac{K_{sp}}{a^a \cdot b^b}\right)^{1/(a+b)} \tag{26}$$

and assuming $[A] = a{\cdot}s^*$, $[B] = b{\cdot}s^*$, $[C] = c{\cdot}s^*$, from Eq. (2), we have

$$s^* = \left(\frac{K_{sp}}{a^a \cdot b^b \cdot c^c}\right)^{1/(a+b+c)} \tag{27}$$

As a rule, the formulas (26) and (27) are invalid for different reasons, indicated in this chapter. This invalidity results, among others, from inclusion of the simplest/minor species in Eq. (26) or (27) and omission of hydroxo-complexes + other soluble complexes formed by A, and proto-complexes + other soluble complexes, formed by B. In other words, not only the species entering the expression for the related solubility product are present in the solution considered. Then the concentrations: [A], [B] or [A], [B], and [C] are usually minor species relative to the other species included in the respective balances, considered from the viewpoint of GATES [8].

4.2. Dissolution of hydroxides

We refer first to the simplest two-phase systems, with insoluble hydroxides as the solid phases. In all instances, s^* denotes the solubility obtained from stoichiometric considerations, whereas s relates to the solubility calculated on the basis of full/attainable physicochemical knowledge related to the system in question where, except the solubility product (K_{sp}), other physicochemical data are also involved.

Applying formula (26) to hydroxides ($B = OH^{-1}$): **$Ca(OH)_2$** ($pK_{sp1} = 5.03$) and **$Fe(OH)_3$** ($pK_{sp2} = 38.6$), we have [35]

$$\mathbf{Ca(OH)_2} = Ca^{+2} + 2OH^{-1} (K_{sp1} = [Ca^{+2}][OH^{-1}]^2,\ s^* = (K_{sp1}/4)^{1/3} = 0.0133\ \text{mol/L}) \tag{28}$$

$$\mathbf{Fe(OH)_3} = Fe^{+3} + 3OH^{-1} (K_{sp2} = [Fe^{+3}][OH^{-1}]^3,\ s^* = (K_{sp2}/27)^{1/4} = 0.98 \times 10^{-10}\ \text{mol/L}) \tag{29}$$

respectively. However, Ca^{+2} and Fe^{+3} form the related hydroxo-complexes: $[CaOH^{+1}] = 10^{1.3}{\cdot}[Ca^{+2}][OH^{-1}]$ and: $[FeOH^{+2}] = 10^{11.0}{\cdot}[Fe^{+3}][OH^{-1}]$, $[Fe(OH)_2^{+1}] = 10^{21.7}{\cdot}[Fe^{+3}][OH^{-1}]^2$; $[Fe_2(OH)_2^{+4}] = 10^{25.1}{\cdot}[Fe^{+3}]^2[OH^{-1}]^2$ [31]. The corrected expression for the solubility of **$Ca(OH)_2$** is as follows

$$s = [Ca^{+2}] + [CaOH^{+1}] \tag{30}$$

Inserting $[Ca^{+2}] = K_{sp1}/[OH^{-1}]^2$ and $[OH^{-1}] = K_W/[H^{+1}]$, $[H^{+1}] = 10^{-pH}$ ($pK_W = 14.0$ for ionic product of water, K_W) into the charge balance

$$2[Ca^{+2}] + [CaOH^{+1}] + [H^{+1}] - [OH^{-1}] = 0 \tag{31}$$

we get, by turns,

$$2 \cdot 10^{-5.03}/[OH^{-1}]^2 + 10^{1.3} \cdot 10^{-5.03}/[OH^{-1}][H^{+1}] - [OH^{-1}] = 0$$
$$\Rightarrow 2 \cdot 10^{-5.03+28-2pH} + 10^{1.3} \cdot 10^{-5.03+14-pH} + 10^{-pH} - 10^{pH-14} = 0$$

$$y(pH) = 2 \cdot 10^{22.97-2pH} + 10^{10,17-pH} + 10^{-pH} - 10^{pH-14} = 0 \tag{32}$$

where $pH = -\log[H^{+1}]$. Applying the zeroing procedure to Eq. (30), we get pH_0 = 12.453 (**Table 1**), where: $[Ca^{+2}]$ = 0.0116, $[CaOH^{+1}]$ = 0.00656, s = 0.0182 mol/L (Eq. (28)). As we see, $[CaOH^{+1}]$ is comparable with $[Ca^{+2}]$, and there are none reasons to omit $[CaOH^{+1}]$ in Eq. (28).

The alkaline reaction in the system with $\mathbf{Ca(OH)_2}$ results immediately from Eq. (29): $[OH^{-1}] - [H^{+1}] = 2[Ca^{+2}] + [CaOH^{+1}] > 0$.

Analogously, for the system with $\mathbf{Fe(OH)_3}$, we have the charge balance

$$3[Fe^{+3}] + 2[FeOH^{+2}] + [Fe(OH)_2^{+1}] + 4[Fe_2(OH)_2^{+4}] + [H^{+1}] - [OH^{-1}] = 0 \tag{33}$$

and then

$$y(pH) = 3 \cdot 10^{3.4-3pH} + 2 \cdot 10^{0.4-2pH} + 10^{-2.9-pH} + 4 \cdot 10^{3.9-4pH} + 10^{-pH} - 10^{pH-14} = 0 \tag{34}$$

Eq. (32) zeroes at pH_0 = 7.0003 (**Table 2**), where the value

$$s = [Fe^{+3}] + [FeOH^{+2}] + [Fe(OH)_2^{+1}] + 2[Fe_2(OH)_2^{+4}] \tag{35}$$

is close to $s \cong [Fe(OH)_2^{+1}] = 10^{-9.9}$. Alkaline reaction for this system, i.e., $[OH^{-1}] > [H^{+1}]$, results immediately from Eq. (30), and pH_0 = 7.0003 (>7).

At pH = 7, $Fe(OH)_2^{+1}$ (not Fe^{+3}) is the predominating species in the system, $[Fe(OH)_2^{+1}]/[Fe^{+3}] = 10^{21.7-14} = 5 \cdot 10^7$, i.e., the equality/assumption $s^* = [Fe^{+3}]$ is extremely invalid. Moreover, the value $[OH^{-1}] = 3 \cdot s^* = 2.94 \cdot 10^{-10} = 10^{-9.532}$, i.e., pH = 4.468; this pH-value is contradictory with the inequality $[OH^{-1}] > [H^{+1}]$ resulting from Eq. (31). Similarly, extremely invalid result was

pH	y(pH)	$[OH^{-1}]$	$[Ca^{+2}]$	$[CaOH^{+1}]$
12.451	0.000377	0.02825	0.01169	0.006592
12.452	0.000193	0.02831	0.01164	0.006577
12.453	**8.30E-06**	**0.02838**	**0.01159**	**0.006561**
12.454	−0.000176	0.02844	0.01153	0.006546
12.455	−0.000359	0.02851	0.01148	0.006531

Table 1. Zeroing the function (30) for the system with $\mathbf{Ca(OH)_2}$ precipitate introduced into pure water (copy of a fragment of display).

http://dx.doi.org/10.5772/67840

pH	y(pH)	$[Fe^{+3}]$	$[FeOH^{+2}]$	$[Fe(OH)_2^{+1}]$	$[Fe_2(OH)_2^{+4}]$
7.0001	7.99E-11	2.510E-18	2.511E-14	1.259E-10	7.936E-25
7.0002	3.38E-11	2.508E-18	2.510E-14	1.258E-10	7.929E-25
7.0003	**−1.23E-11**	**2.507E-18**	**2.508E-14**	**1.258E-10**	**7.921E-25**
7.0004	−5.84E-11	2.505E-18	2.507E-14	1.258E-10	7.914E-25
7.0005	−1.04E-10	2.503E-18	2.506E-14	1.257E-10	7.907E-25

Table 2. Zeroing the function (32) for the system with $\mathbf{Fe(OH)_3}$ precipitate introduced into pure water (copy of a fragment of display).

obtained in Ref. [36], where the strong hydroxo-complexes were totally omitted, and weak chloride complexes of Fe^{+3} ions were included into considerations.

Taking only the main dissociating species formed in the solution saturated with respect to $\mathbf{Fe(OH)_3}$, we check whether the reaction $\mathbf{Fe(OH)_3} = Fe(OH)_2^{+1} + OH^{-1}$ with $K_{sp1} = [Fe(OH)_2^{+1}][OH^{-1}] = 10^{21.7}\cdot 10^{-38.6} = 10^{-16.9}$ can be used for calculation of solubility $s' = (K_{sp1})^{1/2}$ for $\mathbf{Fe(OH)_3}$; the answer is also negative. Simply, the main part of OH^{-1} ions originates here from dissociation of water, where the precipitate has been introduced, and then $Fe(OH)_2^{+1}$ and OH^{-1} differ significantly. As we see, the diversity in K_{sp} value related to a precipitate depends on its dissociation reaction notation, which disqualifies the calculation of s^* based solely on the K_{sp} value. This fact was not stressed in the literature issued hitherto.

Concluding, the application of the option 1, based on the stoichiometry of the reaction (29), leads not only to completely inadmissible results for s^+, but also to a conflict with one of the fundamental rules of conservation obligatory in electrolytic systems, namely the law of charge conservation.

Similarly, critical/disqualifying remarks can be related to the series of formulas considered in the chapter [37], e.g., $K_{sp} = 27(s^*)^4$ for precipitates of $\mathbf{A_3B}$ and $\mathbf{AB_3}$ type, and $K_{sp} = 108(s^*)^5$ for $\mathbf{A_2B_3}$ and $\mathbf{A_3B_2}$. For $\mathbf{Ca_5(PO_4)_3OH}$, the formula $K_{sp} = 84375(s^*)^9$ (!) was applied [38].

As a third example let us take a system, where an excess of $\mathbf{Zn(OH)_2}$ precipitate is introduced into pure water. It is usually stated that $\mathbf{Zn(OH)_2}$ dissociates according to the reaction

$$\mathbf{Zn(OH)_2} = Zn^{+2} + 2\,OH^{-1} \tag{36}$$

applied to formulate the expression for the solubility product

$$K_{sp3} = [Zn^{+2}][OH^{-1}]^2 (pK_{sp3} = 15.0) \tag{37}$$

The soluble hydroxo-complexes $Zn(OH)_i^{+2-i}$ (i=1,...,4), with the stability constants, K_i^{OH}, expressed by the values $\log K_i^{OH}$ = 4.4, 11.3, 13.14, 14.66, are also formed in the system in question. The charge balance (ChB) has the form

$$2[Zn^{+2}] + [ZnOH^{+1}]–[Zn(OH)_3^{-1}] - 2[Zn(OH)_4^{-2}] + [H^{+1}] - [OH^{-1}] = 0 \quad (38)$$

i.e., $2{\cdot}10^{-15}/[OH^{-1}]^2 + 10^{4.4}{\cdot}10^{-15}/[OH^{-1}] - 10^{13.14}{\cdot}10^{-15}{\cdot}[OH^{-1}] - 2{\cdot}10^{14.66}{\cdot}10^{-15}{\cdot}[OH^{-1}]^2 = 0$

$$y(pH) = 2 \cdot 10^{13-2pH} + 10^{3.4-pH}–10^{-15.86+pH}–2 \cdot 10^{-28.34+2pH} + 10^{-pH}–10^{pH-14} = 0 \quad (39)$$

The function (39) zeroes at $pH_0 = 9.121$ (see **Table 3**). The basic reaction of this system is not immediately stated from Eq. (38) (there are positive and negative terms in expression for $[OH^{-1}] - [H^{+1}]$). The solubility s value

$$s = [Zn^{+2}] + [ZnOH^{+1}] + [Zn(OH)_2] + [Zn(OH)_3^{-1}] + [Zn(OH)_4^{-2}] = 2.07 \cdot 10^{-4}$$

calculated at this point is different from $s^* = (K_{so3}/4)^{1/3} = 6.3{\cdot}10^{-6}$, and $[OH^{-1}]/[Zn^{+2}] \neq 2$; such incompatibilities contradict application of this formula.

4.3. Dissolution of MeL_2-type salts

Let us refer now to dissolution of precipitates $\mathbf{MeL_2}$ formed by cations Me^{+2} and anions L^{-1} of a strong acid HL, as presented in **Table 4**. When an excess of $\mathbf{MeL_2}$ is introduced into pure water, the concentration balances and charge balance in two-phase system thus formed are as follows:

pH	$[OH^{-1}]$	$[Zn^{+2}]$	$[ZnOH^{+1}]$	$[Zn(OH)_2]$	$[Zn(OH)_3^{-1}]$	$[Zn(OH)_4^{-2}]$	y(pH)	s [mol/L]
9.118	1.3122E-05	5.8076E-06	1.9143E-06	0.0002	1.8113E-07	7.8705E-11	2.2702E-07	0.00020743
9.119	1.3152E-05	5.7810E-06	1.9099E-06	0.0002	1.8155E-07	7.9068E-11	1.3858E-07	0.00020740
9.120	1.3183E-05	5.7544E-06	1.9055E-06	0.0002	1.8197E-07	7.9433E-11	5.0322E-08	0.00020737
9.121	**1.3213E-05**	**5.7280E-06**	**1.9011E-06**	**0.0002**	**1.8239E-07**	**7.9800E-11**	**−3.7750E-08**	**0.00020734**
9.122	1.3243E-05	5.7016E-06	1.8967E-06	0.0002	1.8281E-07	8.0168E-11	−1.2564E-07	0.00020731
9.123	1.3274E-05	5.6755E-06	1.8923E-06	0.0002	1.8323E-07	8.0538E-11	−2.1335E-07	0.00020728

Table 3. Zeroing the function (39) for the system with $\mathbf{Zn(OH)_2}$ precipitate introduced into water; $pK_W = 14$.

Me^{+2}	$MeOH^{+1}$	$Me(OH)_2$	$Me(OH)_3^{-1}$	L^{-1}	MeL^{+1}	MeL_2	MeL_3^{-1}	MeL_4^{-2}	MeL_2
	$\log K_1^{OH}$	$\log K_2^{OH}$	$\log K_3^{OH}$		$\log K_1$	$\log K_2$	$\log K_3$	$\log K_4$	pK_{sp}
Hg^{+2}	10.3	21.7	21.2	I^{-1}	12.87	23.82	27.60	29.83	28.54
Pb^{+2}	6.9	10.8	13.3	I^{-1}	1.26	2.80	3.42	3.92	8.98
				Cl^{-1}	1.62	2.44	2.04	1.0	4.79

Table 4. $\log K_i^{OH}$ and $\log K_i$ values for the stability constants K_i and K_j of soluble complexes $Me(OH)_i^{+2-i}$ and MeL_j^{+2-j} and pK_{sp} values for the precipitates $\mathbf{MeL_2}$; $[MeL_i^{+2-i}] = K_i[Me^{+2}][L^{-1}]^i$, $K_{sp} = [Me^{+2}][L^{-1}]^2$.

http://dx.doi.org/10.5772/67840

$$[\mathbf{MeL_2}] + [\mathrm{Me}^{+2}] + \sum_{i=1}^{I}[\mathrm{Me(OH)}_i^{+2-i}] + \sum_{j=1}^{J}[\mathrm{MeL}_j^{+2-j}] = C_{\mathrm{Me}} \tag{40}$$

$$2[\mathbf{MeL_2}] + [\mathrm{L}^{-1}] + \sum_{j=1}^{J} j[\mathrm{MeL}_j^{+2-j}] = C_{\mathrm{L}} \tag{41}$$

$$[\mathrm{H}^{+1}] - [\mathrm{OH}^{-1}] + 2[\mathrm{Me}^{+2}] + \sum_{i=1}^{I}(2-i)[\mathrm{Me(OH)}_i^{+2-i}] + \sum_{j=1}^{J}(2-j)[\mathrm{MeL}_j^{+2-j}] - [\mathrm{L}^{-1}] = 0 \tag{42}$$

where $[\mathbf{MeL_2}]$ denotes the concentration of the precipitate $\mathbf{MeL_2}$. At $C_{\mathrm{L}} = 2C_{\mathrm{Me}}$, we have

$$2[\mathrm{Me}^{+2}] + 2\sum_{i=1}^{I}[\mathrm{Me(OH)}_i^{+2-i}] + \sum_{j=1}^{J}(2-j)[\mathrm{MeL}_j^{+2-j}] = [\mathrm{L}^{-1}] \tag{43}$$

From Eqs. (40) and (41)

$$\alpha = [\mathrm{H}^{+1}] - [\mathrm{OH}^{-1}] = \sum_{i=1}^{I} i[\mathrm{Me(OH)}_i^{+2-i}] \tag{44}$$

i.e., reaction of the solution is acidic, $[\mathrm{H}^{+1}] > [\mathrm{OH}^{-1}]$. Applying the relations for the equilibrium constants:

$[\mathrm{Me}^{+2}][\mathrm{L}^{-1}]^2 = K_{sp}$, $[\mathrm{Me(OH)}_i^{+2-i}] = K_i^{\mathrm{OH}}[\mathrm{Me}^{+2}][\mathrm{OH}^{-1}]^i$ ($i = 1,\ldots, I$), $[\mathrm{MeL}_j^{+2-j}] = K_j[\mathrm{Me}^{+2}][\mathrm{L}^{-1}]^j$ ($j = 1,\ldots, J$)

from Eqs. (43) and (44) we have

$$2[\mathrm{Me}^{+2}]^{3/2} \cdot (1 + (1 + \sum_{i=1}^{I} x_i) + K_{sp}{}^{1/2} \cdot [\mathrm{Me}^{+2}] \cdot \sum_{j=1}^{J}(2-j)K_j[\mathrm{L}^{-1}] - K_{sp}{}^{1/2} = 0 \tag{45}$$

where

$$[\mathrm{Me}^{+2}] = \frac{\alpha}{\sum_{i=1}^{I} i \cdot x_i}; \alpha = [\mathrm{H}^{+1}] - [\mathrm{OH}^{-1}] = 10^{-\mathrm{pH}} - 10^{\mathrm{pH}-pK_{\mathrm{W}}}; [\mathrm{L}^{-1}]$$
$$= \left(\frac{K_{sp}}{[\mathrm{Me}^{+2}]}\right)^{1/2}; x_i = K_i^{\mathrm{OH}} \cdot (K_{\mathrm{W}}/[\mathrm{H}^{+1}])^i$$

In particular, for $I = 3$, $J = 4$ (**Table 4**), we have

$$2 \cdot \left(1+\sum_{i=1}^{3} x_i\right) \cdot \frac{[\mathrm{Me}]^2}{K_{\mathrm{sp}}{}^{1/2}} + K_1 \cdot [\mathrm{Me}]^{3/2} - (K_3 \cdot K_{\mathrm{sp}} + 1) \cdot [\mathrm{Me}]^{1/2} - 2 \cdot K_4 \cdot K_{\mathrm{sp}}{}^{3/2} = 0 \qquad (46)$$

Applying the zeroing procedure to Eq. (46) gives the pH = pH_0 of the solution at equilibrium. At this pH_0 value, we calculate the concentrations of all species and solubility of this precipitate recalculated on s_{Me} and s_L. When zeroing Eq. (46), we calculate pH = pH_0 of the solution in equilibrium with the related precipitate. The solubilities are as follows:

$$s = s_{\mathrm{Me}} = [\mathrm{Me}^{+2}] + \sum_{i=1}^{I} [\mathrm{Me(OH)}_i{}^{+2-i}] + \sum_{j=1}^{J} [\mathrm{MeL}_j{}^{+2-j}] \qquad (47)$$

$$s = s_{\mathrm{L}} = [\mathrm{L}^{-1}] + \sum_{j=1}^{4} j[\mathrm{MeL}_j{}^{+2-j}] \qquad (48)$$

The calculations of s_{Me} and s_L for the precipitates specified in **Table 4** can be realized with use of Excel spreadsheet, according to zeroing procedure, as suggested above (**Table 1**).

For **PbI_2**: $pH_0 = 5.1502$, $s_{Pb} = 6.5276 \cdot 10^{-4}$, $s_I = 1.3051 \cdot 10^{-3}$, see **Table 6**. The difference between s_I and $2s_{Pb} = 1.3055 \cdot 10^{-3}$ results from rounding the pH_0-value.

For **HgI_2**: pH_0 = 6.7769, s_{Hg} = $1.91217 \cdot 10^{-5}$, s_I = $3.82435 \cdot 10^{-5}$, see **Table 7**. The difference between s_I and $2s_{Hg}$ = $3.82434 \cdot 10^{-5}$ results from rounding the pH-value. The concentration $[HgI_2]$ = $K_2 K_{sp}$ = $1.90546 \cdot 10^{-5}$ is close to the s_{Hg} value. For comparison, $4(s^*)^3 = K_{sp} \Longrightarrow s^* = 1.93 \cdot 10^{-10}$, i.e., $s^*/s \approx 10^{-5}$.

pH	$[Pb^{+2}]$	$[PbOH^{+1}]$	$[Pb(OH)_2]$	$[Pb(OH)_3{}^{-1}]$	$[PbCl^{+1}]$	$[PbCl_2]$	$[PbCl_3{}^{-1}]$	$[PbCl_4{}^{-2}]$	$[Cl^{-1}]$	y
4.5343	0.010749606	2.92208E-05	7.94315E-11	8.59592E-18	0.017405892	0.004466836	6.90723E-05	2.44685E-07	0.038842191	0.000138249
4.5344	0.010744657	2.92141E-05	7.94315E-11	8.5979E-18	0.017401884	0.004466836	6.90882E-05	2.44798E-07	0.038851136	7.7139E-05
4.5345	**0.01073971**	**2.92074E-05**	**7.94315E-11**	**8.59988E-18**	**0.017397878**	**0.004466836**	**6.91041E-05**	**2.44911E-07**	**0.038860083**	**1.60945E-05**
4.5346	0.010734765	2.92007E-05	7.94315E-11	8.60186E-18	0.017393872	0.004466836	6.912E-05	2.45023E-07	0.038869032	-4.48848E-05
4.5347	0.010729823	2.91939E-05	7.94315E-11	8.60384E-18	0.017389867	0.004466836	6.91359E-05	2.45136E-07	0.038877983	-0.000105799

Table 5. Fragment of display for **$PbCl_2$**.

pH	$[Pb^{+2}]$	$[PbOH^{+1}]$	$[Pb(OH)_2]$	$[Pb(OH)_3{}^{-1}]$	$[PbI^{+1}]$	$[PbI_2]$	$[PbI_3{}^{-1}]$	$[PbI_4{}^{-2}]$	$[I^{-1}]$	y
5.15	0.000630817	7.07789E-06	7.94152E-11	3.54735E-17	1.47894E-05	6.60693E-07	3.54853E-09	1.44576E-11	0.001288393	0.000138249
5.1501	0.000630527	7.07626E-06	7.94152E-11	3.54816E-17	1.4786E-05	6.60693E-07	3.54935E-09	1.44643E-11	0.001288689	7.7139E-05
5.1502	**0.000630236**	**7.07463E-06**	**7.94152E-11**	**3.54898E-17**	**1.47826E-05**	**6.60693E-07**	**3.55016E-09**	**1.44709E-11**	**0.001288986**	**1.60945E-05**
5.1503	0.000629946	7.073E-06	7.94152E-11	3.5498E-17	1.47792E-05	6.60693E-07	3.55098E-09	1.44776E-11	0.001289283	-4.48848E-05
5.1504	0.000629656	7.07137E-06	7.94152E-11	3.55061E-17	1.47758E-05	6.60693E-07	3.5518E-09	1.44843E-11	0.00128958	-0.000105799

Table 6. Fragment of display for **PbI_2**.

http://dx.doi.org/10.5772/67840

pH	$[Hg^{+2}]$	$[HgOH^{+1}]$	$[Hg(OH)_2]$	$[Hg(OH)_3^{-1}]$	$[HgI^{+1}]$	$[HgI_2]$	$[HgI_3^{-1}]$	$[HgI_4^{-2}]$	$[I^{-1}]$	y
6.7767	2.99681E-15	3.57569E-12	5.37106E-08	1.01569E-15	2.17936E-09	1.90546E-05	1.12634E-08	1.87646E-13	9.81003E-08	1.35932E-10
6.7768	2.99398E-15	3.57313E-12	5.36844E-08	1.01543E-15	2.17833E-09	1.90546E-05	1.12688E-08	1.87824E-13	9.81467E-08	7.72021E-11
6.7769	**2.99114E-15**	**3.57056E-12**	**5.36583E-08**	**1.01517E-15**	**2.1773E-09**	**1.90546E-05**	**1.12741E-08**	**1.88002E-13**	**9.81932E-08**	1.8567E-11
6.777	2.98831E-15	3.568E-12	5.36322E-08	1.0149E-15	2.17627E-09	1.90546E-05	1.12794E-08	1.88181E-13	9.82398E-08	-3.99731E-11
6.7771	2.98548E-15	3.56544E-12	5.3606E-08	1.01464E-15	2.17524E-09	1.90546E-05	1.12848E-08	1.88359E-13	9.82863E-08	-9.84182E-11

Table 7. Fragment of display for $\mathbf{HgI_2}$.

4.4. Dissolution of $CaCO_3$ in the presence of CO_2

The portions 0.1 g of calcite $\mathbf{CaCO_3}$ ($M = 100.0869$ g/mol, $d = 2.711$ g/cm^3) are inserted into 100 mL of: pure water (task A) or aqueous solutions of CO_2 specified in the tasks: B1, B2, B3, and equilibrated. Denoting the starting (t = 0) concentrations [mol/L]: C^o for $\mathbf{CaCO_3}$ and C_{CO_2} for CO_2 in the related systems, on the basis of equilibrium data collected in **Table 8**:

(A) we calculate pH = pH_{01} and solubility s = s(pH_{01}) of $\mathbf{CaCO_3}$ at equilibrium in the system;

(B1) we calculate pH = pH_{02} and solubility s = s(pH_{02}) of $\mathbf{CaCO_3}$ in the system, where C_{CO_2} refers to saturated (at 25 °C) solution of CO_2, where 1.45 g CO_2 dissolves in 1 L of water [39].

(B2) we calculate minimal C_{CO_2} in the starting solution needed for complete dissolution of $\mathbf{CaCO_3}$ in the system and the related pH = pH_{03} value, where s = s(pH_{03}) = C^o;

(B3) we plot the $\log s_{Ca}$ versus V, pH versus V and $\log s_{Ca}$ versus pH relationships for the system obtained after addition of V mL of a strong base MOH (C_b = 0.1) into V_0 = 100 mL of the system with $\mathbf{CaCO_3}$ presented in (B1). The *quasistatic* course of the titration is assumed.

The volume 0.1/2.711 = 0.037 cm^3 of introduced $\mathbf{CaCO_3}$ is negligible when compared with V_0 at the start (t = 0) of the dissolution. Starting concentration of $\mathbf{CaCO_3}$ in the systems: A, B1, B2, B3 is $C^o = (0.1/100)/0.1 = 10^{-2}$ mol/L. At $t > 0$, concentration of $\mathbf{CaCO_3}$ is c^o mol/L. The balances are as follows:

No.	Reaction	Expression for the equilibrium constant	Equilibrium data
1	$\mathbf{CaCO_3} = Ca^{+2} + CO_3^{-2}$	$[Ca^{+2}][CO_3^{-2}] = K_{sp}$	$pK_{sp} = 8.48$
2	$Ca^{+2} + OH^{-1} = CaOH^{+1}$	$[CaOH^{+1}] = K_{10}[Ca^{+2}][OH^{-1}]$	$\log K_{10} = 1.3$
3	$H_2CO_3 = H^{+1} + HCO_3^{-1}$	$[H^{+1}][HCO_3^{-1}] = K_1[H_2CO_3]$	$pK_1 = 6.38$
4	$HCO_3^{-1} = H^{+1} + CO_3^{-2}$	$[H^{+1}][CO_3^{-2}] = K_2[HCO_3^{-1}]$	$pK_2 = 10.33$
5	$Ca^{+2} + HCO_3^{-1} = CaHCO_3^{+1}$	$[CaHCO_3^{+1}] = K_{11}[Ca^{+2}][HCO_3^{-1}]$	$\log K_{11} = 1.11$
6	$Ca^{+2} + CO_3^{-2} = CaCO_3$	$[CaCO_3] = K_{12}[Ca^{+2}][CO_3^{-2}]$	$\log K_{12} = 3.22$
7	$\mathbf{Ca(OH)_2} = Ca^{+2} + 2OH^{-1}$	$[Ca^{+2}][OH^{-1}]^2 = K_{sp1}$	$pK_{sp1} = 5.03$
8	$H_2O = H^{+1} + OH^{-1}$	$[H^{+1}][OH^{-1}] = K_W$	$pK_W = 14.0$

Table 8. Equilibrium data.

$$C^o = c^o + [Ca^{+2}] + [CaOH^{+1}] + [CaHCO_3^{+1}] + [CaCO_3] \text{ (for A, B1, B2, B3)} \quad (49)$$

$$C^o = c^o + [CaHCO_3^{+1}] + [CaCO_3] + [H_2CO_3] + [HCO_3^{-1}] + CO_3^{-2}] \text{ (for A)} \quad (50)$$

$$C^o + C_{CO2} = c^o + [CaHCO_3^{+1}] + [CaCO_3] + [H_2CO_3] + [HCO_3^{-1}] + [CO_3^{-2}] \text{(for B1, B2, B3)} \quad (51)$$

$$[H^{+1}] - [OH^{-1}] + 2[Ca^{+2}] + [CaOH^{+1}] + [CaHCO_3^{+1}] - [HCO_3^{-1}] - 2[CO_3^{-2}] = 0 \text{(for A, B1, B2)} \quad (52)$$

$$[H^{+1}] - [OH^{-1}] + [M^{+1}] + 2[Ca^{+2}] + [CaOH^{+1}] + [CaHCO_3^{+1}] - [HCO_3^{-1}] - 2[CO_3^{-2}] = 0 \text{(for B3)} \quad (52a)$$

where $[M^{+1}] = C_bV/(V_0+V)$.

- For (A)

  From Eqs. (49) and (50), we have

$$[Ca^{+2}] + [CaOH^{+1}] = [H_2CO_3] + [HCO_3^{-1}] + [CO_3^{-2}] \quad (53)$$

Considering the solution saturated with respect to **$CaCO_3$** and denoting: $f_1 = 10^{16.71-2pH} + 10^{10.33-pH} + 1$, $f_2 = 1 + 10^{pH-12.7}$, from Eq. (53) and **Table 1**, we have the relations:

$[Ca^{+2}]f_2 = [CO_3^{-2}] \cdot f_1 \Rightarrow [Ca^{+2}] = 10^{-4.24} \cdot (f_1/f_2)^{0.5}$; $[CO_3^{-2}] = 10^{-4.24} \cdot (f_2/f_1)^{0.5}$; $[CaOH^{+1}] = 10^{pH-16.94} \cdot (f_1/f_2)^{0.5}$;
$[CaCO_3] = 10^{-5.26}$; $[CaHCO_3^{+1}] = 10^{2.96-pH}$; $[HCO_3^{-1}] = 10^{6.09-pH} \cdot (f_2/f_1)^{0.5}$; $[H^{+1}] = 10^{-pH}$; $[OH^{-1}] = 10^{pH-14}$.

Inserting them into the charge balance (52), rewritten into the form

$$z = z(pH) = 10^{-pH} - 10^{pH-14} + 2 \cdot 10^{-4.24} \cdot (f_1/f_2)^{0.5} + 10^{pH-16.94} \cdot (f_1/f_2)^{0.5} + 10^{2.96-pH} - 10^{6.09-pH} \cdot (f_2/f_1)^{0.5} - 2 \cdot 10^{-4.24} \cdot (f_2/f_1)^{0.5} \quad (54)$$

and applying the zeroing procedure to the function (54), we find $pH_{01} = 9.904$, at $z = z(pH_{01}) = 0$. The solubility s = s(pH) of **$CaCO_3$**, resulting from Eq. (49), is

$$s = [Ca^{+2}] + [CaOH^{+1}] + [CaHCO_3^{+1}] + [CaCO_3] \quad (55)$$

$$= 10^{-4.24} \cdot (f_1/f_2)^{0.5} + 10^{pH-16.94} \cdot (f_1/f_2)^{0.5} + 10^{2.96-pH} + 10^{-5.26} \quad (55a)$$

We have $s = s(pH = pH_{01}) = 1.159 \cdot 10^{-4}$ mol/L.

- For (B1)

  Subtraction of Eq. (49) from Eq. (51) gives

$$[H_2CO_3] + [HCO_3^{-1}] + [CO_3^{-2}] - ([Ca^{+2}] + [CaOH^{+1}])$$
$$= C_{CO2} \Rightarrow [CO_3^{-2}] \cdot f_1 - [Ca^{+2}] \cdot f_2 - C_{CO2} = 0 \Rightarrow [Ca^{+2}]^2 \cdot f_2 + C_{CO2} \cdot [Ca^{+2}] - K_{sp} \cdot f_1 = 0$$

In this case,

$$[Ca^{+2}] = \frac{\sqrt{(C_{CO_2})^2 + 4 \cdot K_{sp} \cdot f_1 \cdot f_2} - C_{CO_2}}{2 \cdot f_2} \tag{56}$$

where C_{CO_2} = 1.45/44 = 0.0329 mol/L. Eq. (55) has the form

$$s = [Ca^{+2}] \cdot f_2 + 10^{2.96-pH} + 10^{-5.26} \tag{57}$$

and the charge balance is transformed into the zeroing function

$$z = z(pH) = 10^{-pH} - 10^{pH-14} + [Ca^{+2}] \cdot (2 + 10^{pH-12.7}) + 10^{2.96-pH} - [CO_3^{-2}] \cdot (10^{10.33-pH} + 2) \tag{58}$$

where $[CO_3^{\ -2}] = 10^{-8.48}/[Ca^{+2}]$, and $[Ca^{+2}]$ is given by Eq. (56). Eq. (58) zeroes at pH = pH_{02} = 6.031. Then from Eq. (57) we calculate s = s(pH_{02}) = $6.393 \cdot 10^{-3}$ mol/L, at pH = pH_{02} = 6.031.

- For (B2)

  At pH = pH_{03}, where $c^o = 0$, i.e., s = C^o, the solution (a monophase system) is saturated toward $\mathbf{CaCO_3}$, i.e., the relation $[Ca^{+2}][CO_3^{\ -2}] = K_{sp}$ is still valid. Applying Eqs. (56) and (57), we find pH values zeroing Eq. (58) at different, preassumed C_{CO_2} values. Applying these pH-values in Eq. (57), we calculate the related s = s(pH, C_{CO_2}) values (Eq. (57), **Table 9**). Graphically, C_{CO_2} = 0.100 is found at pH_{03} = 5.683, as the abscissa of the point of intersection of the lines: s = s(pH) and s = C^o = 0.01. **Table 9** shows other, preassumed s = C^o values.

- For (B3)

  We apply again the formulas used in (B1) and (B2), and the charge balance (Eq. (52a)), which is transformed there into the function

C_{CO_2}	0.090	0.091	0.092	0.093	0.094	0.095	0.096	0.097	0.098	0.099	0.100	0.101	0.102
pH	5.716	5.712	5.709	5.706	5.702	5.699	5.696	5.693	5.690	5.687	**5.683**	5.680	5.577
s	9.58E-3	9.64E-3	9.67E-3	9.70E-3	9.77E-3	9.80E-3	9.84E-3	9.87E-3	9.91E-3	9.94E-3	**10.01E-3**	10.06E-3	10.10E-3

Table 9. The set of points used for searching the C_{CO_2} value at s = C^o = 0.01; at this point, we have pH_{03} = 5.683.

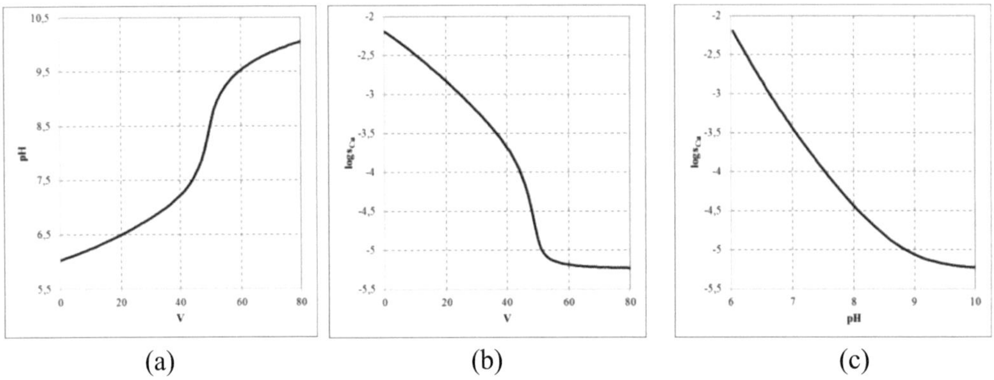

Figure 2. Graphical presentation of the data considered in (b3): (a) pH versus V, (b) log s_{Ca} versus V, (c) log s_{Ca} versus pH relationships.

$$z = z(\mathrm{pH}, \mathrm{V}) = 10^{-\mathrm{pH}} - 10^{\mathrm{pH}-14} + C_b V/(V_0 + V) + [\mathrm{Ca}^{+2}] \cdot (2 + 10^{\mathrm{pH}-12.7}) \\ +10^{2.96-\mathrm{pH}} - [\mathrm{CO}_3^{-2}] \cdot (10^{10.33-\mathrm{pH}} + 2) \quad (59)$$

applied for zeroing purposes, at different V values. The data thus obtained are presented graphically in Figures 2a–c. The data presented in the dynamic solubility diagram (**Figure 2**b), illustrating the solubility changes affected by pH changes (**Figure 2**a) resulting from addition of a base, MOH; **Figure 2**c shows a synthesis of these changes. Solubility product of **$Ca(OH)_2$** is not crossed in this system.

5. Nonequilibrium solid phases in aqueous media

Some solids when introduced into aqueous media (e.g., pure water) may appear to be nonequilibrium phases in these media.

5.1. Silver dichromate ($Ag_2Cr_2O_7$)

The equilibrium data related to the system, where **$Ag_2Cr_2O_7$** is introduced into pure water, were taken from Refs. [33, 40, 41], and presented in **Table 10**. A large discrepancy between pK_{sp2} values (6.7 and 10) in the cited literature is taken here into account. We prove that **$Ag_2Cr_2O_7$** changes into **Ag_2CrO_4**.

On the dissociation step, each dissolving molecule of **$Ag_2Cr_2O_7$** gives two ions Ag^{+1} and 1 ion $Cr_2O_7^{-2}$, where two atoms of Cr are involved; in the contact with water, these ions are hydrolyzed, to varying degrees. In the initial step of the dissolution, before the saturation of the solution with respect to an equilibrium solid phase (not specified at this moment), we can write the concentration balances

Reaction	Equilibrium data
$H_2O = H^{+1} + OH^{-1}$	$pK_w = 14.0$
$H_2CrO_4 = H^+ + HCrO_4^{-1}$	$pK_1 = 0.8$
$HCrO_4^{-1} = H^+ + CrO_4^{-2}$	$pK_2 = 6.5$
$HCr_2O_7^{-1} = H^{+1} + Cr_2O_7^{-2}$	$\log K_3 = 0.07$
$2HCrO_4^{-1} = Cr_2O_7^{-2} + H_2O$	$\log K_4 = 1.52$
$Ag^{+1} + OH^{-1} = AgOH$	$\log K_1^{OH} = 2.3$
$Ag^{+1} + 2OH^{-1} = Ag(OH)_2^{-1}$	$\log K_2^{OH} = 3.6$
$Ag^{+1} + 3OH^{-1} = Ag(OH)_3^{-2}$	$\log K_3^{OH} = 4.8$
$\mathbf{Ag_2CrO_4} = 2Ag^{+1} + CrO_4^{-2}$	$pK_{sp1} = 11.9$
$\mathbf{Ag_2Cr_2O_7} = 2Ag^{+1} + Cr_2O_7^{-2}$	$pK_{sp2} = 6.7$
$\mathbf{AgOH} = Ag^{+1} + OH^{-1}$	$pK_{sp3} = 7.84$

Table 10. Physicochemical equilibrium data relevant to the $\mathbf{Ag_2Cr_2O_7}$ + H_2O system ($pK = -\log K$), at "room" temperatures.

$$2[\mathbf{Ag_2Cr_2O_7}] + [Ag^{+1}] + [AgOH] + [Ag(OH)_2^{-1}] + [Ag(OH)_3^{-2}] = 2C_0 \tag{60}$$

$$2[\mathbf{Ag_2Cr_2O_7}] + [H_2CrO_4] + [HCrO_4^{-1}] + [CrO_4^{-2}] + 2[HCr_2O_7^{-1}] + 2[Cr_2O_7^{-2}] = 2C_0 \tag{61}$$

where $2C_0$ is the total concentration of the solid phase in the system, at the moment ($t = 0$) of introducing this phase into water, $[\mathbf{Ag_2Cr_2O_7}]$ is the concentration of this phase at a given moment of the intermediary step. As previously, we assume that addition of the solid phase (here: $\mathbf{Ag_2Cr_2O_7}$) does not change the volume of the system in a significant degree, and that $\mathbf{Ag_2Cr_2O_7}$ is added in a due excess, securing the formation of a solid (that is not specified at this moment), as an equilibrium solid phase. The balances in Eqs. (60) and (61) are completed by the charge balance

$$\begin{aligned}[H^{+1}] - [OH^{-1}] + [Ag^{+1}] - [Ag(OH)_2^{-1}] - 2[Ag(OH)_3^{-2}] - [HCrO_4^{-1}] \\ -2[CrO_4^{-2}] - [HCr_2O_7^{-1}] - 2[Cr_2O_7^{-2}] = 0\end{aligned} \tag{62}$$

used, as previously, to formulation of the zeroing function, y = y(pH), and the set of relations for equilibrium data specified in **Table 10**. From these relations, we get

$$\begin{aligned}[H_2CrO_4] = 10^{7.3-2pH} \cdot [CrO_4^{-2}]; \quad [HCrO_4^{-1}] = 10^{6.5-pH} \cdot [CrO_4^{-2}]; \\ [HCr_2O_7^{-1}] = 10^{14.59-3pH} \cdot [CrO_4^{-2}]^2;\end{aligned} \tag{63}$$

$$[Cr_2O_7^{-2}] = 10^{14.52-2pH} \cdot [CrO_4^{-2}]^2 \tag{63a}$$

Denoting by $2c_0$ ($< 2C_0$) the total concentration of dissolved Ag and Cr species formed, in a transition stage, from $\mathbf{Ag_2Cr_2O_7}$, we can write

$$[Ag^{+1}] + [AgOH] + [Ag(OH)_2^{-1}] + [Ag(OH)_3^{-2}] = 2c_0 \tag{64}$$

$$[H_2CrO_4] + [HCrO_4^{-1}] + [CrO_4^{-2}] + 2[HCr_2O_7^{-1}] + 2[Cr_2O_7^{-2}] = 2c_0 \tag{65}$$

From **Table 10** and formulas (63)–(65) we get the relations:

$$\text{(a) } [Ag^{+1}] = 2c_0/g_0;\ 2g_2[CrO_4^{-2}]^2 + g_1[CrO_4^{-1}]\text{–}2c_0 = 0 \Rightarrow (b)\ [CrO_4^{-2}] = \frac{(g_1^2 + 16 \cdot c_0 g_2)^{0.5} - g_1}{4 \cdot g_2} \tag{66}$$

where $g_0 = 1 + 10^{pH-11.7} + 10^{2pH-24.4} + 10^{3pH-37.2}$; $g_1 = 10^{7.3-2pH} + 10^{6.5-pH} + 1$; $g_2 = 10^{14.59-3pH} + 10^{14.52-2pH}$. Applying them in Eq. (62), we get the zeroing function

$$y = y(pH) = 10^{-pH} – 10^{pH-14} + g_3 \cdot [Ag^{+1}] – g_4 \cdot [CrO_4^{-2}] - g_5 \cdot [CrO_4^{-2}]^2 \tag{67}$$

where $g_3 = 1 - 10^{2pH-24.4} - 2 \cdot 10^{3pH-37.2}$; $g_4 = 10^{6.5-pH} + 2$; $g_5 = 10^{14.59-3pH} + 2 \cdot 10^{14.52-2pH}$, and $[Ag^{+1}]$ and $[CrO_4^{-2}]$ are defied above, as functions of pH.

The calculation procedure, realizable with use of Excel spreadsheet, is as follows. We assume a sequence of growing numerical values for $2c_0$. At particular $2c_0$ values, we calculate pH = pH($2c_0$) value zeroing the function (67), and then calculate the values of the products: $q_1 = [Ag^{+1}]^2[CrO_4^{-2}]/K_{sp1}$ and $q_2 = [Ag^{+1}]^2[Cr_2O_7^{-2}]/K_{sp2}$, where: $[Ag^{+1}]$, $[CrO_4^{-2}]$, and $[Cr_2O_7^{-2}]$ are presented above (Eqs. (66a), (66b) and (63a), resp.), pK_{sp1} = 11.9, pK_{sp2} = 6.7. As results from **Figure 3**, where $\log q_1$ and $\log q_2$ are plotted as functions of $2c_0$; $\log q_1 = 0 \Leftrightarrow q_1 = 1 \Leftrightarrow [Ag^{+1}]^2[CrO_4^{-2}] = K_{sp1}$ at lower $2c_0$ value, whereas $\log q_2 < 0 \Leftrightarrow q_2 < 1 \Leftrightarrow [Ag^{+1}]^2[Cr_2O_7^{-2}] < K_{sp2}$, both for pK_2 = 6.7 and 10, cited in the literature. The x_1=1

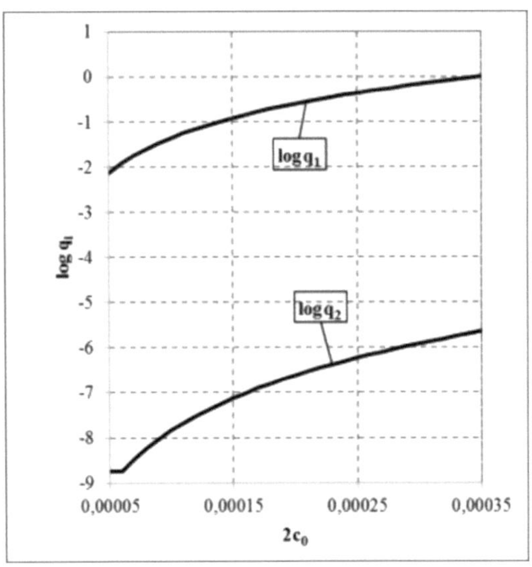

Figure 3. The convergence of $\log q_1$ and $\log q_2$ to 0 value; K_{sp1} is attained at lower $2c_0$ value.

http://dx.doi.org/10.5772/67840

value is attained at $2c_0 = 3.5 \cdot 10^{-4} \Longrightarrow c_0 = 1.75 \cdot 10^{-4}$; then $\mathbf{Ag_2CrO_4}$ precipitates as the new solid phase, i.e., total depletion of $\mathbf{Ag_2Cr_2O_7}$ occurs. It means that $\mathbf{Ag_2Cr_2O_7}$ is not the equilibrium solid phase in this system. This fact was confirmed experimentally, as stated in [42], i.e., $\mathbf{Ag_2Cr_2O_7}$ is transformed into $\mathbf{Ag_2CrO_4}$ upon boiling with H_2O; at higher temperatures, this transformation proceeds more effectively. Concluding, the formula $s^* = (K_{sp2}/4)^{1/3}$ applied for $K_{sp2} = [Ag^{+1}]^2[Cr_2O_7^{-2}]$ is not "the best answer," as stated in Ref. [43].

The system involved with $\mathbf{Ag_2CrO_4}$ was also considered in context with the Mohr's method of Cl^{-1} determination [44–46]. As were stated there, the systematic error in Cl^{-1} determining according to this method, expressed by the difference between the equivalence (eq) volume ($V_{eq} = C_0V_0/C$) and the volume V_{end} corresponding to the end point where the K_{sp1} for $\mathbf{Ag_2CrO_4}$ is crossed, equals to

$$V_{eq} - V_{end} = \frac{K_{sp}}{C} \cdot \left(\frac{C_{01}V_0}{K_{sp1}}\right)^{0.5} \cdot (V_0 + V_{end})^{0.5} - \frac{1}{C} \cdot \left(\frac{K_{sp1}}{C_{01}V_0}\right)^{0.5} \cdot (V_0 + V_{end})^{1.5}$$

where $K_{sp} = [Ag^{+1}][Cl^{-1}]$ ($pK_{sp} = 9.75$), V_0 is the volume of titrant with NaCl (C_0) + K_2CrO_4 (C_{01}) titrated with $AgNO_3$ (C) solution; $V_{end} = V_{eq}$ at $C_{01} = (1 + V_{end}/V_0) \cdot K_{sp1}/K_{sp}$.

All calculations presented above were realized using Excel spreadsheets. For more complex nonequilibrium two-phase systems, the use of iterative computer programs, e.g., ones offered by MATLAB [8, 47], is required. This way, the *quasistatic* course of the relevant processes under isothermal conditions can be tested [48].

5.2. Dissolution of struvite

The fact that NH_3 evolves from the system obtained after leaving pure struvite pr1 in contact with pure water, e.g., on the stage of washing this precipitate, has already been known at the end of nineteenth century [49]. It was noted that the system obtained after mixing magnesium, ammonium, and phosphate salts at the molar ratio 1:1:1 gives a system containing an excess of ammonium species remaining in the solution and the precipitate that "*was not struvite, but was probably composed of magnesium phosphates*" [50]. This effect can be explained by the reaction [20]

$$3\mathbf{MgNH_4PO_4} = \mathbf{Mg_3(PO_4)_2} + HPO_4^{-2} + NH_3 + 2NH_4^{+1} \qquad (68)$$

Such inferences were formulated on the basis of X-ray diffraction analysis, the crystallographic structure of the solid phase thus obtained. It was also stated that the precipitation of struvite requires a significant excess of ammonium species, e.g., Mg:N:P = 1:1.6:1. Struvite (pr1) is the equilibrium solid phase only at a due excess of one or two of the precipitating reagents. This remark is important in context with gravimetric analysis of magnesium as pyrophosphate. Nonetheless, also in recent times, the solubility of struvite is calculated from the approximate formula $s^* = (K_{sp1})^{1/3}$ based on an assumption that it is the equilibrium solid phase in such a system.

Struvite is not the equilibrium solid phase also when introduced into aqueous solution of CO_2 (C_{CO_2}, mol/L), modified (or not) by free strong acid HB (C_a, mol/L) or strong base MOH (C_b, mol/L).

The case of struvite requires more detailed comments. The reaction (68) was proved theoretically [20], on the basis of simulated calculations performed by iterative computer programs, with use of all attainable physicochemical knowledge about the system in question. For this purpose, the fractions

$$q_1 = [Mg^{+2}][NH_4^{+1}][PO_4^{-3}]/K_{sp1}, q_2 = [Mg^{+2}]^3[PO_4^{-3}]^2/K_{sp2}, q_3 = [Mg^{+2}][HPO_4^{-2}]/K_{sp3}, q_4[Mg^{+2}][OH^{-1}]^2/K_{sp4} \tag{69}$$

were calculated for: pr1 = **$MgNH_4PO_4$** (pK_{sp1} = 12.6), pr2 = **$Mg_3(PO_4)_2$** (pK_{sp2} = 24.38), pr3 = **$MgHPO_4$** (pK_{sp3} = 5.5), pr4 = **$Mg(OH)_2$** (pK_{sp4} = 10.74) and are presented in **Figure 4**, at an initial concentration of pr1, equal $C^0 = [pr1]_{t=0} = 10^{-3}$ mol/L ($pC^0 = (ppr1)_{t=0} = 3$); ppr1 = −log [pr1]. As we see, the precipitation of pr2 (Eq. (68)) starts at ppr1 = 3.088; other solubility products are not crossed. The changes in concentrations of some species, resulting from dissolution of pr1, are indicated in **Figure 5**, where s is defined by equation [20]

$$s = s_{Mg} = [Mg^{+2}] + [MgOH^{+1}] + [MgH_2PO_4^{+1}] + [MgHPO_4] + [MgPO_4^{-1}] + [MgNH_3^{+2}] + [Mg(NH_3)_2^{+2}] + [Mg(NH_3)_3^{+2}] \tag{70}$$

involving all soluble magnesium species are identical in its form, irrespective of the equilibrium solid phase(s) present in this system. Moreover, it is stated that pH in the solution equals

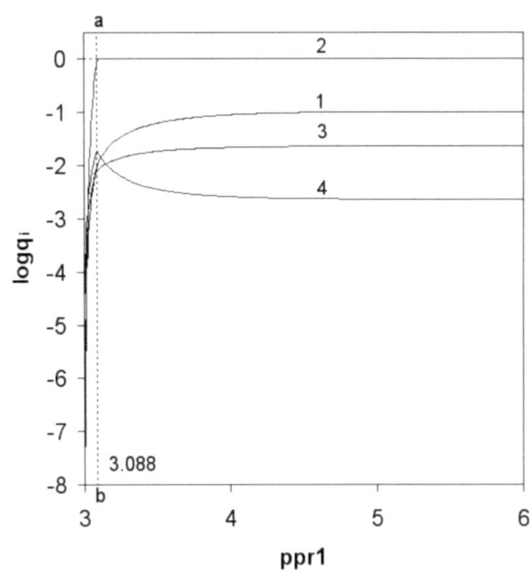

Figure 4. Plots of $\log q_i$ versus ppr1 = −log[pr1] relationships, at $(ppr1)_{t=0} = 3$; i = 1,2,3,4 refer to pr1, pr2, pr3 and pr4, respectively.

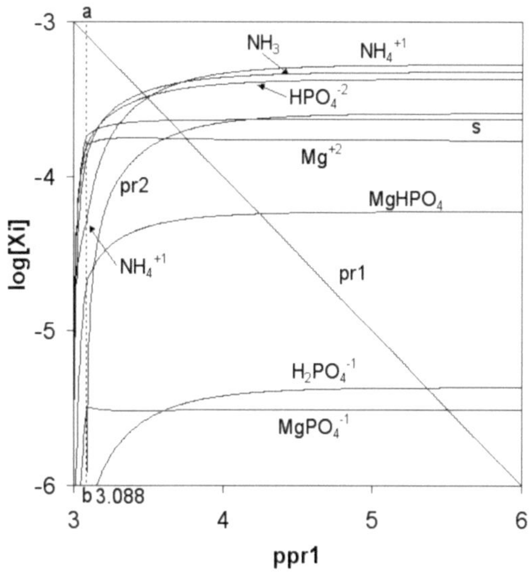

Figure 5. The speciation curves for indicated species resulting from dissolution of pr1 at $(ppr1)_{t=0} = 3$.

ca. 9–9.5 (**Figure 6**); this pH can be affected by the presence of CO_2 from air. Under such conditions, NH_4^{+1} and NH_3 occur there at comparable concentrations $[NH_4^{+1}] \approx [NH_3]$, but $[HPO_4^{-2}]/[PO_4^{-3}] = 10^{12.36-pH} \approx 10^3$. This way, the scheme (10) would be more advantageous, provided that struvite is the equilibrium solid phase; but it is not the case, see Eq. (68). The reaction (68) occurs also in the presence of CO_2 in water where struvite was introduced.

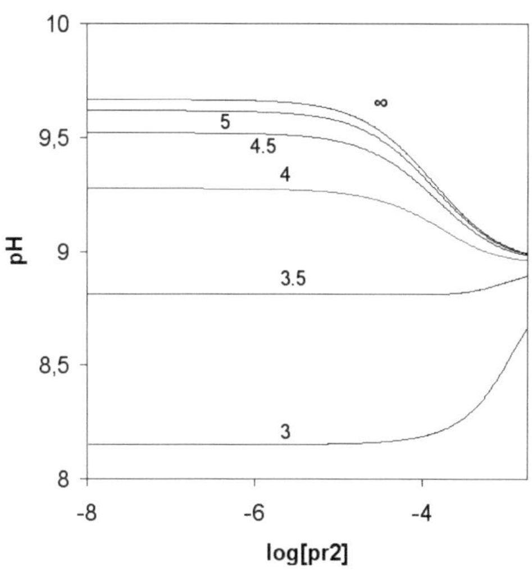

Figure 6. The pH versus log[pr2] relationship; pr2 = $\mathbf{Mg_3(PO_4)_2}$, at $[ppr1]_{t=0} = 3$. The numbers at the corresponding lines indicate $pCO_2 = -\log C_{CO_2}$ values; $pCO_2 = \infty \Leftrightarrow C_{CO_2} = 0$.

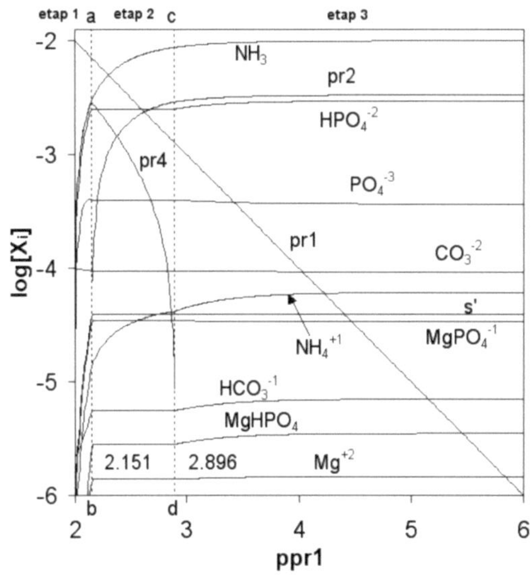

Figure 7. The speciation curves for indicated species $X_i^{z_i}$, resulting from dissolution of pr1 = $\mathbf{MgNH_4PO_4}$, at (pC_0, pCO_2, pC_b) = (2, 4, 2); s′ is defined by Eq. (71).

After introducing struvite pr1 (at $pC_0 = [ppr1]_{t=0} = 2$) into alkaline ($C_b = 10^{-2}$ mol/L KOH, $pC_b = 2$) solution of CO_2 ($pCO_2 = 4$), the dissolution is more complicated and proceeds in three steps, see **Figure 7**.

In step 1, pr4 precipitates first, pr1 + $2OH^{-1}$ = pr4 + NH_3 + HPO_4^{-2}, nearly from the very start of pr1 dissolution, up to ppr1 = 2.151, where K_{sp2} is attained. Within step 2, the solution is saturated toward pr2 and pr4. In this step, the reaction expressed by the notation 2pr1 + pr4 = pr2 + $2NH_3$ + $2H_2O$ occurs up to total depletion of pr4 (at ppr1 = 2.896). In this step, the reaction 3pr1 + $2OH^{-1}$ = pr2 + $3NH_3$ + HPO_4^{-1} + $2H_2O$ occurs up to total depletion of pr1, i.e., the solubility product K_{sp1} for pr1 is not crossed. The curve s′ (**Figure 7**) is related to the function

$$s' = s + [MgHCO_3^{+1}] + [MgCO_3] \tag{71}$$

where s is expressed by Eq. (70).

6. Solubility of nickel dimethylglyoximate

The precipitate of nickel dimethylglyoximate, $\mathbf{NiL_2}$, has soluble counterpart with the same formula, i.e., NiL_2, in aqueous media. If $\mathbf{NiL_2}$ is in equilibrium with the solution, concentration of the soluble complex NiL_2 assumes constant value: $[NiL_2] = K_2 \cdot [Ni^{2+}][L^-]^2 = K_2 \cdot K_{sp}$, where $K_2 = 10^{17.24}$, $K_{sp} = [Ni^{2+}][L^-]^2 = 10^{-23.66}$ [14, 17, 18], and then $[NiL_2] = 10^{-6.42}$ (i.e., $\log[NiL_2] = -6.42$). The concentration $[NiL_2]$ is the constant, limiting component in expression for solubility $s = s_{Ni}$ of nickel dimethylglyoximate, $\mathbf{NiL_2}$. Moreover, it is a predominant component in

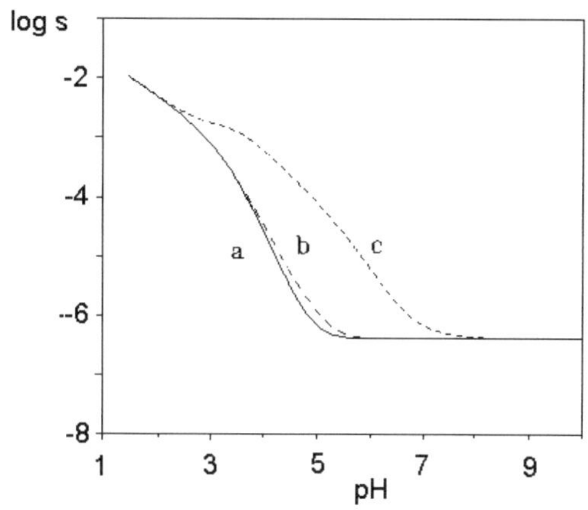

Figure 8. Solubility curves for nickel dimethylglyoximate $\mathbf{NiL_2}$ in (a) ammonia, (b) acetate+ammonia, and (c) citrate +acetate+ ammonia media at total concentrations [mol/L]: $C_{Ni} = 0.001$, $C_L = 0.003$, $C_N = 0.5$, $C_{Ac} = 0.3$, $C_{Cit} = 0.1$ [14].

expression for s in alkaline media, see **Figure 8**. This pH range involves pH of ammonia buffer solutions, where $\mathbf{NiL_2}$ is precipitated from $NiSO_4$ solution during the gravimetric analysis of nickel; the expression for solubility

$$s = s_{Ni} = [Ni^{+2}] + [NiOH^{+1}] + [NiSO_4] + \sum_{i=1}^{6}[Ni(NH_3)_i{}^{+2}] + [NiL_2] \tag{72}$$

The effect of other, e.g., citrate (Cit) and acetate (Ac) species as complexing agents can also be considered for calculation purposes, see the lines b and c in **Figure 8**. The presence of citrate does not affect significantly the solubility of $\mathbf{NiL_2}$ in ammonia buffer media, i.e., at pH ≈ 9, where $s_{Ni} \cong [NiL_2]$.

Calculations of $s = s_{Ni}$ were made at $C_{Ni} = 0.001$ mol/L and $C_L = 0.003$ mol/L HL, i.e., at the excessive HL concentration equal $C_L - 2C_{Ni} = 0.001$ mol/L. Solubility of HL in water, equal 0.063 g HL/100 mL H_2O (25°C) [51], corresponds to concentration 0.63/116.12 = 0.0054 mol/L of the saturated HL solution, 0.003 < 0.0054. Applying higher C_L values needs the HL solution in ethanol, where HL is fairly soluble. However, the aqueous-ethanolic medium is thus formed, where equilibrium constants are unknown. To avoid it, lower C_{Ni} and C_L values were applied in calculations. The equilibrium data were taken from Ref. [31].

The soluble complex having the formula identical to the formula of the precipitate occurs also in other, two-phase systems. In some pH range, concentration of this soluble form is the dominant component of the expression for the solubility s. As stated above, such a case occurs for $\mathbf{NiL_2}$. Then one can assume the approximation

$$s = K_2 K_{sp} \tag{73}$$

Similar relationship exists also for other precipitates. By differentiation of Eq. (73) with respect to temperature T at p = const, and application of van't Hoff's isobar equation for K_2 and K_{sp}, we obtain

$$\frac{1}{s}\cdot\left(\frac{\partial s}{\partial T}\right)_p = \frac{1}{RT^2}\cdot(\Delta G_1^o + \Delta G_2^o) \tag{74}$$

where

$$\Delta G_1^o = RT^2\cdot\left(\frac{\partial \ln K_{sp}}{\partial T}\right)_p \quad \text{and} \quad \Delta G_2^o = RT^2\cdot\left(\frac{\partial \ln K_2}{\partial T}\right)_p$$

Because, as a rule,

$$\left(\frac{\partial K_{sp}}{\partial T}\right)_p > 0 \quad \text{and} \quad \left(\frac{\partial K_2}{\partial T}\right)_p < 0$$

then $\Delta G_1^o > 0$ and $\Delta G_2^o < 0$, and Eq. (74) can be rewritten into the form

$$\frac{1}{s}\cdot\left(\frac{\partial s}{\partial T}\right)_p = \frac{1}{RT^2}\cdot(|\Delta G_1^o| - |\Delta G_2^o|) \tag{75}$$

If $|\Delta G_1^o| \approx |\Delta G_2^o|$ within the temperature range (T_0, T), the value of s is approximately constant. Let T_0 denote the room temperature (at which,as a rule—all the equilibrium constants are determined) and $T \neq T_0$ is the temperature at which the precipitate is filtered and washed. In this case, the solubility s and then theoretical accuracy of gravimetric analysis does not change with temperature.

7. Calculation of solubility in dynamic redox systems

7.1. Preliminary information

The redox system presented in this section is resolvable according to generalized approach to redox systems (GATES), formulated by Michałowski (1992) [8]. According to GATES principles, the algebraic balancing of any electrolytic system is based on the rules of conservation of particular elements/cores Y_g (g = 1,…, G), and on charge balance (ChB), expressing the rule of electroneutrality of this system; the terms element and core are then distinguished. The core is a cluster of elements with defined composition (expressed by its chemical formula) and external charge that remains unchanged during the chemical process considered, e.g., titration. For ordering purposes, we assume: Y_1 = H, Y_2 = O,…. For modeling purposes, the closed systems, composed of condensed phases separated from its environment by diathermal (freely permeable by heat) walls, are considered; it enables the heat exchange between the system and its environment. Any chemical process, such as titration, is carried out under isothermal conditions, in a *quasistatic* manner; constant temperature is one of the conditions securing constancy of equilibrium constants values. An exchange of the matter (H_2O, CO_2, O_2,…) between the system and its environment is thus forbidden, for modeling purposes. The elemental/core balance $F(Y_g)$ for the g-th element/core (Y_g) (g = 1,…, G) is expressed by an equation interrelating the numbers of Y_g-atoms or cores in components of the system with the numbers of Y_g-atoms/cores in the species of the system thus formed; we have F(H) for Y_1 = H, F(O) for Y_2 = O, etc.

The key role in redox systems is due to generalized electron balance (GEB) concept, discovered by Michałowski as the Approach I (1992) and Approach II (2006) to GEB; both approaches are equivalent:

$$\text{Therefore, Approach II to GEB} \leftrightarrow \text{Approach I to GEB} \tag{76}$$

GEB is fully compatible with charge balance (ChB) and concentration balances $F(Y_g)$, formulated for different elements and cores. The primary form of GEB, pr-GEB, obtained according to Approach II to GEB is the linear combination

$$\text{pr} - \text{GEB} = 2 \cdot F(\text{O}) – F(\text{H}) \tag{77}$$

Both approaches (I and II) to GEB were widely discussed in the literature [7–12, 14, 15, 17, 18, 34, 52–74], and in three other chapters in textbooks [75–79] issued in 2017 within InTech. The GEB is perceived as a law of nature [9, 10, 17, 67, 71, 73, 74], as the hidden connection of physicochemical laws, as a breakthrough in the theory of electrolytic redox systems. The GATES refers to mono- and polyphase, redox, and nonredox, equilibrium and metastable [20, 21–23, 78, 79] static and dynamic systems, in aqueous, nonaqueous, and mixed-solvent media [69, 72], and in liquid-liquid extraction systems [53]. Summarizing, Approach II to GEB needs none prior information on oxidation numbers of all elements in components forming a redox system and in the species in the system thus formed. The Approach I to GEB, considered as the "short" version of GEB, is useful if all the oxidation numbers are known beforehand; such a case is obligatory in the system considered below. The terms "oxidant" and "reductant" are not used within both approaches. In redox systems, $2{\cdot}F(\text{O}) - F(\text{H})$ is linearly independent on CHB and $F(Y_g)$ ($g \geq 3,\ldots, G$); in nonredox systems, $2{\cdot}F(\text{O}) - F(\text{H})$ is dependent on those balances. This property distinguishes redox and nonredox systems of any degree of complexity. Within GATES, and GATES/GEB in particular, the terms: "stoichiometry," "oxidation number," "oxidant," "reductant," "equivalent mass" are considered as redundant, old-fashioned terms. The term "mass action law" (MAL) was also replaced by the equilibrium law (EL), fully compatible with the GATES principles. Within GATES, the law of charge conservation and law of conservation of all elements of the system tested have adequate importance/significance.

A detailed consideration of complex electrolytic systems requires a collection and an arrangement of qualitative (particular species) and quantitative data; the latter ones are expressed by interrelations between concentrations of the species. The interrelations consist of material balances and a complete set of expressions for equilibrium constants. Our further considerations will be referred to a titration, as a most common example of dynamic systems. The redox and nonredox systems, of any degree of complexity, can be resolved in analogous manner, without any simplifications done, with the possibility to apply all (prior, preselected) physicochemical knowledge involved in equilibrium constants related to a system in question. This way, one can simulate (imitate) the analytical prescription to any process that may be realized under isothermal conditions, in mono- and two-phase systems, with liquid-liquid extraction systems included.

7.2. Solubility of CuI in a dynamic redox system

The system considered in this section is related to iodometric, indirect analysis of an acidified (H_2SO_4) solution of $CuSO_4$ [14, 64]. It is a very interesting system, both from analytical and physicochemical viewpoints. Because the standard potential $E_0 = 0.621$ V for (I_2, I^{-1}) exceeds $E_0 = 0.153$ V for (Cu^{+2}, Cu^{+1}), one could expect (at a first sight) the oxidation of Cu^{+1} by I_2. However, such a reaction does not occur, due to the formation of sparingly soluble **CuI** precipitate ($pK_{sp} = 11.96$).

This method consists of four steps. In the preparatory step (step 1), an excess of H_2SO_4 is neutralized with NH_3 (step 1) until a blue color appears, which is derived from $Cu(NH_3)_i^{+2}$ complexes. Then the excess of CH_3COOH is added (step 2), to attain a pH ca. 3.6. After subsequent introduction of an excess of KI solution (step 3), the mixture with **CuI** precipitate and dissolved iodine formed in the reactions: $2Cu^{+2} + 4I^{-1} = 2\mathbf{CuI} + I_2$, $2Cu^{+2} + 5I^{-1} = 2\mathbf{CuI} + I_3^{-1}$ is titrated with $Na_2S_2O_3$ solution (step 4), until the reduction of iodine: $I_2 + 2S_2O_3^{-2} = 2I^{-1} + S_4O_6^{-2}$, $I_3^{-1} + 2S_2O_3^{-2} = 3I^{-1} + S_4O_6^{-2}$ is completed; the reactions proceed quantitatively in mildly acidic solutions (acetate buffer), where the thiosulfate species are in a metastable state. In strongly acidic media, thiosulfuric acid disproportionates according to the scheme $H_2S_2O_3 = H_2SO_3 + \mathbf{S}$ [80].

7.3. Formulation of the system

We assume that V mL of C mol/L $Na_2S_2O_3$ solution is added into the mixture obtained after successive addition of: V_N mL of NH_3 (C_1) (step 1), V_{Ac} mL of CH_3COOH (C_2) (step 2), V_{KI} mL of KI (C_3) (step 3), and V mL of $Na_2S_2O_3$ (C) (step 4) into V_0 mL of titrand D composed of $CuSO_4$ (C_0) + H_2SO_4 (C_{01}). To follow the changes occurring in particular steps of this analysis, we assume that the corresponding reagents in particular steps are added according to the titrimetric mode, and the assumption of the volumes additivity is valid.

In this system, three electron-active elements are involved: Cu (atomic number $Z_{Cu} = 29$), I ($Z_I = 53$), S ($Z_S = 16$). Note that sulfur in the core SO_4^{-2} is not involved here in electron-transfer equilibria between $S_2O_3^{-2}$ and $S_4O_6^{-2}$; then the concentration balance for sulfate species can be considered separately.

The balances written according to Approach I to GEB, in terms of molar concentrations, are as follows:

- Generalized electron balance (GEB)

$$(Z_{Cu}-2)([Cu^{+2}] + [CuOH^{+1}] + [Cu(OH)_2] + [Cu(OH)_3^{-1}] + [Cu(OH)_4^{-2}] + [CuNH_3^{+2}] + [Cu(NH_3)_2^{+2}] + [Cu(NH_3)_3^{+2}] + [Cu(NH_3)_4^{+2}] + [CuCH_3COO^{+1}] + [Cu(CH_3COO)_2]) + (Z_{Cu} - 2 + Z_I - 5)[CuIO_3^{+1}] + (Z_{Cu} - 1)([Cu^{+1}] + [CuNH_3^{+1}] + [Cu(NH_3)_2^{+1}]) + (Z_{Cu} + Z_I)[\mathbf{CuI}_{(s)}] + (Z_{Cu} + 2Z_I + 1)[CuI_2^{-1}] + (Z_I + 1)[I^{-1}] + (3Z_I + 1)[I_3^{-1}] + 2Z_I([I_2] + a \cdot [\mathbf{I}_{2(s)}]) + (Z_I - 1)([HIO] + [IO^{-1}]) + (Z_I - 5)([HIO_3] + [IO_3^{-1}]) + (Z_I - 7)([H_5IO_6] + [H_4IO_6^{-1}] + [H_3IO_6^{-2}]) + 2 \cdot (Z_S - 2)([H_2S_2O_3] + [HS_2O_3^{-1}]) + [S_2O_3^{-2}]) + 4 \cdot (Z_S-2.5)[S_4O_6^{-2}] + (Z_{Cu}-1 + 2 \cdot (Z_S-2))[CuS_2O_3^{-1}] + (Z_{Cu}-1 + 4 \cdot (Z_S-2))[Cu(S_2O_3)_2^{-3}] + (Z_{Cu}-1 + 6 \cdot (Z_S-2))[Cu(S_2O_3)_3^{-5}] - ((Z_{Cu} - 2)C_0V_0 + (Z_I + 1)C_3V_{KI} + 2 \cdot (Z_S - 2)CV)/(V_0+V_N+V_{Ac}+V_{KI}+V) = 0 \quad (78)$$

http://dx.doi.org/10.5772/67840

- CHB

$$[H^{+1}]–[OH^{-1}] + [K^{+1}] + [Na^{+1}] - [HSO_4^{-1}] - 2[SO_4^{-2}] + [NH_4^{+1}] - [CH_3COO^{-1}] + 2[Cu^{+2}] +[CuOH^{+1}] - [Cu(OH)_3^{-1}]–2[Cu(OH)_4^{-2}] + 2[Cu(NH_3)^{+2}] + 2[Cu(NH_3)_3^{+2}] + 2[Cu(NH_3)_3^{+2}] +2Cu(NH_3)_4^{+2}] + [CuCH_3COO^{+1}] - [I^{-1}] - [I_3^{-1}]–[IO^{-1}] - [IO_3^{-1}] - [H_4IO_6^{-1}] - 2[H_3IO_6^{-2}] +[Cu^{+1}] + [CuNH_3^{+1}] + [Cu(NH_3)_2^{+1}] - [CuI_2^{-1}] + [CuIO_3^{+1}] - [HS_2O_3^{-1}] - 2[S_2O_3^{-2}] -2[S_4O_6^{-2}] - [CuS_2O_3^{-1}] - 3[Cu(S_2O_3)_2^{-3}] - 5[Cu(S_2O_3)_3^{-5}] = 0 \quad (79)$$

- F(Cu)

$$[Cu^{+2}] + [CuOH^{+1}] + [Cu(OH)_2] + [Cu(OH)_3^{-1}] + [Cu(OH)_4^{-2}] + [CuSO_4] + [CuNH_3^{+2}] +[Cu(NH_3)_2^{+2}] + [Cu(NH_3)_3^{+2}] + [Cu(NH_3)_4^{+2}] + [CuCH_3COO^{+1}] + [Cu(CH_3COO)_2] +[Cu^{+1}] + [CuNH_3^{+1}] + [Cu(NH_3)_2^{+1}] + [CuI_2^{-1}] + [\mathbf{CuI}_{(s)}] + [CuIO_3^{+1}] + [CuS_2O_3^{-1}] +[Cu(S_2O_3)_2^{-3}] + [Cu(S_2O_3)_3^{-5}]–C_0V_0/(V_0 + V_N + V_{Ac} + V_{KI} + V) = 0 \quad (80)$$

- F(SO$_4$)

$$[HSO_4^{-1}] + [SO_4^{-2}] + [CuSO_4]–C_{01}V_0/(V_0 + V_N + V_{Ac} + V_{KI} + V) = 0 \quad (81)$$

- F(NH$_3$)

$$[NH_4^{+1}] + [NH_3] + [CuNH_3^{+2}] + 2[Cu(NH_3)_2^{+2}] + 3[Cu(NH_3)_3^{+2}] + 4[Cu(NH_3)_4^{+2}] +[CuNH_3^{+1}] + 2[Cu(NH_3)_2^{+1}]–C_1V_N/(V_0 + V_N + V_{Ac} + V_{KI} + V) = 0 \quad (82)$$

- F(CH$_3$COO)

$$[CH_3COOH] + [CH_3COO^{-1}] + [CuCH_3COO^{+1}] + 2[Cu(CH_3COO)_2] –C_2V_{Ac}/(V_0 + V_N + V_{Ac} + V_{KI} + V) = 0 \quad (83)$$

- F(K)

$$[K^{+1}] = C_3V_{KI}/(V_0 + V_N + V_{Ac} + V_{KI} + V) \quad (84)$$

- F(I)

$$[I^{-1}] + 3[I_3^{-1}] + 2([I_2] + [\mathbf{I}_{2(s)}]) + [HIO] + [IO^{-1}] + [HIO_3] + [IO_3^{-1}] + [H_5IO_6] + [H_4IO_6^{-1}] +[H_3IO_6^{-2}] + 2[CuI_2^{-1}] + [\mathbf{CuI}_{(s)}] + [CuIO_3^{+1}]–C_3V_{KI}/(V_0 + V_N + V_{Ac} + V_{KI} + V) = 0 \quad (85)$$

- F(S)

$$2[H_2S_2O_3] + 2[HS_2O_3^{-1}] + 2[S_2O_3^{-2}] + 4[S_4O_6^{-2}] + 2[CuS_2O_3^{-1}] + 4[Cu(S_2O_3)_2{}^{-3}]$$
$$+6[Cu(S_2O_3)_3{}^{-5}]–2CV/(V_0 + V_N + V_{Ac} + V_{KI} + V) = 0 \Rightarrow [H_2S_2O_3] + [HS_2O_3^{-1}] + [S_2O_3^{-2}]$$
$$+2[S_4O_6^{-2}] + [CuS_2O_3^{-1}] + 2[Cu(S_2O_3)_2{}^{-3}] + 3[Cu(S_2O_3)_3{}^{-5}]$$
$$–CV/(V_0 + V_N + V_{Ac} + V_{KI} + V) = 0 \quad (86)$$

- F(Na)

$$[Na^{+1}] = 2CV/(V_0 + V_N + V_{Ac} + V_{KI} + V) \quad (87)$$

The GEB is presented here in terms of the Approach I to GEB, based on the "card game" principle, with Cu (Eq. (80)), I (Eq. (85)) as S (Eq. (86)) as "players," and H, O, S (Eq. (81)), C (from Eq. (83)), N (from Eq. (82)), K, Na as "fans." There are together 47 species involved in 2 + 6 = 8, Eqs. (78)–(83), (85), (86) and two equalities; $[K^{+1}]$ (Eq. (84)) and $[Na^{+1}]$ (Eq. (87)) are not involved in expressions for equilibrium constants, and then are perceived as numbers (not variables), at a particular *V*-value. Concentrations of the species in the equations are interrelated in 35 independent equilibrium constants:

$$[H^{+1}] = 10^{-pH}, [OH^{-1}] = 10^{pH-14}(pK_W = 14), [CuOH^{+1}] = 10^{7.0} \cdot [Cu^{+2}][OH^{-1}], [Cu(OH)_2]$$
$$= 10^{13.68} \cdot [Cu^{+2}][OH^{-1}]^2,$$
$$[Cu(OH)_3{}^{-1}] = 10^{17.0} \cdot [Cu^{+2}][OH^{-1}]^3, [Cu(OH)_4^{-2}] = 10^{18.5} \cdot [Cu^{+2}][OH^{-1}]^4, [CuNH_3^{+2}]$$
$$= 10^{3.39} \cdot [Cu^{+2}][NH_3],$$
$$[Cu(NH_3)_2^{+2}] = 10^{7.33} \cdot [Cu^{+2}][NH_3]^2, [Cu(NH_3)_3^{2+}] = 10^{10.06} \cdot [Cu^{+2}][NH_3]^3, [Cu(NH_3)_4^{+2}]$$
$$= 10^{12.03} \cdot [Cu^{+2}][NH_3]^4,$$
$$[CuSO_4] = 10^{2.36} \cdot [Cu^{+2}][SO_4^{-2}], [NH_4^{+1}] = 10^{9.35} \cdot [H^{+1}][NH_3], [HSO_4^{-1}] = 10^{1.8} \cdot [H^{+1}][SO_4^{-2}],$$
$$[CH_3COOH] = 10^{4.65} \cdot [H^{+1}][CH_3COO^{-1}], [Cu^{+1}][I^{-1}] = 10^{-11.96}(\text{solubitltiy product for } \mathbf{CuI}),$$
$$[CuI_2^{-1}] = 10^{8.85} \cdot [Cu^{+1}][I^{-1}]^2, [CuIO_3^{+1}] = 10^{0.82} \cdot [Cu^{+2}][IO_3^{-1}], [CuCH_3COO^{+1}]$$
$$= 10^{2.24} \cdot [Cu^{+2}][CH_3COO^{-1}],$$
$$[Cu(CH_3COO)_2] = 10^{3.30} \cdot [Cu^{+2}][CH_3COO^{-1}]^2, [HS_2O_3^{-1}] = 10^{1.72} \cdot [H^{+1}][S_2O_3^{-2}], [H_2S_2O_3]$$
$$= 10^{2.32} \cdot [H^{+1}]^2[S_2O_3^{-2}],$$
$$[CuS_2O_3^{-1}] = 10^{3.30} \cdot [Cu^{+1}][S_2O_3^{-2}], [Cu(S_2O_3)_2^{-3}] = 10^{12.2} \cdot [Cu^{+1}][S_2O_3^{-2}]^2, [Cu(S_2O_3)_3^{-5}]$$
$$= 10^{13.8} \cdot [Cu^{+1}][S_2O_3^{-2}]^3$$

Applying *A* = 16.92 [16], we have

$$[Cu^{+2}] = [Cu^{+1}] \cdot 10^{A(E-0.153)}; [I_2] = [I^{-1}]^2 \cdot 10^{2A(E-0.621)}, s = 1.33 \cdot 10^{-3}\ \mathrm{mol/L(solubility\ of\ } \mathbf{I}_{2(s)}),$$
$$[I_3^{-1}] = [I^{-1}]^3 \cdot 10^{2A(E-0.545)}, [IO^{-1}] = [I^-] \cdot 10^{2A(E-0.49)+2pH-28}, [HIO] = 10^{10.6} \cdot [H^{+1}][IO^{-1}], [IO_3^{-1}]$$
$$= [I^{-1}] \cdot 10^{6A(E-1.08)+6pH},$$
$$[HIO_3] = 10^{0.79} \cdot [H^{+1}][IO_3^{-1}], [H_5IO_6] = [I^{-1}] \cdot 10^{8A(E-1.24)+7pH}, [H_4IO_6^{-1}]$$
$$= [H_5IO_6] \cdot 10^{-3.3+pH}, [H_3IO_6^{-2}] = [I^{-1}] \cdot 10^{8A(E-0.37)+9pH-126}.$$

In the calculations made in this system according to the computer programs attached to Ref. [64], it was assumed that $V_0 = 100$, $C_0 = 0.01$, $C_{01} = 0.01$, $C_1 = 0.25$, $C_2 = 0.75$, $C_3 = 2.0$, $C_4 = C = 0.1$; $V_N = 20$, $V_{Ac} = 40$, $V_K = 20$. At each stage, the variable V is considered as a volume of the solution added, consecutively: NH_3, CH_3COOH, KI, and $Na_2S_2O_3$, although the true/factual titrant in this method is the $Na_2S_2O_3$ solution, added in stage 4.

The solubility s [mol/L] of **CuI** in this system (**Figures 8a** and **b**) is put in context with the speciation diagrams presented in **Figure 9**. This precipitate appears in the initial part of titration with KI (C_3) solution (**Figure 8a**) and further it accompanies the titration, also in stage 4 (**Figure 8b**). Within stage 3, at $V \geq C_0V_0/C_3$, we have

$$s = s_3 = [Cu^{+2}] + \sum_{i=1}^{4}[Cu(OH)_i^{+2-i}] + \sum_{i=1}^{4}[Cu(NH_3)_i^{+2}] + [CuSO_4] + [CuIO_3^{+1}] \\ +\sum_{i=1}^{2}[Cu(CH_3COO)_i^{+2-i}] + [Cu^{+1}] + [CuI_2^{-1}] + \sum_{i=1}^{2}[Cu(NH_3)_i^{+1}] \tag{88}$$

and in stage 4

$$s = s_4 = s_3 + \sum_{i=1}^{3}[Cu(S_2O_3)_i^{+1-2i}] \tag{89}$$

The small concentration of Cu^{+1} (**Figure 9**, stage 3) occurs at a relatively high total concentration of Cu^{+2} species, determining the potential ca. 0.53–0.58 V, $[Cu^{+2}]/[Cu^{+1}] = 10^{A(E-0.153)}$, see **Figure 10a**. Therefore, the concentration of Cu^{+2} species determine a relatively high solubility s in the initial part of stage 3. The decrease in the s value in further parts of stage 3 is continued in stage 4, at $V < V_{eq} = C_0V_0/C = 0.01 \cdot 100/0.1 = 10$ mL. Next, a growth in the solubility s_4 at $V > V_{eq}$ is involved with formation of thiosulfate complexes, mainly $CuS_2O_3^{-1}$ (**Figure 9**, stage 4). The species I_3^{-1} and I_2 are consumed during the titration in stage 4 (**Figure 9d**). A sharp drop of E value at $V_{eq} = 10$ mL (**Figure 10b**) corresponds to the fraction titrated $\Phi_{eq} = 1$.

The course of the E versus V relationship within the stage 3 is worth mentioning (**Figure 10a**). The corresponding curve initially decreases and reaches a "sharp" minimum at the point corresponding to crossing the solubility product for **CuI**. Precipitation of **CuI** starts after

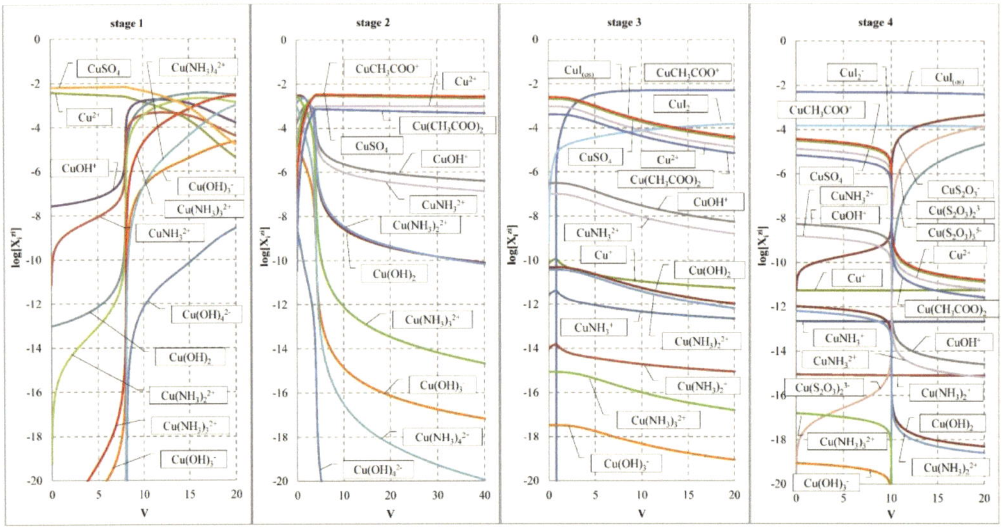

Figure 9. The speciation plots for indicated Cu-species within the successive stages. The *V*-values on the abscissas correspond to successive addition of V mL of: 0.25 mol/L NH_3 (stage 1); 0.75 mol/L CH_3COOH (stage 2); 2.0 mol/L KI (stage 3); and 0.1 mol/L $Na_2S_2O_3$ (stage 4). For more details see text.

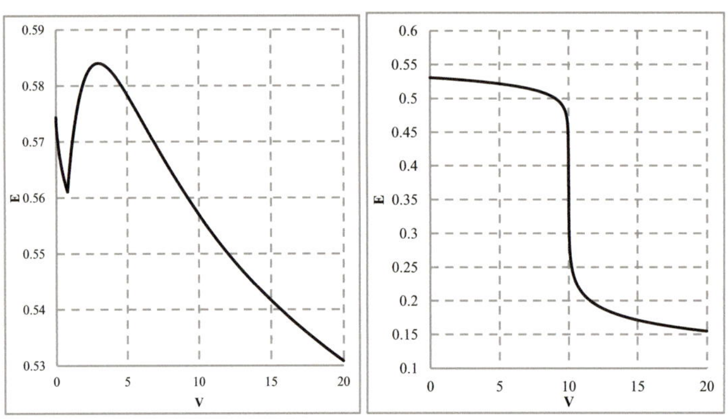

(a) (b)

Figure 10. Plots of *E* versus *V* for (a) stage 3 and (b) stage 4.

addition of 0.795 mL of 2.0 mol/L KI (**Figure 11a**). Subsequently, the curve in **Figure 10**a increases, reaches a maximum and then decreases. At a due excess of the KI (C_3) added on the stage 3 (V_K = 20 mL), solid iodine ($\mathbf{I_{2(s)}}$, of solubility 0.00133 mol/L at 25°C) is not precipitated.

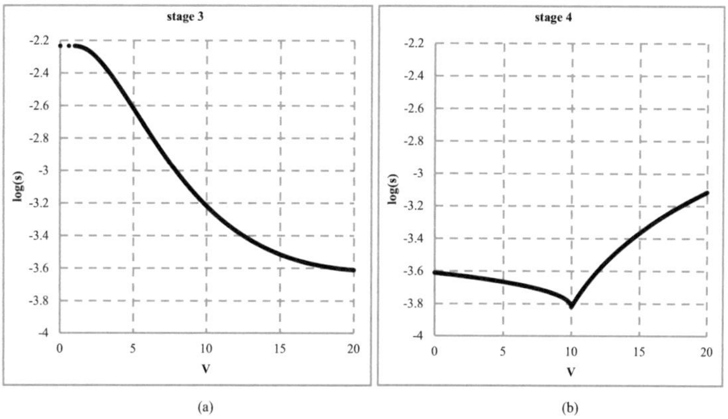

Figure 11. Solubility s of **CuI** within stage 3 (a) and stage 4 (b).

8. Final comments

The solubility and dissolution of sparingly soluble salts in aqueous media are among the main educational topics realized within general chemistry and analytical chemistry courses. The principles of solubility calculations were formulated at a time when knowledge of the two-phase electrolytic systems was still rudimentary. However, the earlier arrangements persisted in subsequent generations [81], and little has changed in the meantime [82]. About 20 years ago, Hawkes put in the title of his article [83] a dramatic question, corresponding to his statement presented therein that "the simple algorithms in introductory texts usually produce dramatic and often catastrophic errors"; it is hard not to agree with this opinion.

In the meantime, Meites et al. [84] stated that "It would be better to confine illustrations of the solubility product principle to 1:1 salts, like silver bromide (…), in which the (…) calculations will yield results close enough to the truth." The unwarranted simplifications cause confusion in teaching of chemistry. Students will trust us enough to believe that a calculation we have taught must be generally useful.

The theory of electrolytic systems, perceived as the main problem in the physicochemical studies for many decades, is now put on the side. It can be argued that the gaining of quantitative chemical knowledge in the education process is essentially based on the stoichiometry and proportions.

Overview of the literature indicates that the problems of dissolution and solubility calculation are not usually resolved in a proper manner; positive (and sole) exceptions are the studies and practice made by the authors of this chapter. Other authors, e.g., [13, 85], rely on the simplified schemes (ready-to-use formulas), which usually lead to erroneous results, expressed by dissolution denoted as s^* [mol/L]; the values for s^* are based on stoichiometric reaction notations and expressions for the solubility product values, specified by Eqs. (1) and (2). The calculation of s^* contradicts the common sense principle; this was clearly stated in the example with **$Fe(OH)_3$** precipitate. Equation (27) was applied to struvite [50] and dolomite [86], although these precipitates are nonequilibrium solid phases when introduced into pure water, as were proved in Refs. [20–23]. The fact of the struvite instability was known at the end of nineteenth century [49]; nevertheless, the formula $s^* = (K_{sp})^{1/3}$ for struvite may be still encountered in almost all textbooks and learning materials; this problem was raised in Ref. [15]. In this chapter, we identified typical errors involved with s^* calculations, and indicated the proper manner of resolution of the problem in question.

The calculations of solubility s^*, based on stoichiometric notation and Eq. (3), contradict the calculations of s, based on the matter and charge preservation. In calculations of s, all the species formed by defined element are involved, not only the species from the related reaction notation. A simple zeroing method, based on charge balance equation, can be applied for the calculation of $pH = pH_0$ value, and then for calculation of concentrations for all species involved in expression for solubility value.

The solubility of a precipitate and the pH-interval where it exists as an equilibrium-solid phase in two-phase system can be accurately determined from calculations based on charge and concentration balances, and complete set of equilibrium constant values referred to the system in question.

In the calculations performed here we assumed *a priori* that the K_{sp} values in the relevant tables were obtained in a manner worthy of the recognition, i.e., these values are true. However, one should be aware that the equilibrium constants collected in the relevant tables come from the period of time covering many decades; it results from an overview of dates of references contained in some textbooks [31, 85] relating to the equilibrium constants. In the early literature were generally presented the results obtained in the simplest manner, based on K_{sp} calculation from the experimentally determined s* value, where all soluble species formed in solution by these ions were included on account of simple cations and anions forming the expression for K_{sp}. In many instances, the K_{sp}^* values should be then perceived as conditional equilibrium constants [87]. Moreover, the differences between the equilibrium constants obtained under different physicochemical conditions in the solution tested were credited on account of activity coefficients, as an antidote to any discrepancies between theory and experiment.

First dissociation constants for acids were published in 1889. Most of the stability constants of metal complexes were determined after the announcement 1941 of Bjerrum's works, see Ref. [88], about ammine-complexes of metals, and research studies on metal complexes were carried out intermittently in the twentieth century [89]. The studies of complexes formed by simple ions started only from the 1940s; these studies were related both to mono- and two-phase systems. It should also be noted that the first mathematical models used for determination of equilibrium constants were adapted to the current computing capabilities. Critical comments in this regard can be found, among others, in the Beck [90] monograph; the variation between the values obtained by different authors for some equilibrium constants was startling, and reaching 20 orders of magnitude. It should be noted, however, that the determination of a set of stability constants of complexes as parameters of a set of suitable algebraic equations requires complex mathematical models, solvable only with use of an iterative computer program [91–93].

http://dx.doi.org/10.5772/67840

The difficulties associated with the resolution of electrolytic systems and two-phase systems, in particular, can be perceived today in the context of calculations using (1°) spreadsheets (2°) iterative calculation methods. In (1°), a calculation is made by the zeroing method applied to the function with one variable; both options are presented in this chapter.

The expression for solubility products, as well as the expression of other equilibrium constants, is formulated on the basis of mass action law (MAL). It should be noted, however, that the underlying mathematical formalism contained in MAL does not inspire trust, to put it mildly. For this purpose, the equilibrium law (EL) based on the Gibbs function [94] and the Lagrange multipliers method [95–97] with laws of charge and elements conservation was suggested lately by Michałowski.

From semantic viewpoint, the term "solubility product" is not adequate, e.g., in relation to Eq. (8). Moreover, K_{sp} is not necessarily the product of ion concentrations, as indicated in formulas (4), (5), and (11). In some (numerous) instances of sparingly soluble species, e.g., sulfur, solid iodine, 8-hydroxyquinoline, dimethylglyoxime, the term solubility product is not applied. In some instances, e.g., for $\mathbf{MnO_2}$, this term is doubtful.

One of the main purposes of the present chapter is to familiarize GEB within GATES as GATES/GEB to a wider community of analysts engaged in electrolytic systems, also in aspect of solubility problems.

In this context, owing to large advantages and versatile capabilities offered by GATES/GEB, it deserves a due attention and promotion. The GATES is perceived as a step toward reductionism [19, 71] of chemistry in the area of electrolytic systems and the GEB is considered as a general law of nature; it provides the real proof of the world harmony, harmony of nature.

Author details

Anna Maria Michałowska-Kaczmarczyk[1], Aneta Spórna-Kucab[2] and Tadeusz Michałowski[2]*

*Address all correspondence to: michalot@o2.pl

1 Department of Oncology, The University Hospital in Cracow, Cracow, Poland

2 Faculty of Chemical Engineering and Technology, Cracow University of Technology, Cracow, Poland

References

[1] Clever HL, Gevantman LH. Obituary: A.S. Kertes. Pure and Applied Chemistry. 1989;**61**: 121. http://dx.doi.org/10.1351/pac198961020121

[2] Clever L, The IUPAC solubility data project. A brief history: 1972–2001. Chemistry International. 2004;**26**(3). http://www.iupac.org/publications/ci/2004/2603/4_clever.html

[3] Gamsjäger H, Lorimer JW, Salomon M, Shaw DG, Tomkins RPT. The IUPAC-NIST solubility data series: A guide to preparation and use of compilations and evaluations (IUPAC Technical Report). Pure and Applied Chemistry. 2010;**82**(5):1137-1159. DOI: 10.1351/PAC-REP-09-10-33

[4] Dirkse TP, Michałowski T, Akaiwa H, Izumi F. Copper, Silver, Gold and Zinc, Cadmium, Mercury Oxides and Hydroxides, Solubility Data Series. Vol. 23. Oxford: Pergamon; 1986. http://www.worldcat.org/title/copper-silver-gold-and-zinc-cadmium-mercury-oxides--and-hydroxides/oclc/12945958

[5] Yalkowsky SH. Handbook of Aqueous Solubility Data (e-Book Google). CRC Press, LLC, Boca Raton, FL. 2003

[6] Kobayashi T, Sasaki T, Takagi I, Moriyama H. Solubility of zirconium (IV) hydrous oxides. Journal of Nuclear Science and Technology. 2007;**44**(1):90-94. http://dx.doi.org/10.1080/18811248.2007.9711

[7] Michałowski T, Lesiak A. Formulation of generalized equations for redox titration curves. Chemia Analityczna (Warsaw). 1994;**39**:623-637

[8] Michałowski T. Application of GATES and MATLAB for resolution of equilibrium, metastable and non-equilibrium electrolytic systems. In: Michałowski T, editor. Applications of MATLAB in Science and Engineering. Rijeka: InTech-Open Access Publisher in the Fields of Science, Technology and Medicine; 2011. pp. 1-34. Chapter 1. http://www.intechopen.com/books/show/title/applications-of-matlab-in-science-and-engineering. ISBN 978-953-307-708-6

[9] Michałowski T, Toporek M, Michałowska-Kaczmarczyk AM, Asuero AG. New trends in studies on electrolytic redox systems. Electrochimica Acta. 2013;**109**:519-531. DOI: http://dx.doi.org/10.1016/j.electacta.2013.07.125

[10] Michałowska-Kaczmarczyk AM, Toporek M, Michałowski T. Speciation diagrams in dynamic iodide + dichromate system. Electrochimica Acta. 2015;**155**:217-227. DOI: http://dx.doi.org/10.1016/j.electacta.2015.01.003

[11] Meija J, Michałowska-Kaczmarczyk AM, Michałowski T. Redox titration challenge. Analytical and Bioanalytical Chemistry. 2017;**409**(1):11-13. DOI 10.1007/s00216-016-0020-0

[12] Michałowski T, Michałowska-Kaczmarczyk AM, Meija J, Solution of redox titration challenge, Analytical and Bioanalytical Chemistry, be issued in 409(17).

[13] Kotrlý S, Šůcha L, Handbook of Chemical Equilibria in Analytical Chemistry, Ellis Horwood Limited, Halsted Press, a division of John Wiley & Sons, New York 1985.

[14] Michałowski T. Calculations in Analytical Chemistry with Elements of Computer Programming (in Polish). Cracow: PK; 2001 http://suw.biblos.pk.edu.pl/resourceDetails&rId=3974

[15] Michałowska-Kaczmarczyk AM, Asuero AG, Michałowski T. "Why not stoichiometry" versus "Stoichiometry—why not?" Part I. General context. Critical Reviews in Analytical Chemistry. 2015;**45**(2):166-188. DOI: 10.1080/10408347.2014.937852

[16] Michałowska-Kaczmarczyk AM, Michałowski T. Calculation of solubility of oxyquinolinates. Journal of Analytical Sciences, Methods and Instrumentation. 2014;**4**(3):71-79. DOI: 10.4236/jasmi.2014.43010

[17] Michałowska-Kaczmarczyk AM, Michałowski T. Solubility product challenge. Analytical and Bioanalytical Chemistry. 2015;**407**(6):1789-1791

[18] Michałowska-Kaczmarczyk AM, Michałowski T. Solution to the solubility product challenge. Analytical and Bioanalytical Chemistry. 2015;**407**(17):4877-4878

[19] Michałowski T. Constructivist perception of chemical knowledge. Implications in terms of electrolytic systems (in Polish). In: Use of Information Technology in Academic Teaching of Chemistry (Maciejowska I, Ruszak M, Witkowski S, eds.), Jagiellonian University, Cracow, 2007. pp. 130–136. http://www2.chemia.uj.edu.pl/~ictchem/book.html

[20] Michałowski T, Pietrzyk A. A thermodynamic study of struvite+water system. Talanta. 2006;**68**:594-601

[21] Michałowski T, Pietrzyk A. Quasistatic processes in non-equilibrium two-phase systems with ternary salts: I. Struvite + aqueous solution (CO_2 + KOH). Chemia Analityczna (Warsaw). 2008;**53**:33-46

[22] Michałowski T, Borzęcka M, Toporek M, Wybraniec S, Maciukiewicz P, Pietrzyk A. Quasistatic processes in non-equilibrium two-phase systems with ternary salts: II. Dolomite + aqueous media. Chemia Analityczna (Warsaw). 2009;**54**:1203-1217

[23] Michałowski T, Asuero AG, Thermodynamic modelling of dolomite behavior in aqueous media. Journal of Thermodynamics. 2012;(2012):12. Article ID 723052. DOI: 10.1155/2012/723052. http://www.hindawi.com/journals/jtd/2012/723052/cta/

[24] http://www.thefreedictionary.com/ternary

[25] Bartecki A. On the chemistry of MeO_2^{n+} oxycation compounds. Chemické Zvesti. 1965;**19**: 161-166

[26] Pauling L. The Nature of the Chemical Bond and the Structure of Molecules and Crystals: An Introduction to Modern Structural Chemistry. 3rd ed. Ithaca, New York: Cornell University Press; 1960

[27] López FA, Alguacil FJ, Roman CP, Tayibi H, López-Delgado A. Disposal of elemental mercury via sulphur reaction by milling; 1 st International Conference on "Hazardous Waste Management", 1–3 Octubre, Chaina, Grete, Greece 2008, p. 49. http://digital.csic.es/bitstream/10261/7692/1/DISPOSAL%20ELEMENTALHg.pdf

[28] Chadwick BM, Sharpe AG. Transition metal cyanides and their complexes. In: Emeleus HJ, Sharpe AG, editors. Advances in Inorganic Chemistry and Radiochemistry. Vol. 8. New York: Academy Press Inc.; 1966. p. 118

[29] Michałowski T, Nizińska-Pstrusińska M, Sztark W, Baterowicz A. Laboratory Trainings in Analytical Chemistry (in Polish). Kraków: Wyd. PK; 2002. http://bc.biblos.pk.edu.pl/

bc/resources/MD/MichalowskiT/CwiczeniaLaboratoryjne/pdf/MichalowskiT_CwiczeniaLaboratoryjne.pdf

[30] http://en.wikipedia.org/wiki/Solubility_equilibrium

[31] Inczédy J, Analytical applications of complex equilibria, Chichester, E. Horwood, New York, Halsted Press, 1976.

[32] Ruetschi P, Giovanoli R. The behaviour of MnO_2 in strongly acidic solutions. Journal of Applied Electrochemistry. 1982;**12**(1):109-114. DOI: 10.1007/BF01112071

[33] Nikolsky BP, editors. Chemist's Handbook (in Russian). Vol. III. Moscow: Khimia; 1964

[34] Michałowska-Kaczmarczyk AM, Rymanowski M, Asuero AG, Toporek M, Michałowski T. Formulation of titration curves for some redox systems. American Journal of Analytical Chemistry. 2014;**5**:861-878. http://www.scirp.org/journal/PaperInformation.aspx?paperID=50016#.VGOZ92dvHFw.

[35] https://answers.yahoo.com/question/index?qid=20070415234848AA54Gus

[36] Hulanicki, A., Reactions of acids and bases in analytical chemistry, Chichester, West Sussex, England : E. Horwood ; New York : Halsted Press ; Warsaw : PWN-Polish Scientific Publishers, 1987. https://iucat.iu.edu/iun/3772927

[37] Gordus AA. Chemical equilibrium, VIII. Precipitates. Journal of Chemical Education. 1991;**68**(11):927-930. http://dx.doi.org/10.1021/ed068p927.

[38] https://answers.yahoo.com/question/index?qid=20090316151751AAyF87o

[39] https://en.wikipedia.org/wiki/Carbon_dioxide

[40] Lurie YY. Handbook of Analytical Chemistry (in Russian). Moscow: Izd. Khimia; 1971. s. 105

[41] deCogan D. Transmission Line Matrix (TLM) Techniques for Diffusion Applications. Australia: Gordon and Breach Science Publishers; 1998. p. 134

[42] Considine DM, Considine GD. Van Nostrand's Scientific Encyclopedia. 8th ed. New York: Springer Science+Business Media, LLC; 1995. p. 673

[43] https://answers.yahoo.com/question/index?qid=20080518113447AAg17u3

[44] Michałowski T. Evaluation of the systematic error related to determination of chloride according to Mohr method. Chemia Analityczna (Warsaw). 1989;**34**:461

[45] Meija J, Michałowska-Kaczmarczyk AM, Michałowski T. Mohr's method Challenge. Analytical and Bioanalytical Chemistry. 2016;**408**(7):1721-1722. http://paperity.org/p/75256979/mohrs-method-challenge

[46] Meija J, Michałowska-Kaczmarczyk AM, Michałowski T. Solution to Mohr's method challenge. Analytical and Bioanalytical Chemistry. 2016;**408**(17):4469-4471. DOI: 10.1007/s00216-015-9273-2

[47] E.P. Leite (Ed.), Matlab - Modelling, Programming and Simulations, Sciyo, Published by Sciyo, Rijeka, Croatia 2010. ISBN 978-953-307-125-1. http://www.intechopen.com/books/matlab-modelling-programming-and-simulations

[48] Michałowska-Kaczmarczyk AM, Michałowski T. Evaluation of transition points between different solid phases in aqueous media. Journal of Analytical Sciences, Methods and Instrumentation. 2014;**4**(3):87-94. http://www.scirp.org/journal/PaperInformation.aspx?PaperID=49567#.VGOZo2dvHFw

[49] Beilstein F, Grosset Th. Ueber die Bestimmung der freien Schwefelsaure in der Schwefelsauren Thonerde. Zeitschrift für Analytische Chemie. 1890;**29**:73-78

[50] Doyle JD, Parsons SA. Struvite formation, control and recovery. Water Research. 2002;**36**(16):3925-3940

[51] http://chemlab.truman.edu/CHEM222manual/pdf/nickelgrav.pdf

[52] Michalowski T. Calculation of pH and potential E for bromine aqueous solution. Journal of Chemical Education. 1994;**71**:560-562. DOI: 10.1021/ed071p560

[53] Michałowski T, Lesiak A. Acid-base titration curves in disproportionating redox systems. Journal of Chemical Education. 1994;**71**:632-636. DOI: 10.1021/ed071p632

[54] Michałowski T, Wajda N, Janecki D. A unified quantitative approach to electrolytic systems. Chemia Analityczna (Warsaw). 1996;**41**:667-685. http://www.chem.uw.edu.pl/chemanal/PDFs/1996/CHAN1996V0041P00667.pdf

[55] Michałowski T, Baterowicz A, Madej A, Kochana J. An extended Gran method and its applicability for simultaneous determination of Fe(II) and Fe(III). Analytica Chimica Acta. 2001;**442**(2):287-293. http://www.sciencedirect.com/science/article/pii/S0003267001011722

[56] Michałowski T, Toporek M, Rymanowski M. Overview on the Gran and other linearization methods applied in titrimetric analyses. Talanta. 2005;**65**(5):1241-1253. http://pl.scribd.com/doc/173699711/Talanta-65-2005-1241-1253

[57] Michałowski T, Rymanowski M, Pietrzyk A. Non-typical Brönsted's acids and bases. Journal of Chemical Education. 2005;**82**(3):470-472. http://pubs.acs.org/doi/abs/10.1021/ed082p470

[58] Michałowski T, Kupiec K, Rymanowski M. Numerical analysis of the Gran methods. A comparative study. Analytica Chimica Acta. 2008;**606**(2):172-183. http://www.sciencedirect.com/science/article/pii/S0003267007018673

[59] Ponikvar M, Michałowski T, Kupiec K, Wybraniec S, Rymanowski M. Experimental verification of the modified Gran methods applicable to redox systems. Analytica Chimica Acta. 2008;**628**(2):181-189. http://pl.scribd.com/doc/173699262/ACA-2008-2

[60] Michałowski T, Pietrzyk A. The generalized electron balance concept. Derivation based on elementary rules of the matter conservation. In: Zuba D, Parczewski A, editors.

Chemometrics: Methods and Applications. Institute of Forensic Research, Kraków; 2006. pp. 415-422. https://pbn.nauka.gov.pl/sedno-webapp/works/210292

[61] Michałowski T. Electron Balance as the Basis of Calculations in Redox Systems (in Polish). In: Use of Information Technology in Academic Teaching of Chemistry (Maciejowska I, Ruszak M, Witkowski S, eds.), Jagiellonian University, Cracow, 2007. pp. 162-169. http://www.chemia.uj.edu.pl/~ictchem/book.html

[62] Michałowski T, Pietrzyk A. Complementarity of physical and chemical laws of preservation in aspect of electrolytic systems (in Polish). Wiadomości Chemiczne. 2007;**61**:625-640

[63] Michałowska-Kaczmarczyk AM, Asuero AG, Toporek M, Michałowski T. "Why not stoichiometry" versus "Stoichiometry – why not?" Part II. GATES in context with redox systems. Critical Reviews in Analytical Chemistry. 2015;**45**(3):240-268. http://www.tandfonline.com/doi/full/10.1080/10408347.2014.937853

[64] Michałowska-Kaczmarczyk AM, Michałowski T, Toporek M, Asuero AG. "Why not stoichiometry" versus "Stoichiometry – why not?" Part III. Extension of GATES/GEB on Complex Dynamic Redox Systems. Critical Reviews in Analytical Chemistry. 2015;**45**(4):348-366. DOI: 10.1080/10408347.2014.953673. http://www.researchgate.net/publication/274401037_Why_Not_Stoichiometry_Versus_Stoichiometry_-_Why_Not_Part_III_Extension_of_GatesGeb_on_Complex_Dynamic_Redox_Systems

[65] Michałowska-Kaczmarczyk AM, Michałowski T, Toporek M. Formulation of dynamic redox systems according to GATES/GEB principles. International Journal of Electrochemical Science. 2016;**11**:2560-2578. DOI:10.20964/e110340

[66] Michałowski T, Ponikvar-Svet M, Asuero AG, Kupiec K. Thermodynamic and kinetic effects involved with pH titration of As(III) with iodine in a buffered malonate system. Journal of Solution Chemistry. 2012;**41**(3):436-446. http://www.springerlink.com/content/p2m73068h2q5u174/. DOI: 10.1007/s10953-012-9815-6

[67] Michałowska-Kaczmarczyk AM, Michałowski T. Comparative balancing of non-redox and redox electrolytic systems and its consequences. American Journal of Analytical Chemistry. 2013;**4**(10):46-53. http://www.scirp.org/journal/PaperInformation.aspx?PaperID=38569

[68] Toporek M, Michałowska-Kaczmarczyk AM, Michałowski T. Disproportionation reactions of HIO and NaIO in static and dynamic systems. American Journal of Analytical Chemistry. 2014;**5**:1046-1056. http://www.scirp.org/journal/PaperInformation.aspx?PaperID=51637#.VHXKcWfpt74

[69] Michałowska-Kaczmarczyk AM, Michałowski T. Generalized electron balance for dynamic redox systems in mixed-solvent media. Journal of Analytical Sciences, Methods and Instrumentation. 2014;**4**(4):102-109. http://www.scirp.org/Journal/PaperInformation.aspx?PaperID=52018#.VH1N5Gfpt74

[70] Michałowska-Kaczmarczyk AM, Michałowski T. Compact formulation of redox systems according to GATES/GEB principles. Journal of Analytical Sciences, Methods and Instrumentation. 2014;**4**(2):39-45. http://www.scirp.org/journal/PaperInformation.aspx?PaperID=46335

[71] Michałowska-Kaczmarczyk AM, Michałowski T. GATES as the unique tool for simulation of electrolytic redox and non-redox systems. Journal of Analytical & Bioanalytical Techniques. http://omicsonline.org/open-access/gates-as-the-unique-tool-for-simulation-of-electrolytic-redox-and-non-redox-systems-2155-9872.1000204.pdf

[72] Michałowski T, Pilarski B, Asuero AG, Michałowska-Kaczmarczyk AM. Modelling of acid-base properties in binary-solvent systems. In: Wypych G, editor. Handbook of Solvents, Vol. 1. Properties. Toronto: ChemTec Publishing; 2014. pp. 623-648. Chapter 9.4

[73] Michałowski T, Michałowska-Kaczmarczyk AM, Toporek M. Formulation of general criterion distinguishing between non-redox and redox systems. Electrochimica Acta. 2013;**112**:199-211. http://www.sciencedirect.com/science/article/pii/S0013468613016836

[74] Toporek M, Michałowska-Kaczmarczyk AM, Michałowski T. Symproportionation versus disproportionation in bromine redox systems. Electrochimica Acta. 2015;**171**:176-187. http://www.sciencedirect.com/science/article/pii/S0013468615011238

[75] Michałowska-Kaczmarczyk AM, Spórna-Kucab A, Michałowski T. A Distinguishing Feature of the Balance $2{\cdot}f(O) - f(H)$ in Electrolytic Systems. The Reference to Titrimetric Methods of Analysis, in: "Titration", InTech 2017

[76] Michałowska-Kaczmarczyk AM, Spórna-Kucab A, Michałowski T. Principles of Titrimetric Analyses According to Generalized Approach to Electrolytic systems (GATES), in: "Titration", InTech 2017

[77] AM, Spórna-Kucab A, Michałowski T. Generalized Electron Balance (GEB) as the Law of Nature in Electrolytic Redox Systems, in: "Redox: Principles and Advanced Applications", InTech 2017.

[78] Pietrzyk A, Michałowski T. Computer simulation of the process of struvite $MgNH_4PO_4{\cdot}6H_2O$ dissolution in aqueous solutions, (in Polish). Use of Information Technology in Academic Teaching of Chemistry. Kraków: UJ; 2007. pp.137-146. http://www.chemia.uj.edu.pl/~ictchem/book.html

[79] Pietrzyk A, Michałowski T, Conditions of struvite formation in aqueous systems. In: Zuba D, Parczewski A, editors. Chemometrics: Methods and Applications. Kraków; 2006. pp. 423-427. https://pbn.nauka.gov.pl/works/210292

[80] Steudel R, editor. Elemental Sulfur and Sulfur-Rich Compounds. II. Topics in Current Chemistry. Berlin, Heidelberg: Springer-Verlag; 2003

[81] Clark RW, Bonicamp JM. The K_{sp}-solubility conundrum. Journal of Chemical Education. 1998;**75**(9):1182-1185. DOI: 10.1021/ed075p1182

[82] Umland JB, Bellama JM. General Chemistry. 2nd ed. Minneapolis, MN: West; 1996. pp. 621-622

[83] Hawkes SJ. What should we teach beginners about solubility and solubility products? Journal of Chemical Education. 1998;**75**(9):1179-1181. DOI: 10.1021/ed075p1179

[84] Meites L, Pode JSF, Thomas HC. Are solubilities and solubility products related? Journal of Chemical Education. 1966;**43**:667-672

[85] Šůcha L, Kotrlý S. Solution Equilibrium in Analytical Chemistry. London: Van Nostrand Reinhold Comp.; 1975

[86] Chen G, Tao D. Effect of solution chemistry on flotability of magnesite and dolomite. International Journal of Mineral Processing. 2004;**74**(1–4):343-357

[87] Ringbom A. Complexation in Analytical Chemistry. New York: Interscience Publishers; 1963

[88] Rossotti H. The study of ionic equilibria. An introduction. London: Longman; 1978

[89] Sillén LG, Martell AE, Stability Constants of Metal-Ion Complexes, The Chemical Society, London, 1964; Supplement No. 1, 1971.

[90] Beck MT. Chemistry of Complex Equilibria. New York: Van Nostrand Reinhold Co.; 1970

[91] Janecki D, Doktór K, Michałowski T. Determination of stability constants of complexes of $M_iK_jH_kL$ type in concentrated solutions of mixed salts. Talanta. 1999;**48**:1191-1197 http://www.sciencedirect.com/science/article/pii/S0039914098003452

[92] Janecki D, Doktór K, Michałowski T. Erratum to "Determination of stability constants of complexes of $M_jK_jH_kL$ type in concentrated solutions of mixed salts" [48 (1999) 1191]. Talanta. 1999;**49**:943. http://www.sciencedirect.com/science/article/pii/S0039914099001745

[93] Janecki D, Styszko-Grochowiak K, Michałowski T. The catenation and isomerisation effects on stability constants of complexes formed by some diprotic acids. Talanta. 2000;**52**:555-562. http://www.ncbi.nlm.nih.gov/pubmed/18968016

[94] https://en.wikipedia.org/wiki/Gibbs_free_energy

[95] Zeidler E. Nonlinear Functional Analysis and its Applications: III: Variational Methods and Optimization. New York: Springer Science +Business Media; 1985

[96] Sieniutycz S. Thermodynamic Approaches in Engineering Systems. Amsterdam: Elsevier; 2016

[97] Bertsekas DP. Nonlinear Programming. 2nd ed. Cambridge, MA: Athena Scientific; 1999

Chapter 6

Design of Novel Classes of Building Blocks for Nanotechnology: Core-Modified Metalloporphyrins and Their Derivatives

Aleksey E. Kuznetsov

Additional information is available at the end of the chapter

http://dx.doi.org/10.5772/67728

Abstract

Metalloporphyrins and related macrocycles have been of great interest due to their role in biology and their numerous technological applications. Engineering of the porphyrins by replacing pyrrole nitrogens with other elements is a highly promising approach for tuning properties of porphyrins. To date, numerous efforts have been made to the modification of the porphyrin core with main-group elements, such as chalcogens (O, S, Se) and phosphorus. Thus, the modification of the porphyrin core by incorporation of heteroatoms instead of nitrogens is a very promising strategy for obtaining novel compounds with unusual optical, electrochemical and coordinating properties as well as reactivity. These novel compounds can be used as building blocks in various nanotechnological applications. Within the framework of this research, the following questions can be formulated: (i) what structures will core-modified porphyrins adopt? (ii) How will electronic properties of core-modified porphyrins differ from those of common tetrapyrroles? (iii) Will the core-modified porphyrins be able to form stacks and other arrays like regular porphyrins? (iv) Can core-modified porphyrins form complexes with fullerenes? (v) Can core-modified porphyrins activate small molecules, e.g. O_2 or N_2? (vi) Will the core-modified porphyrins be able to form complexes with nanoparticles?

Keywords: metalloporphyrins, core modification, chalcogens, phosphorus, structural changes

1. Introduction

Various (metallo)tetrapyrrole compounds, for example, porphyrins (P), porphyrazines (Pz) and phthalocyanines (Pc) (**Figure 1**), are representatives of the huge class of π-conjugated

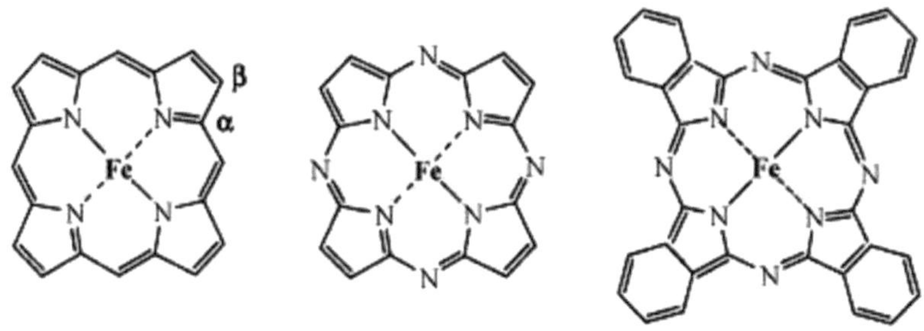

Figure 1. Molecular structures of some tetrapyrrole macrocycles. Reprinted (adapted) with permission from Liao et al. [20]. Copyright 2005 American Chemical Society.

(aromatic) organic heterocycles [1–5]. They can be found as cofactors in numerous enzymes: as hemes in various cytochromes, catalases, peroxidases, etc.; as chlorophyll and pheophytin in photosynthetic proteins and as corrin and corphin in other proteins [1–3, 5]. The metalloporphyrins have numerous biological functions such as: (i) O_2 transport and storage, (ii) oxidative metabolism, (iii) gas sensing, (iv) antibactericides/microbicides, (v) collection and transport of light energy, (vi) conversion of solar energy to chemical energy, (vii) electron transfer and (viii) NO scavenging and a significant number of other functions [1–3, 5–10]. Numerous technological applications of porphyrins include: catalysis [1, 2, 4, 11, 12], molecular photonic devices [4, 13, 14], medicine [1, 2, 4, 15], artificial photosynthesis [16, 17], sensitizers for dye-sensitized solar cells [18] and sensor devices [19].

The size, shape, electronic properties and binding ability of porphyrins can be broadly tuned by replacing one or more pyrrole nitrogens with other elements [21–24]. This type of the porphyrin core modification is a highly promising approach for tuning the various properties of porphyrin species. It brings to life the following questions:

(i) What structures will core-modified porphyrins adopt? (ii) How will atomic charges and other electronic properties (frontier orbital energies, HOMO/LUMO and optical gaps, ionization potentials, electron affinities, etc.) in core-modified porphyrins differ from regular tetrapyrroles? How can we tune these properties? (iii) What novel properties will core-modified porphyrins possess?

In recent years, there has been increasing interest in porphyrin core modification with the chalcogens (O, S, Se), which resulted in numerous experimental and computational works in this extremely promising area. Core modification of tetrapyrroles by P has been of long-lasting interest as well. Of course, it would not be possible to cover all the studies on core modification of porphyrins in this review. Thus, this chapter will cover the most significant and interesting works devoted to the core modification of porphyrins and derivatives with the principal focus on *completely core-modified* compounds. The important works on *partially core-modified* compounds will be considered as well.

2. Core modification with different main-group elements

2.1. Core modification with chalcogens

The first porphyrins *fully modified* by the chalcogens, the tetraoxaporphyrin dication **1** [26, 27] and tetrathiaporphyrin dication **2** [27], were reported by Vogel et al. in 1988 and 1989, respectively (**Figure 2**). The X-ray structure and ^{1}H NMR and electronic absorption spectroscopy data for the compounds **1** and **2** were consistent with 18 π-electron aromatic cycles [26]. However, S_4P^{2+} was found to exhibit low solubility and was shown to be unstable in common organic solvents. To resolve these issues, the octaethyltetrathiaporphyrin dication (S_4OEP^{2+}) was subsequently prepared [28], but its spectroscopic studies and use in practical applications was found to be problematic. The UV/Vis spectrum of the perchlorate salt of **1** was found to have a sharp, high-intensity B (Soret) band and a series of Q bands in 96% H_2SO_4, whereas the UV/Vis spectrum of the perchlorate salt of the compound 2 showed strongly broadened and red-shifted bands. This difference was ascribed to the planarity of the compound 1 and distorted structure of 2 [27].

Also, research interest was focused on the core modification of the 20 π-electron *N,N'*-dihydroporphyrins (isophlorins). The possible formation of an isophlorin was first noted during the total synthesis of chlorophyll by Woodward [29] who proposed the 18 π porphyrin **4** – 20 π isophlorin 5 redox system (**Figure 2**). The first synthesis of the isophlorin, 21,22,23,24-tetramethyloctaethylisophlorin 6 (**Figure 2**), was realized in 1991 [30]. Isophlorin 6 was found to have saddle shape with the *syn,anti,syn,anti*-conformation of the N-attached methyl groups in the solid state. Its UV/vis spectrum was shown to have a band at 356 nm with a shoulder at 516 nm. *Six* was reported to be readily oxidized into the dication and was found to be moderately stable upon exposure to air. Several relatively stable isophlorins were obtained by employing strong electron-withdrawing substituents. Thus, in 2007 Chen et al. reported the synthesis of β-tetrakis(trifluoromethyl)-meso-tetraphenyl-isophlorin 7 [31]. Isophlorins **6** and

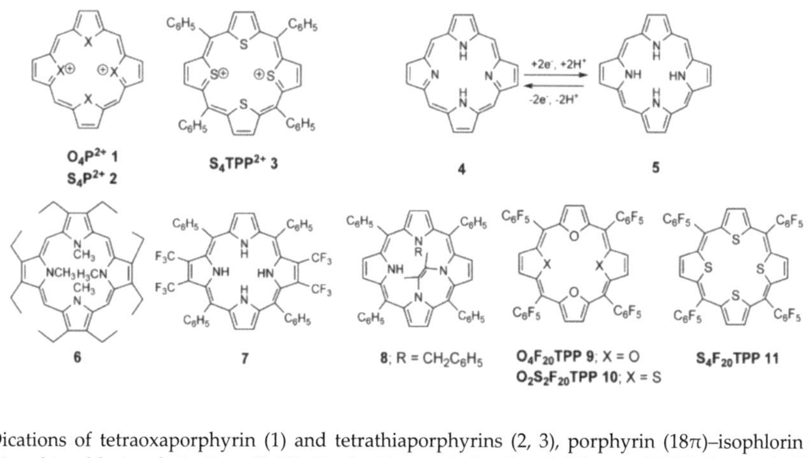

Figure 2. Dications of tetraoxaporphyrin (1) and tetrathiaporphyrins (2, 3), porphyrin (18π)–isophlorin (20π) redox system (4, 5) and isophlorine derivatives (6–11). Used with permission from Mishra et al. [25]. Copyright 2016 Wiley.

7 were *nonaromatic*. *Antiaromatic* almost planar isophlorin derivative 21,22-bridged-23-alkylporphyrin 8 was reported by Setsune et al. in 1999 [32]. In 2008, Anand and Reddy reported fully core-modified tetra*oxa*- and dithiadioxaisophlorines 9 and 10 [33], with planar structures and *antiaromatic* properties [33].

In 2012, Kon-no et al. reported the synthesis, structures, optical properties, and electronic structures of the *fully core-modified* 18 π-electron 5,10,15,20-tetraphenyl-21,22,23,24-tetrathiaporphyrin dication, S_4TPP^{2+} (**6**) and the 20 π-electron 5,10,15,20-tetrakis-(penta-fluorophenyl)-21,22,23,24-tetra*thia*isophlorin, S_4F_{20}-TPP (7) [34]. The tetraaryl tetrathiaisophlorin was supposed to be *non-planar owing to the larger size of the S-atom* [34]. The X-ray analysis of the $[S_4TPP^{2+}][B(C_6F_5)_4^-]_2$ and compound **7** showed the following: (i) The thiophene moieties located on the *y*-axis of $[S_4TPP^{2+}][B(C_6F_5)_4^-]_2$ were tilted above and below the plane formed by the four *meso*-carbons so that the four S-atoms could be accommodated inside the central cavity, and the two thiophene rings on the *x*-axis were located within the x/y plane. Thus, S_4TPP^{2+} had a wave conformation. The structure of S_4TPP^{2+} inside $[S_4TPP^{2+}][B(C_6F_5)_4^-]_2$ was considered as being very similar to that of the tetrathiaporphyrin dication S_4P^{2+} (2) [27]. (ii) In the structure of 7, two thiophene rings were shown to be tilted out of the plane of the four *meso*-carbons along one axis in a disordered manner with an occupancy factor of 0.50 above and below the plane, whilst the other two thiophene rings were found to be almost coplanar with the rest of the π-system (saddle conformation). The bond-alternation pattern for S_4F_{20}TPP was found to be consistent with that normally anticipated for an isophlorin. (iii) The B3LYP/6-31G(d) optimized geometries of S_4TPP^{2+} and S_4F_{20}TPP were shown to be very similar to their X-ray structures.

The results of time-dependent DFT and ZINDO/s calculations were compared to the observed magnetic circular dichroism (MCD) spectra and the electronic absorption spectra to study the effects of core modification on the electronic structures of S_4TPP^{2+} and S_4F_{20}TPP. For $[S_4TPP^{2+}][ClO_4^-]_2$, the MCD spectrum showed correspondence with the weaker bands at 948 and 733 nm in the near-IR region of the electronic absorption spectrum and a more intense band in the visible region at 491 nm. These bands were assigned as Q_{00}, Q_{01} and B_{00} bands, respectively. In contrast to 21- and 21,23-core-modified porphyrinoids [35], a marked red-shift of the Q bands into the near-IR region was observed owing to a narrowing of the HOMO–LUMO gap. *Full core modification was shown to result in a marked destabilization of the HOMO.* The ring current calculations were carried out for model compounds S_4P^{2+} in C_s and D_{4h} symmetries and S_4P in C_{2v} symmetry. The optimized S_4P^{2+} C_s structure was comparable to the B3LYP-optimized structure of S_4TPP^{2+} and the X-ray structure for $[S_4TPP^{2+}][B(C_6F_5)_4^-]_2$. The calculations for the S_4P^{2+} C_s and S_4P^{2+} D_{4h} model structures predicted the *aromatic* character of S_4P^{2+}. The current density map of the S_4P in C_{2v} symmetry showed *nonaromatic* character of the S_4P and, therefore, S_4F_{20}TPP, π-system. In general, current density map calculations for the model structures predicted the core modification and non-planarity of the macrocycles to modify patterns of the ring currents significantly. Nuclear independent chemical shift (NICS) values obtained for the structure of S_4TPP^{2+} (6) were consistent with a *diamagnetic* ring current and an *aromatic* π-system. The NICS values computed for the compound 7 were essentially nonaromatic. The greater stability of tetraaryl tetrathiaporphyrins, as stated by the authors, makes these species potentially suitable for use as organic devices in practical applications.

Also in 2012, Rurack and coworkers reported the synthesis of novel *partially core-modified* and fused-ring-expanded tetraphenyldiphenanthroporphyrins [36], denoted as N_2O_2, N_2S_2, N_2Se_2 and N_2Te_2. In these compounds, chalcogens replaced the pyrrole NH-groups along the *y*-axis. Peripheral-fused phenanthrene rings were substituted onto the pyrroles on the *x*-axis. Trends in the optical properties and electronic structures were explored and the suitability of these compounds for near IR region dye applications was studied. The terminology N_2Y_2 (Y = O, S, Se, Te) used in this publication referred to the four core atoms on the inner perimeter of the porphyrin π-system that can coordinate a central metal atom (**Figure 4**). Other notations used in the paper were as follows: *P* = porphyrin; *TPP* = 5,10,15,20-tetraphenylporphyrin; P_1, P_1-Bz_y, and P_3, P_3-Bz_y = core-modified diphenanthro- and diacenaphthoporphyrins, respectively, with fused bicyclo[2.2.2]octadiene (BCOD) and benzene rings along the *y*-axis; *TPTPhenPn* = *meso*-tetraphenyltetraphenanthroporphyrin; *TPhenP* = tetraphenanthroporphyrin; 2Phen$_x N_4$ = non-core-modified tetraphenyldiphenanthroporphyrin. The partially chalcogen core-modified macrocycles have a potential to stabilize metals in unusual oxidation states. Thus, 5,10,15,20-tetraphenyl-21-oxaporphyrin was shown to stabilize Ni(I) [37], which was not possible for *TPP*, due to the presence of an N_3O, rather than an N_4, core. The effects of the heteroatoms on the electronic structures and optical properties of the porphyrinoids were examined using TD-DFT calculations and MCD and fluorescence measurements. MCD measurements were carried out on a series of compounds with n-hexadecyloxy groups attached at the *para*-positions of the phenyl substituents ($N_2Y_2OC_{16}H_{33}$): these groups enhanced the solubility of the compounds in optically transparent solvents. To examine the extent to which the aryl substituents could be used to fine-tune the optical properties of core-modified porphyrins, electron-donating OMe- and $N(CH_3)_2$-groups (giving compounds denoted as N_2Se_2-OMe and N_2Se_2-NMe_2) and electron-withdrawing F-atoms (giving compounds denoted as N_2O_2-F and N_2S_2-F) were employed.

To compare the relative effects of the heavy atoms and core modification on the emission properties of the core-modified porphyrins, Cl, Br and I were used to generate the compounds denoted as N_2O_2-*Cl*, N_2O_2-*Br* and N_2O_2-*I*. The effects of steric crowding and peripheral-fused-ring expansion were explored by synthesizing core-modified diphenanthro- (P_1 and P_1-Bz_y) and diacenaphthoporphyrins (P_3 and P_3-Bz_y), with fused BCOD and benzene rings along the *y*-axis. The results of the research might be summarized as follows. (i) Steric hindrance between the *meso*-phenyl substituents and the peripheral-fused-ring moieties resulted in a significant saddling distortion of the π-systems. According to the results of computational studies, when the porphyrin cores were modified by introducing furan, thiophene, selenophene, or tellurophene moieties along the *y*-axis of the porphyrin core, the obtained core-modified porphyrin structures remained highly non-planar but the saddling distortion of the π-system steadily diminished as the heteroatoms became progressively larger on going from O to Te to form N_2O_2, N_2S_2, N_2Se_2 and N_2Te_2. One of the selenophene or tellurophene moieties was found to tilt out of the saddled structure that was formed by the rest of the π-system, due to a marked increase in the length of the C-Se and C-Te bonds. However, the twofold axis of symmetry was retained in the X-ray structure of N_2Se_2 and also when the structure was optimized with the 3–21G** rather than the 6–31G(d) basis sets. (ii) Upon core modification with chalcogens, a slight increase of the average HOMO–LUMO gap was predicted from O to S, with slight decrease from S to Se to Te, primarily based on a slight relative destabilization of the so-called *s MOs* (the MOs with nodal planes along the *y*-axis of the compound were referred to as 'a' and '-a' MOs, whereas MOs with large coefficients were referred to as 's' and '-s' MOs [36]). The TD-DFT predicted average HOMO-LUMO gaps for the N_2Y_2 compounds were found to vary from ca. 2.15 to ca. 2.4 eV [36]. Core modification was found to result in a red shift of the lowest-energy Q and B bands on moving from $N_2O_2OC_{16}H_{33}$ to $N_2Se_2OC_{16}H_{33}$ and then to N_2Te_2, as well as from $N_2S_2OC_{16}H_{33}$ to $N_2O_2OC_{16}H_{33}$. N_2Te_2 was found to be unstable to oxidation due to the destabilization of the HOMO and thus considered as not useful for optoelectronic or photodynamic applications. The absorption spectra of N_2O_2 and N_2Te_2 were shown to be markedly different from those of N_2S_2 and N_2Se_2 and the bands in the UV region were found to increase in intensity relative to the Q and B bands. This increase was explained by the effect of changes in the degree of saddling of the phenanthrene moieties and, where N_2Te_2 is concerned, by the effect on the vibrational bands of tilting one of the tellurophene moieties out of the saddled C_{2v} symmetry structure. The Q_{00} bands of N_2O_2, N_2S_2, N_2Se_2 *and* N_2Te_2 were found to be relatively weak. (iii) Incorporation of electron donating -NMe_2 groups at the *para*-positions of *meso*-attached benzene rings was shown to introduce a strong mesomeric interaction with the main porphyrin macrocycle which resulted in a significant intensification and red-shift of the Q bands. (iv) The diphenanthro- and diacenaphtho-fused N_2S_2 compounds containing two benzo-fused thiophene moieties were shown to exhibit a narrowing of the HOMO–LUMO gap relative to TPP due to primarily a stabilization of the LUMO rather than a destabilization of the HOMO and enhanced absorption intensity in the NIR region. (v) The differing effects of incorporating benzene, phenanthrene and acenaphthalene fused ring moieties along the *x* and *y* axes were shown to substantially modify the relative energies of the four frontier π-MOs of the compounds studied. Some of the core-modified compounds studied were shown to be promising candidates for use in photodynamic therapy.

In 2016, Goto, Shinmyozu and coworkers reported the synthesis, optical and redox properties, and electronic structure of the completely core-modified tetrakis(pentafluorophenyl)tetrathiaisophlorin dioxide (12) [25]. After the synthesis of the fully core-modified 5,10,15,20-tetrakis (pentafluorophenyl)-21,22,23,24-tetrathiaisophlorin (11) (**Figure 2**) [25], the authors aimed to oxidize the S-atoms of the thiophene moieties of **11** to reveal its reactivity toward oxidation, compared to that of simple thiophene derivatives and to elucidate the structure and electronic properties of the oxidized products. Earlier, Bongini et al. reported that oxidation of thiophene to the corresponding 1-oxide led to only a minor change in the ionization potential, but to a dramatic change in electron affinity [38]. The product of oxidation of the compound **11** was found to be the 20 π-electron tetrathiaisophlorin dioxide **12**, stable at room temperature. The thiophene moieties and S-atoms of the thiophene 1-oxide moieties of **12** were found to be tilted above and below the plane formed by the four meso-carbons. Cyclic voltammetry measurements indicated for **12** a significant stabilization of the HOMO, but the LUMO energy remained essentially unaltered. This corresponded to the significant blue shift of the λ_{max} of the absorption band (348 nm), compared with that of the parent compound **11**. This result was also supported by MCD spectra and molecular orbital calculations (B3LYP/6-31G* level).

http://dx.doi.org/10.5772/67728

The MCD spectrum of **12** was interpreted as that of a 4n π-antiaromatic system. Based on the computed NICS values and ^1H NMR spectroscopy data, compound **12** was assigned more antiaromatic character than **11** (which was aromatic). This study demonstrated, for the first time, the following: (i) a tetrathiaporphyrin can be oxidized to the dioxide stable at room temperature; (ii) an attachment of O-atoms to the S-atoms of a tetrathiaporphyrin could modify its redox potentials and optical and electronic properties, along with its aromaticity properties.

Within this chapter, it is also of interest to mention the work of Sukumaran, Detty, and coworkers who in 2002 reported their studies of Te-containing 21- and 21,23-core-modified porphyrins [39]. Ono and coworkers who also studied the partial core modification of tetrabenzoporphyrins and tetraphenyltetrabenzoporphyrins with O and S observed only minor changes in the optical spectra of 21- and 21,23-core-modified tetrabenzoporphyrins [35].

Very recently, Anand and coworkers reported extremely interesting synthesis and characterization of the meso-meso linked antiaromatic tetraoxaisophlorin dimer [40]. It should be noted that antiaromatic units are seldom used as components of functional π-materials [41], although they can be employed in organic electronics due to their noticeable paramagnetic properties [42]. The chemistry of antiaromatic systems is severely hindered by the very small number of stable antiaromatic compounds. The 4nπ isophlorins offer a rare opportunity to explore novel antiaromatic organic materials for potential applications in optoelectronics. The 20 π-electrons isophlorin derivatives of thiophene and furan represent the simplest of the stable and planar antiaromatic compounds. Isophlorin can non-covalently bind to C_{60} through conventional π-π interactions, as was shown by the same research group in 2015, thus highlighting the utility of isophlorin as a synthon for supramolecular chemistry [43]. It was found that the compound **3** formed the co-crystallized product **3**·C_{60} along with the unexpected *meso–meso* linked dimer, **4**, bound non-covalently to C_{60}. The formation of the dimer was confirmed by MALDI TOF-TOF mass spectrometry and by ^1H NMR spectroscopy. The compound **3** was found to exhibit a flat geometry (as observed for other tetraoxaisophlorins) with very close contacts (2.58 and 2.61 Å) between its π-surface and the surface of C_{60}. The macrocycles in the compound **4** were found to make an angle of 35.57°, which supported the single bond nature of the meso-meso link between these two macrocycles. The macrocycles in the compound **4** were found to bind the fullerene through uncommon short π–π interactions (2.70, 2.78 and 2.93 Å) between their surface and the surface of the C_{60}. The antiaromatic character of **4** was further supported by NICS calculations. The estimated NICS values in the centre of the macrocycle, NICS(0), of +30.38 and +12.90 for **3** and **4**, respectively, showed *antiaromaticity* of these compounds. The reduced antiaromaticity of **4** was attributed to the loss of planar structure upon binding the fullerene C_{60} at the centre of the macrocycle. The electronic absorption spectrum of the dimer **4** in dichloromethane displayed a red shift with respect to the monomer **3**. An intense absorption similar to the Soret-like band at 372 nm, and Q-like bands at 436, 466 and 503 nm were considered as suggesting electronic coupling between the macrocyclic units in spite of the non-coplanar orientation of the both the macrocycles. The compound **4** was found to exhibit a strong resistivity towards oxidation as was observed for tetraoxaisophlorins by Reddy and Anand before [65]. Moreover, **4** was also found to resist the formation of β-β links upon action of strong oxidants to yield the completely fused macrocyclic dimer.

Also, it is worthwhile to mention the following several works on *partially core-modified* porphyrins. The 2009 micro-review on aza-deficient porphyrins considered briefly, among other compounds, 21-heteroporphyrins containing O, S, Se, Te or P, and 21,23-ditellura-porphyrin, which possessed significant distortion due to the size of Te-atom [44]. The 2015 report by Maeda et al., considered the synthesis and photophysical properties of cyano and ethynyl-substituted carbazole-based chalcogen-porphyrins containing either two S- or the Se-atoms [45]. One year earlier, Maeda et al. reported the synthesis of carbazole-based hetero-core-modified (by S and Se) porphyrins [46]. In 2015, Zhang and coworkers reported the DFT study of the magnetically induced current strengths as well as NICS of aromatic heteroporphyrins and antiaromatic 22,24-dideazaheteroporphyrins [47]. Heteroporphyrins were shown to sustain a diatropic induced current while 22,24-dideazaheteroporphyrins revealed paratropic ring current. The substitution of pyrrole NH groups by O and S atoms was shown to not change the total induced current strengths and total $NICS(0)_{\pi zz}$ values.

2.2. Core modification with phosphorus

In this section, we will first address the studies on *partially core-modified* porphyrins and their derivatives reported by Matano et al. [53–62] for the series of *mono-phosphorus*-substituted porphyrins, calixphyrins and calixphospholes [48] and then will proceed to the *fully core-modified* porphyrins recently studied by Kuznetsov.

Partial core modification of tetrapyrroles by P has been of long-lasting interest [48]. The *mono-P-modified* Matano/Imahori structures showed interesting redox chemistry both in their coordination compounds with Pd, Pt, Rh, Zn, Au and Ni and as free ligands, along with the catalytic activity (see the discussion below). Therefore, more heavily P-modified metalloporphyrins should possess intriguing structural, electronic and optical properties. Stepwise syntheses of tri- and tetraphosphaporphyrinogens with numerous groups attached to the porphyrin core were reported by Carmichael et al. [48].

Now a few words should be said about the phosphole, C_5H_5P, as the phosphorus isologue of pyrrole, C_5H_5N. C_5H_5P has much lower aromaticity than pyrrole due to insufficient π-conjugation between the cis-dienic π-system and the lone electron pair of the P-atom [49, 50]. The phosphole species possesses the following prominent features affecting its structure, electronic properties and reactivities [50]: (1) the P-centre adopts a trigonal pyramidal geometry due to insufficient n-π orbital interaction; (2) the LUMO is located at a lower energy compared to the pyrrole LUMO due to the effective σ^*(P–R) – π^*(1,3-diene) hyperconjugation; (3) orbital energies of the C_5H_5P π-system are easily tunable by chemical modification at the P-centre and (4) the P-bridged 1,3-diene unit is rigid, electron rich and polarizable. These features of phospholes originate from the intrinsic nature of the P 3s and 3p orbitals. Consequently, phospholes behave both as potential building blocks for the π-conjugated materials and as ordinary phosphine ligands [51].

In 2003, Delaere and Nguyen [52] reported the DFT study of the structural and optical properties of the core-modified porphyrins with one or two pyrrole nitrogens replaced by P-atoms. The geometries of the ground states were optimized using the B3LYP/6-31G* approach and energies of the lower-lying excited singlet states of P-modified porphyrins were computed using

http://dx.doi.org/10.5772/67728

the TD-B3LYP/SV(P) method and compared with those of N-porphyrins. The substitution of a NH- by a PH-unit did not distort the carbon skeleton which remains essentially planar, whereas replacement of a N- by a P-atom was found to weakly distort (by 15.3°) the P-containing ring from the porphyrin mean plane. A nearly equal red-shift of both Q- and B-bands was predicted upon substituting NH- by PH-units, whereas the red shift of Q-bands was calculated to be much larger than the red shift of B-bands upon substitution of an N-atom by a P-atom.

Later, Matano et al. [53–62] reported syntheses and characterization of various phosphaporphyrins and their derivatives with only one pyrrole nitrogen replaced by a P-atom. Thus, in their 2010 review [53], the researchers summarized their previous studies on the phosphole-containing porphyrins and their metal complexes. One of the compounds studied, the porphyrin containing trigonal pyramidal P-centre was found to possess a slightly distorted 18π-electron plane, wherein the phosphole and three pyrrole rings were found to be somewhat tilted from the 24-atom mean plane. It was suggested that the porphyrin 18π-electron circuit does not involve the lone electron pair of the trigonal pyramidal P-atom. On the contrary, the 22π-electron porphyrin containing tetrahedral P-centre was shown to have a highly-ruffled structure, with the P-atom deviated significantly from the porphyrin π-plane (1.20 Å) to avoid the steric congestion at the core. The Rh(III) and Pd(II) derivatives of these compounds were also shown to possess significant structural distortions. The metal complexes of these P-modified porphyrins exhibited only a weak antiaromaticity in terms of the magnetic criterion. In the UV/vis absorption spectra of the P-modified porphyrins, the characteristic two transitions of the porphyrin core, B and Q bands, were clearly observed, with significant red shifts. These results showed that the incorporation of a P-atom in the porphyrin core considerably reduced both the S_0–S_2 and S_0–S_1 excitation energies. The 18π-electron Rh-complex also showed characteristic Soret and Q bands, whereas the 20π-electron Pd-complex displayed broad and blue-shifted Soret-like bands and no detectable Q bands, which is typical of highly ruffled, nonaromatic 4nπ porphyrinoids. It was stated that the observed structures, reactivities, and coordinating properties of the studied P-core-modified porphyrins were undoubtedly produced by the P-atom at the core. In this context, the phosphole-containing porphyrins were regarded as metal–affinitive macrocyclic π-systems and could be developed as new classes of metal sensors, sensitizers and catalysts.

Earlier, in the 2009 review [54], Matano and Imahori described the exploration of the utility of phosphole-containing porphyrins and porphyrinogens as macrocyclic, mixed-donor ligands. The convenient methods for the synthesis of calixpyrroles, calixphyrins and porphyrins with P and either O or S substitutions (P,X,N_2-hybrids) were described. Also, the effects of varying the combination of core heteroatoms (P, N, S and O) on the coordination properties of the hybrid macrocycles were investigated. The results were summarized to show that: (i) the P,S,N_2-calixpyrroles behave as monophosphine ligands, (ii) the P,X,N_2-calixphyrins behave as neutral, monoanionic or dianionic tetradentate ligands and (iii) the P,S,N_2-porphyrins behave as a redox-active π-ligand for group 10 metals (Ni, Pd, Pt), affording a novel class of core-modified isophlorin complexes. The incorporation of the phosphole subunit into the macrocyclic framework was proved to provide unprecedented coordinating properties for the porphyrin family.

In 2008, the syntheses, structures and coordination chemistry of phosphole-containing hybrid calixphyrins (P,N_2,X-hybrid calixphyrins) and the catalytic activities of their transition metal complexes were reported [60]. The 5,10-porphodimethene type 14π-P,$(NH)_2$,X- and 16π-P,N_2,X-hybrid calixphyrins (where X = O, S, NH) were prepared. The σ^3-P,$(NH)_2$,S- and σ^3-P,N_2,S-compounds were shown to produce the same Pd(II)-P,N_2,S-hybrid complex. In this complex, the calixphyrin ligand was regarded as a dianion. In the complexation with $[RhCl(CO)_2]_2$ in CH_2Cl_2, the σ^3-P,N_2,S-compound was shown to behave as a neutral ligand producing an ionic Rh(I)-P,N_2,S-hybrid complex. The σ^3-P,N_2,NH-compound was found to behave as an anionic ligand to produce Rh(III)-P,N_3-hybrid complexes. The complexation of $AuCl(SMe_2)$ with the σ^3-P,N_2,X-compounds (X = S, NH) was shown to lead to the formation of the corresponding Au(I)-monophosphine complexes. The calixphyrin-Pd and -Rh complexes were shown to catalyse the Heck reaction and hydrosilylation reaction, respectively, implying that the metal centre in the core was capable of activating the substrates under appropriate reaction conditions. The study results demonstrated the potential utility of the phosphole-containing hybrid calixphyrins as a new class of macrocyclic P,N_2,X-mixed donor ligands for designing highly reactive transition metal complexes.

It is also worthwhile to mention the 2009 theoretical investigation of electronic structure and reactivity for oxidative addition for the Pd-complex of P,S-containing hybrid calixphyrin [62]. Two kinds of valence tautomers were shown for the Pd-complex **1**: (i) with the calixphyrin moiety having -2 charges and the Pd-centre with +2 oxidation state, (ii) with the calixphyrin neutral and the Pd-centre with 0 oxidation state. Complex **1** was shown to take the first form in the ground state. DFT computations clearly showed that the oxidative addition of phenyl bromide (PhBr) to **1** occurred with moderate activation enthalpy, as experimentally proposed. On the other hand, the oxidative additions of PhBr to Pd-complexes of P,S-containing hybrid porphyrin **2** and of conventional porphyrin **3** needed much larger activation enthalpies. The differences in the reactivity among the complexes **1**, **2**, and **3** were theoretically investigated. In **1**, the valence tautomerization was shown to occur with moderate activation enthalpy to afford the form with Pd(0) which was reactive for the oxidative addition. In **2**, the tautomerization from the Pd(+2) form to the Pd(0) form needed very large activation enthalpy. In **3**, such valence tautomerization did not occur at all, indicating that the Pd(+2) must change to the Pd(+4) in the oxidative addition of PhBr to **3**, which is a very difficult process. These differences were interpreted in terms of the π^*-orbital energies of the compounds and the flexibility of their frameworks.

So far, as can be seen, no computational studies (metallo)porphyrins completely core-modified with P-atoms ($P(P)_4$) have been reported, except the 2012 report by Barbee and Kuznetsov on the $NiP(P)_4$ compound [63]. Motivated by the above-listed works on mono-P-core-modified porphyrins and derivatives, Kuznetsov reported the computational studies of the structures and electronic properties of the fully P-core-modified metalloporphyrins, $MP(P)_4$, M = Sc-Zn [64, 65], along with the computational design of the stacks formed by the $ZnP(P)_4$ species [66]. The prominent structural feature of all the $MP(P)_4$ compounds studied was found to be their significant distortion from planarity (**Figure 3**) [63–66].

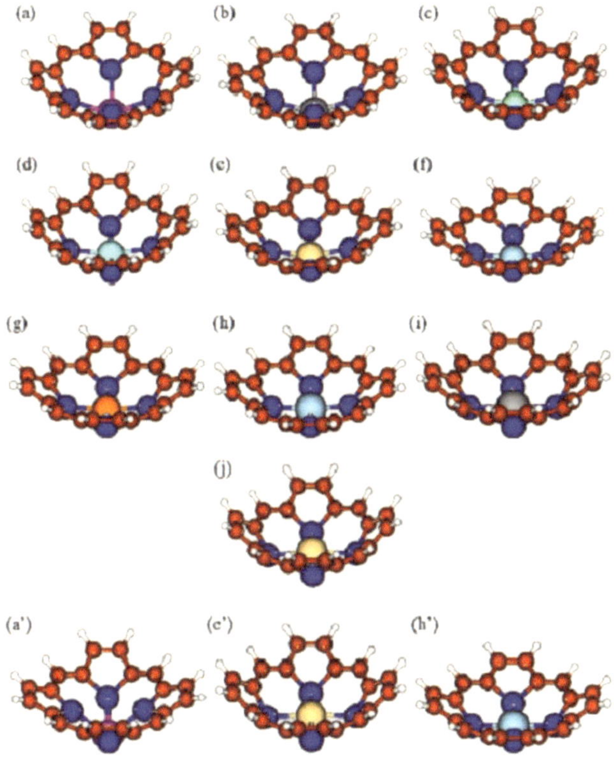

Figure 3. Structures of the $MP(P)_4$ species calculated at the B3LYP/6-31G* level: neutrals: $Sc^{II}P(P)_4$ (a), $Ti^{II}P(P)_4$ (b), $V^{II}P(P)_4$ (c), $Cr^{II}P(P)_4$ (d), $Mn^{II}P(P)_4$ (e), $Fe^{II}P(P)_4$ (f), $Co^{II}P(P)_4$ (g), $Ni^{II}P(P)_4$ (h), $Cu^{II}P(P)_4$ (i) and $Zn^{II}P(P)_4$ (j), and cations: $Sc^{III}P(P)_4$ (a'), $Mn^{IV}P(P)_4$ (e'), and $Ni^{III}P(P)_4$ (h'). Reprinted from Kuznetsov [65]. Copyright (2016), with permission from Elsevier.

In the 2015 work, the first *systematic* DFT study of the $MP(P)_4$ compounds was performed [64]. The $MP(P)_4$ species with increasing number of d-electrons were studied: $3d^14s^2$ (Sc) → $3d^24s^2$ (Ti) → $3d^64s^2$ (Fe) → $3d^84s^2$ (Ni) → $3d^{10}4s^1$ (Cu) → $3d^{10}4s^2$ (Zn). Systematic comparison with the tetrapyrrole MP counterparts was made. As mentioned above, all the $MP(P)_4$ species were calculated to adopt a bowl-like shape, compared to generally planar shapes of their MP counterparts. Significant positive charges were computed to be accumulated on P-atoms in $MP(P)_4$. Positive charges on the metals in $MP(P)_4$ were found to be noticeably lower than in the MP counterparts. The calculated $MP(P)_4$ HOMO-LUMO gaps and optical gaps were noticeably smaller than the corresponding gaps in their MP counterparts, which was explained by stabilization of the $MP(P)_4$ LUMOs.

In the follow-up 2016 work [65], the comparative DFT study, including Natural Bond Orbitals analysis, of the binding energies between all the first-row transition metals M^{n+} (M = Sc–Zn) and two ligands of the similar type, porphine, P^{2-}, and its completely P-modified counterpart, $P(P)_4^{2-}$, was reported. The main findings were as follows: (i) generally, for the $MP(P)_4$ compounds the calculated HOMO-LUMO gaps and optical gaps were shown to be smaller than for their MP counterparts; (ii) the trends in the change of the binding energies between M^{n+}

and $P(P)_4^{2-}/P^{2-}$ were shown to be very similar for both ligands. The full P-modification of the porphyrin core was found to decrease the M^{n+}-ligand binding energies; however, the $MP(P)_4$ compounds studied were shown to be stable according to the E_{bind} values and therefore can be potentially synthesized.

Also in 2016, due to motivation by the phenomenon of formation of stacks by regular metalloporphyrins, the computational check of the stack formation between the $MP(P)_4$ species without any linkers or substituents was performed [66]. Three modes of binding or coordination were found to be possible between the monomeric $ZnP(P)_4$ units (**Figure 4**).

The 'convexity-to-convexity' dimer I was found to be the most stable compound with the highest binding energy. In the dimer I, the strongly convex shape of both monomer units was demonstrated. The Zn–Zn distances in the dimer I, ca. 3.5 Å, were computed to be significantly shorter than in two other dimers. In the dimer I, significant decrease of the charge was found on the Zn-centres.

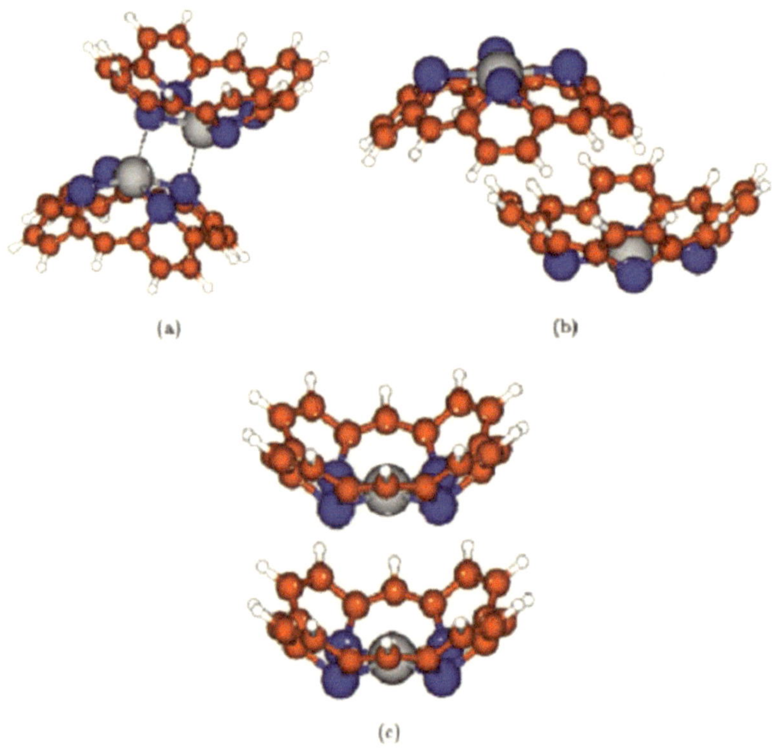

Figure 4. Structures of the $[ZnP(P)_4]_2$ stacks, binding modes I (a), II (b), and III (c), calculated at the B3LYP/6-31G* level of theory. Republished with permission of Journal of Theoretical and Computational Chemistry, Kuznetsov [66]; permission conveyed through Copyright Clearance Center, Inc.

http://dx.doi.org/10.5772/67728

3. Conclusions and perspectives

Thus, as can be seen from this micro-review, core modification of the porphyrins and their derivatives with other elements than N is a very promising and productive approach to modify and fine-tune their structures, electronic and coordination properties and reactivities. The research in this area has already been quite productive and brought for our attention numerous compounds with unusual novel structures and properties, which makes these species great candidates for different fields in chemistry and nanotechnology. Without any doubts, studies in this area will be continued and broadened. Based on the considered studies of porphyrin derivatives core-modified by other elements, we can summarize subareas (or research directions), which would be necessary to focus on to employ the core-modified porphyrins for the design of building blocks for nanotechnology:

i. Ability of core-modified porphyrins to form stacks and other arrays like regular porphyrins. Would be core-modified porphyrins and their derivatives form stacks/arrays without any linkers or substituents?

ii. Ability of core-modified porphyrins to form complexes with fullerenes, similar to regular porphyrins. How stable will be such complexes and what properties and potential applications they will have?

iii. Catalytic properties of core-modified porphyrins. Can they activate, in particular, small molecules, like H_2, O_2, N_2, hydrocarbons?

iv. Ability of core-modified porphyrins to form complexes with various nanoparticles, including semiconductor NPs. What properties and potential applications with these complexes have? How can we tune structures and properties of these complexes?

v. Possibility to develop general synthetic strategies for obtaining the core-modified porphyrins with desired structures and properties. Development of approaches for synthesis of building nanoblocks from these compounds.

vi. Broad and profound application of computational approaches in this area, both to assist the synthesis of novel core-modified porphyrins and to provide insight in their properties. Also, the computational approaches could be broadly used to assist in obtaining various building nanoblocks from these compounds.

Author details

Aleksey E. Kuznetsov

Address all correspondence to: aleksey73kuznets@gmail.com

Instituto de Química, Universidade de São Paulo – SP, São Paulo, SP, Brasil

References

[1] Dolphin D, editor. The Porphyrins. New York: Academic; 1978. Vols. I–VII.

[2] Kadish KM, Smith KM, Guilard R, editors. The Porphyrin Handbook. San Diego, CA: Academic Press; 2000. ISBN: 978-0-08-092390-1

[3] Hill HAO, Sadler PJ, Thomson AJ, editors. Metal Sites in Proteins and Models. Iron Centres. Berlin, Heidelberg, Germany: Springer-Verlag; 1997. 207 p. ISBN 3540628703, 9783540628705

[4] Kadish KM, Smith KM, Guilard R, editors. Handbook of Porphyrin Science with Applications to Chemistry, Physics, Materials Science, Engineering, Biology and Medicine, Singapore: World Scientific; 2010. Vol. III. 539 p. ISBN 978-981-4280-16-7

[5] Severance S, Hamza I. Trafficking of heme and porphyrins in metazoa. Chem. Rev. 2009;**109**:4596-4616. DOI: 10.1021/cr9001116

[6] Ponka P. Cell biology of heme. Am. J. Med. Sci. 1999;**318**:241-256. PMID: 10522552

[7] Rodgers KR. Heme-based sensors in biological systems. Curr. Opin. Chem. Biol. 1999;**3**:158-167. DOI: 10.1016/S1367-5931(99)80028-3

[8] Tsiftsoglou AS, Tsamadou AI, Papadopoulou LC. Heme as key regulator of major mammalian cellular functions: molecular, cellular, and pharmacological aspects. Pharmacol. Ther. 2006;**111**:327-345. DOI: 10.1016/j.pharmthera.2005.10.017

[9] Yin L, Wu N, Curtin JC, Qatanani M, Szwergold NR, Reid RA, Waitt GM, Parks DJ, Pearce KH, Wisely GB, Lazar MA. Rev-erbalpha, a heme sensor that coordinates metabolic and circadian pathways. Science. 2007;**318**:1786-1789. DOI: 10.1126/science.1150179

[10] Faller M, Matsunaga M, Yin S, Loo JA, Guo F. Heme is involved in microRNA processing. Nat. Struct. Mol. Biol. 2007;**14**:23-29. DOI: 10.1038/nsmb1182

[11] Sheldon RA, editor. Metalloporphyrins in Catalytic Oxidation. New York: Marcel Dekker; 1994. 390 p. ISBN 0-8247-9228-9

[12] Lomova TN, Klyueva ME, Klyuev MV. The mechanism of catalytic action of the coordination centres of catalase synthetic models, in: Lomova TN, Zaikov GE, editors. Chemical Processes with Participation of Biological and Related Compounds. Leiden, The Netherlands NV: Koninklijke Brill; 2008, pp. 93-116.

[13] Holten D, Bocian DF, Lindsey JS. Probing electronic communication in covalently linked multiporphyrin arrays. A guide to the rational design of molecular photonic devices. Acc. Chem. Res. 2002;**35**:57-69. DOI: 10.1021/ar970264z

[14] Senge MO, Fazekas M, Notaras EGA, Blau WJ, Zawadzka M, Locos OB, Ni Mhuircheartaigh EM. Nonlinear optical properties of porphyrins. Adv. Mater. 2007;**19**:2737-2774.DOI: 10.1002/adma.200601850

[15] Castano AP, Mroz P, Hamblin MR. Photodynamic therapy and anti-tumour immunity. Nature. 2006;**6**:535-545. DOI: 10.1038/nrc1894

[16] Wasielewski MR. Self-assembly strategies for integrating light harvesting and charge separation in artificial photosynthetic systems. Acc. Chem. Res. 2009;**42**:1910-1921. DOI: 10.1021/ar9001735

[17] Aratani N, Kim D, Osuka A. Discrete cyclic porphyrin arrays as artificial light-harvesting antenna. Acc. Chem. Res. 2009;**42**:1922-1934. DOI: 10.1021/ar9001697

[18] Imanori H, Umeyama T, Ito S. Large π-aromatic molecules as potential sensitizers for highly efficient dye-sensitized solar cells. Acc. Chem. Res. 2009;**42**:1809-1818. DOI: 10.1021/ar900034t

[19] Monti D, Nardis S, Stefanelli M, Paolesse R, Di Natale C, D'Amico A. Porphyrin-based nanostructures for sensing applications. J. Sensors. 2009;**2009**:1-10. Article ID 856053

[20] Liao M-S, Watts JD, Huang M-J. Fe^{II} in different macrocycles: electronic structures and properties. J. Phys. Chem. A. 2005;**109**:7988-8000. DOI: 10.1021/jp0581476

[21] Latos-Grazyński L. Core-modified heteroanalogues of porphyrins and metallo-porphyrins, in: Kadish KM, Smith KM, Guilard R, editors. The Porphyrin Handbook, San Diego, CA: Academic Press; 2000, Vol. 2, Chapter 14.

[22] Chandrashekar TK, Venkatraman S. Core-modified expanded porphyrins: new generation organic materials. Acc. Chem. Res. 2003;**36**:676-691. DOI: 10.1021/ar020284n

[23] Chmielewski PJ, Latos-Grazyński L. Core modified porphyrins—a macrocyclic platform for organometallic chemistry. Coord. Chem. Rev. 2005;**249**:2510-2533. DOI: 10.1016/j.ccr.2005.05.015

[24] Gupta I, Ravikanth M. Recent developments in heteroporphyrins and their analogues. Coord. Chem. Rev. 2006;**250**:468-518. DOI: 10.1016/j.ccr.2005.10.010

[25] Mishra VL, Furuyama T, Kobayashi N, Goto K, Miyazaki T, Yang J.-S, Shinmyozu T. Synthesis, optical properties, and electronic structures of tetrakis(pentafluorophenyl)tetrathiaisophlorin dioxide. Chem. Eur. J. 2016;**22**:9190-9197. DOI: 10.1002/chem.201505079

[26] Vogel E, Haas W, Knipp B, Lex J, Schmickler H. Tetraoxaporphyrin dication. Angew. Chem. Int. Ed. Engl. 1988;**27**:406-409. DOI: 10.1002/anie.198804061

[27] Vogel E, Röhrig P, Sicken M, Knipp B, Herrmann A, Pohl M, Schmickler H, Lex J. The thiophene analogue of porphyrin: tetrathiaporphyrin dication. Angew. Chem. Int. Ed. Engl. 1989;**28**:1651-1655. DOI: 10.1002/anie.198916511

[28] Vogel E, Pohl M, Herrmann A, Wiss T, König C, Lex J, Gross M, Gisselbercht JP. Porphyrinoid macrocycles based on thiophene—the octaethyltetrathiaporphyrin dication. Angew. Chem. Int. Ed. Engl. 1996;**35**:1520-1524. DOI: 10.1002/anie.199615201

[29] Woodward RB. Total synthese des chlorophylls. Angew. Chem. 1960;**72**:651-662. DOI: 10.1002/ange.19600721803

[30] Pohl M, Schmickler H, Lex J, Vogel E. Isophlorins: molecules at the crossroads of porphyrin and annulene chemistry. Angew. Chem. Int. Ed. Engl. 1991;**30**:1693-1697. DOI: 10.1002/anie.199116931

[31] Chao L, Shen D.-M, Chen Q.-Y. Synthesis and reactions of 20 π-electron β-tetrakis-(trifluoromethyl)-meso-tetraphenylporphyrins. J. Am. Chem. Soc. 2007;**129**:5814-5815. DOI: 10.1021/ja070855c

[32] Setsune J.-I, Kashihara K, Wada K-i, Shiozaki H. Photoreduction of N,N′-bridged porphyrins to 20π antiaromatic isophlorins. Chem. Lett. 1999;**28**:847-848. DOI: 10.1246/cl.1999.847

[33] Reddy JS, Anand VG. Planar *meso* pentafluorophenyl core modified isophlorins. J. Am. Chem. Soc. 2008;**130**:3718-3719. DOI: 10.1021/ja710664y

[34] Kon-no M, Mack J, Kobayashi N, Suenaga M, Yoza K, Shinmyozu T. Synthesis, optical properties, and electronic structures of fully core modified porphyrin dications and isophlorins. Chem. Eur. J. 2012;**18**:13361-13371. DOI: 10.1002/chem.201200776

[35] Mack J, Bunya M, Shimizu Y, Uoyama H, Komobuchi N, Okujima T, Uno H, Ito S, Stillman MJ, Ono N, Kobayashi N. Application of MCD spectroscopy and TD-DFT to nonplanar core-modified tetrabenzoporphyrins: effect of reduced symmetry on nonplanar porphyrinoids. Chem. Eur. J. 2008;**14**:5001-5020. DOI: 10.1002/chem.200701611

[36] Xu H.-J, Mack J, Wu D, Xue Z.-L, Descalzo AB, Rurack K, Kobayashi N, Shen Z. Synthesis and properties of fused-ring-expanded porphyrins that were core-modified with group 16 heteroatoms. Chem. Eur. J. 2012;**18**:16844-16867. DOI: 10.1002/chem.201200956

[37] Xu H.-J, Mack J, Descalzo AB, Shen Z, Shen Z, You X.-Z, Rurack K. *meso*-Aryl phenanthroporphyrins: synthesis and spectroscopic properties. Chem. Eur. J. 2011;**17**:8965-8983. DOI:10.1002/chem.201002596

[38] Bongini A, Barbarella G, Zambianchi M, Arbizzani C, Mastragostino M. Thiophene S-oxides: orbital energies and electrochemical properties. Chem. Commun. 2000;439-440. DOI: 10.1039/A909390G

[39] Abe M, Hilmey DG, Stilts CE, Sukumaran DK, Detty MR. 21-Telluraporphyrins. 1. Impact of 21,23-heteroatom interactions on electrochemical redox potentials, ^{125}Te NMR spectra, and absorption spectra. Organometallics. 2002;**21**:2986-2992. DOI: 10.1021/om0202219

[40] Reddy BK, Gadekar SK, Anand VG. The synthesis and characterization of the meso–meso linked antiaromatic tetraoxaisophlorin dimer. Chem. Commun. 2016;**52**:3007-3009. DOI: 10.1039/c5cc10370c

[41] Müller TJJ, Bunz UHF, editors. Functional Organic Materials: Syntheses, Strategies and Applications. Weinheim: Wiley-VCH; 2007. 612 p. ISBN: 978-3-527-31302-0

[42] Garratt PJ. Aromaticity. New York: Wiley; 1986. 318 p. ISBN 0-471-80703-6

[43] Reddy BK, Gadekar SK, Anand VG. Non-covalent composites of antiaromatic isophlorin–fullerene. Chem. Commun. 2015;**51**:8276-8279. DOI: 10.1039/C5CC00771B

[44] Pacholska-Dudziak E, Latos-Grażyński L. Aza-deficient porphyrin as a ligand. Coord. Chem. Rev. 2009;**253**:2036-2048. DOI: 10.1016/j.ccr.2009.01.029

[45] Maeda C, Kurihara K, Masuda M, Yoshioka N. Effects of cyano, ethynyl and ethylenedioxy groups on the photophysical properties of carbazole-based porphyrins. Org. Biomol. Chem. 2015;**13**:11286-11291. DOI: 10.1039/c5ob01824b

[46] Maeda C, Kurihara K, Masuda M, Yoshioka N. Synthesis of carbazole-based hetero-core-modified porphyrins. Org. Biomol. Chem. 2014;**12**:2656-2662. DOI: 10.1039/c3ob42564a

[47] Ren J, Bai F.-Q, Zhang H.-X. The induced current strengths and aromatic pathways of heteroporphyrins and their antiaromatic derivatives. Int. J. Quant. Chem. 2015;**115**:983-988. DOI: 10.1002/qua.24932

[48] Carmichael D, Escalle-Lewis A, Frison G, Le Goffa X, Muller E. Stepwise syntheses of tri- and tetraphosphaporphyrinogens. Chem. Commun. 2012;**48**:302-304. DOI: 10.1039/c1cc15421d

[49] Minkin VI, Glukhovtsev MN, Simkin BY. Aromaticity and Antiaromaticity. Electronic and Structural Aspects. New York: Wiley; 1994. 336 p. ISBN: 978-0-471-59382-9

[50] Matano Y, Imahori H. Design and synthesis of phosphole-based π systems for novel organic materials. Org. Biomol. Chem. 2009;**7**:1258-1271. DOI: 10.1039/b819255n

[51] Baumgartner T, Réau R. Organophosphorus π-conjugated materials. Chem. Rev. 2006;**106**:4681-4727; DOI: 10.1021/cr040179m

[52] Delaere D, Nguyen MT. A density functional study of the ground state electronic structure of phosphorus-porphyrins. Chem. Phys. Lett. 2003;**376**:329-337. DOI: 10.1016/S0009-2614(03)01012-1

[53] Matano Y, Nakabuchi T, Imahori H. Synthesis, structures, and aromaticity of phosphole-containing porphyrins and their metal complexes. Pure Appl. Chem. 2010;**82**:583-593. DOI:10.1351/PAC-CON-09-08-05

[54] Matano Y, Imahori H. Phosphole-containing calixpyrroles, calixphyrins, and porphyrins: synthesis and coordination chemistry. Acc. Chem. Res. 2009;**42**:1193-1204. DOI: 10.1021/ar900075e

[55] Nakabuchi T, Matano Y, Imahori H. Remarkable effects of P-perfluorophenyl group on the synthesis of core-modified phosphaporphyrinoids and phosphadithiasapphyrin. Org. Lett. 2010;**12**:1112-1115. DOI: 10.1021/ol100114j

[56] Nakabuchi T, Nakashima M, Fujishige S, Nakano H, Matano Y, Imahori H. Synthesis and reactions of phosphaporphyrins: reconstruction of π-skeleton triggered by oxygenation of a core phosphorus atom. J. Org. Chem. 2010;**75**:375-389. DOI: 10.1021/jo902060b

[57] Matano Y, Nakashima M, Nakabuchi T, Imahori H, Fujishige S, H. Nakano. Mono-phosphaporphyrins: oxidative π-extension at the peripherally fused carbocycle of the phosphaporphyrin ring. Org. Lett. 2008;**10**:553-556. DOI: 10.1021/ol7029118

[58] Matano Y, Nakabuchi T, Fujishige S, Nakano H, Imahori H. Redox-coupled complexation of 23-phospha-21-thiaporphyrin with group 10 metals: a convenient access to stable core-modified isophorin-metal complexes. J. Am. Chem. Soc. 2008;**130**:16446-16447. DOI: 10.1021/ja807742g

[59] Matano Y, Nakabuchi T, Miyajima T, Imahori H, Nakano H. Synthesis of phosphorus-containing hybrid porphyrin. Org. Lett. 2006;**8**:5713-5716. DOI: 10.1021/ol0622763

[60] Matano Y, Miyajima T, Ochi N, Nakabuchi T, Shiro M, Nakao Y, Sakaki S, Imahori H. Syntheses, structures, and coordination chemistry of phosphole-containing hybrid calixphyrins: promising macrocyclic P,N_2,X-mixed donor ligands for designing reactive transition-metal complexes. J. Am. Chem. Soc. 2008;**130**:990-1002. DOI: 10.1021/ja076709o

[61] Matano Y, Miyajima T, Nakabuchi T, Imahori H, Ochi N, Sakaki S. Phosphorus-containing hybrid calixphyrins: promising mixed-donor ligands for visible and efficient palladium catalysts. J. Am. Chem. Soc. 2006;**128**:11760-11761. DOI: 10.1021/ja0640039

[62] Ochi N, Nakao Y, Sato H, Matano Y, Imahori H, Sakaki S. New palladium(II) complex of P,S-containing hybrid calixphyrin. Theoretical study of electronic structure and reactivity for oxidative addition. J. Am. Chem. Soc. 2009;**131**:10955-10963. DOI: 10.1021/ja901166a

[63] Barbee J, Kuznetsov AE. Revealing substituent effects on the electronic structure and planarity of Ni-porphyrins. Comp. Theor. Chem. 2012;**981**:73-85. DOI: 10.1016/j.comptc.2011.11.049

[64] Kuznetsov AE. Metalloporphyrins with all the pyrrole nitrogens replaced with phosphorus atoms, $MP(P)_4$ (M = Sc, Ti, Fe, Ni, Cu, Zn). Chem. Phys. 2015;**447**:36-45. DOI: 10.1016/j.chemphys.2014.11.018

[65] Kuznetsov AE. How the change of the ligand from L = porphine, P^{2-}, to L = P_4-substituted porphine, $P(P)_4^{2-}$, affects the electronic properties and the M-L binding energies for the first-row transition metals M = Sc-Zn: comparative study. Chem. Phys. 2016;**469-470**:38-48. DOI: 10.1016/j.chemphys.2016.02.010

[66] Kuznetsov AE. Computational design of $ZnP(P)_4$ stacks: three modes of binding. J. Theor. Comp. Chem. 2016;**15**:1650043. DOI: 10.1142/S0219633616500437

Chapter 7

The Mechanisms for the Oxidative Addition of Imidazolium Salts to a Group 9 Transition Metal Atom (Co^0, Rh^0, and Ir^0) and a Group 10 Transition Metal Atom (Ni^0, Pd^0, and Pt^0): A Theoretical Study

Hsin-Yi Liao, Jia-Syun Lu and Ming-Der Su

Additional information is available at the end of the chapter

http://dx.doi.org/10.5772/67567

Abstract

The potential energy surfaces of the oxidative addition reactions, L_2M + imidazolium cation → product and CpM′L + imidazolium cation → product (M = Ni, Pd, Pt; M′ = Co, Rh, Ir; Cp = η^5-C_5H_5; L = 1,3-aryl-N-heterocyclic carbene (NHC), aryl = 2,4,6-trimethylphenyl), are studied at the M06-L/Def2-SVP level of theory. The theoretical findings show that the singlet-triplet splitting ($\Delta E_{st} = E_{triplet} - E_{singlet}$) for the L_2M and CpM′L species can be used to predict the reactivity for their oxidative additions. That is to say, current theoretical evidence suggests that both a 14-electron L_2M complex and a 16-electron CpM′Lcomplex with a better electron-donating ligand L (such as NHC) result in a reduced ΔE_{st} value and facilitate the oxidative addition to the saturated C—H bond. The theoretical results for this study are in good agreement with the obtainable experimental results and allow a number of predictions to be made.

Keywords: oxidative addition reactions, group 9 elements, group 10 elements, imidazolium and density functional theory

1. Introduction

Recent studies involving the chemistry of the heterocyclic nitrogen carbene complexes of transition metals have demonstrated that they can act as precatalysts for a variety of reactions [1–11]. These new species offer many opportunities to advance this field of study [12–30]. The use of palladium carbene complexes for the Heck reaction [31–34] and platinum carbene compounds for the C—H activation reactions [11] has created new opportunities in catalytic chemistry.

Over the last two decades, imidazolium-based ionic liquids have also found many applications in catalysis [35–39], or as nonaqueous alternatives for biphasic catalysis [4, 40–42]. The studies by Cavell and co-workers [43] showed that the reaction of imidazolium-based ionic liquids with low-valence Ni^0 and σ-donor ligands that bear Pd^0 is an easy procedure for the production of unusually stable carbene-metal-hydride complexes (see **Scheme 1**). The major feature of the study was the direct formation of a carbene-metal-hydride, which offers an atom-efficient direct route to an active catalytic species. Besides these experimental facts, it is not surprising that N-heterocyclic carbenes (NHCs) [44–46] have found applications in a series of catalytic reactions, such as amination reactions, the Suzuki-Miyaura and Sonogashira coupling reactions, hydroformylation, hydrosilylation, and polymerization and olefin metathesis [47–50].

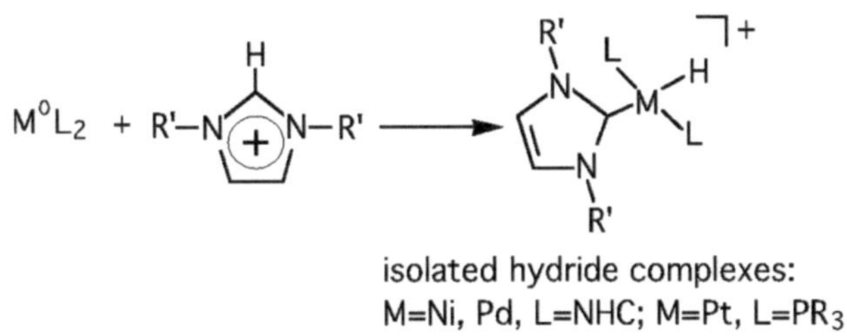

Scheme 1.

The crucial experimental works that are presented in **Scheme 1** inspire this study of the potential energy surfaces of these oxidative addition reactions, using density functional theory (DFT). There have been a number of reports concerning the conventional oxidative additions-reductive eliminations of alkanes to low-valence metals, which has led to an understanding of the factors that affect these reactions [51–56]. These studies have mostly focused on the catalytic reactions of saturated hydrocarbons to zerovalent group 10 elements (i.e., Ni, Pd, and Pt). To the authors' best knowledge, there has been neither experimental nor theoretical study of the catalytic oxidative addition reactions for the group 9 atoms (i.e., Co, Rh, and Ir) or the imidazolium cation. This study gives a thorough understanding of the catalytic reactions for potential transition metal complexes with imidazolium cations (ICs). Accordingly, a study of the important C—H activation reactions, Eqs. (1) and (2), is undertaken:

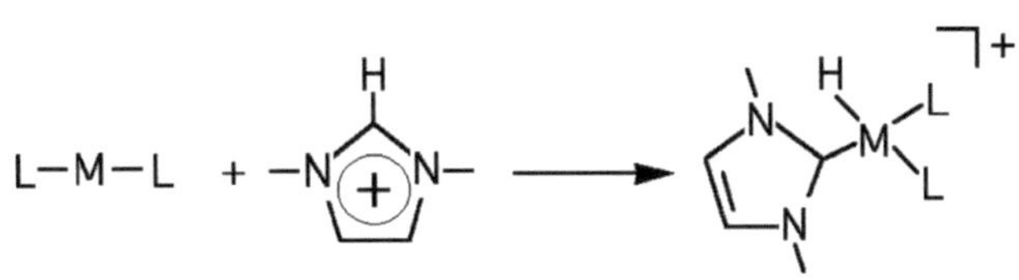

$$(M = Ni, Pd, Pt; L = 1,3\text{-aryl-NHC, aryl} = 2,4,6\text{-trimethylphenyl}) \qquad (1)$$

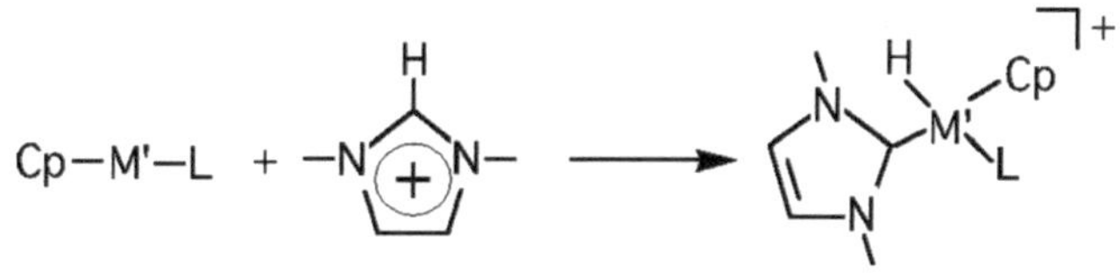

$$\left(M' = Co, Rh, Ir; L = 1,3\text{-aryl-NHC}, aryl = 2,4,6\text{-trimethylphenyl}\right) \quad (2)$$

Since oxidative addition involves charge transfer from the metallic center of both L_2M and CpM′L to the approaching IC, an electron-donating L that increases the electron density on the central metal stabilizes its transition state and lowers the barrier height. That is to say, increasing the electron density on the central metal atom of both L_2M and CpM′L increases the chance of its triplet participating in the oxidative addition reaction (vide infra). Therefore, the reactivity of both substituted 14-electron L_2M and 16-electron CpM′L is verified by the singlet-triplet splitting ($\Delta E_{st} = E_{triplet} - E_{singlet}$), which can result from several factors, such as the effect of the geometrical structure (i.e., linear or bent for the L_2M system) [55], the nature of electron-withdrawal or electron-acceptance for the ancillary ligand, L, and the character of the central transition metal atom. In the organometallic field, the NHC groups are stronger σ-donors and weaker π-acceptor ligands than the traditional PR_3 ligands [47–50]. Therefore, the model systems (both L_2M and CpM′L complexes) that are studied in this work use the NHC as the ancillary ligand L.

Since the transition-metal-catalyzed reactions that contain imidazolium salt are both helpful and novel, a comprehensive understanding of the factors that control the magnitude of the activation barriers and the reaction enthalpies allows a greater understanding of their reactivity. Full realization of the factors that influence the reactivities of transition metal complexes with ICs benefits basic science and a continued expansion of their applications.

2. Theoretical methods

The geometries of all of the stationary points are fully optimized at the M06-L level of theory [57], using the Gaussian 09 program package [58]. These M06-2X calculations are executed using pseudo-relativistic effective core potentials on group 9 and group 10 elements, using the Def2-SVP basis sets [59–63]. These M06-L calculations are denoted as M06-L/Def2-SVP. Frequency computations are accomplished for all structures to verify that the reactants and products have no imaginary frequencies and that the transition states occupy only one imaginary frequency. The relative free energy (ΔG) at 298 K is computated at the M06-L/Def2-SVP level of theory.

3. The origin of the barrier and the reaction enthalpy for the oxidative addition of an imidazolium cation to transition metal complexes

In this section, the valence bond state correlation diagram (VBSCD) model [64–68] that was developed by Shaik and Pross is used to interpret the oxidative addition for an imidazolium cation to transition metal complexes. For the σ-bond insertion reaction, the system must have a number of predetermined states, each of which is approximated by an appropriate electronic configuration [64–68]. In particular, there are two important configurations that contribute significantly to the total wave function, Ψ, and change the shape of the potential energy surface. **Figure 1** shows the qualitative behavior of the two configurations for the insertion of a transition-metal complex (L_nM) into a C(carbenic carbon)—H bond of an IC. The first line shows the reactant ground-state configuration, which connects the excited state for the products, denoted as the reactant configuration (I_R). The second line shows the excited configuration of the reactants, which connects the ground state of the products and is marked as the product configuration (I_P).

From the valence bond (VB) viewpoint, the reactions for the insertion of L_nM fragments into the C—H bond are illustrated in **1** and **2**, as shown in **Figure 1**. In the reactant configuration (I_R), which is labeled $^1[L_nM]^1[IC]$, the two electrons on the L_nM moiety are spin-paired to form a lone pair and the two electrons on the CH moiety are spin-paired to form a C—H σ bond. In the product configuration (I_P), which is labeled $^3[L_nM]^3[IC]$, the electron pairs are coupled to

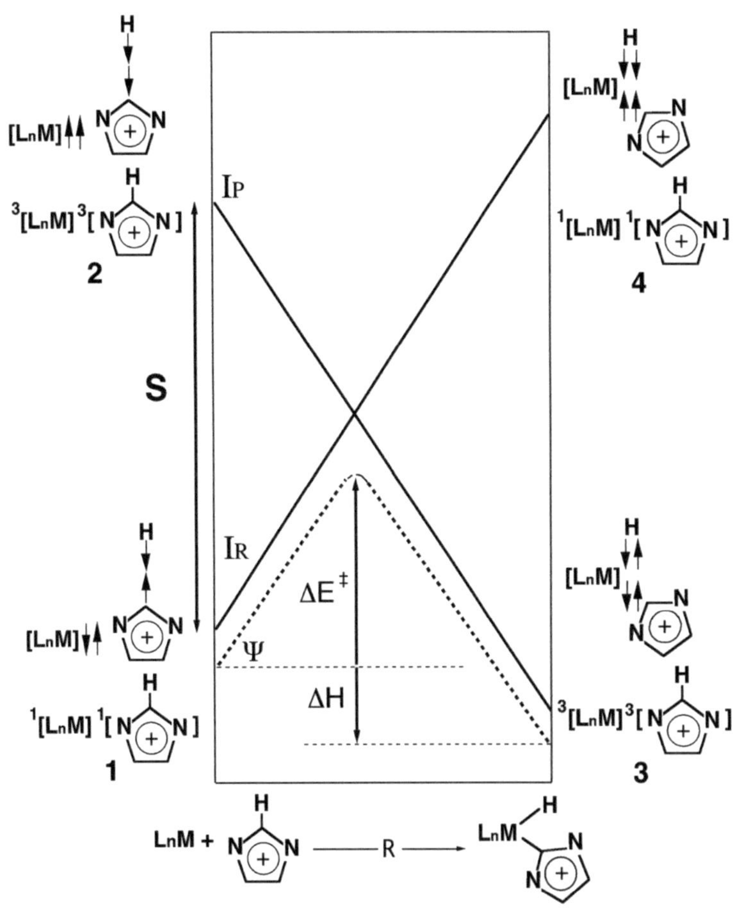

Figure 1. The energy diagram for an oxidative addition reaction, showing the formation of a state curve (Ψ) by mixing two configurations: the reactant configuration (I_R) and the product configuration (I_P). The reactants are separated by an energy gap, S. Configuration mixing near the crossing point causes an avoidance crossing (dotted line). For details see the text.

allow the formation of both an M—C and an M—H bond and the simultaneous breaking of a C—H bond. From the molecular orbital (MO) viewpoint, the representations of VB configurations **1** and **2** are respectively given in **3** and **4**.

It is proposed that the transition state for the reaction that inserts L_nM into a C—H bond is regarded as the respective triplet states of the reactants. It is worthy to note that these individual triplets are coupled to an overall singlet state. Since new M—C and M—H covalent bonds are formed in the product $L_nM(C)(H)$, the bond-prepared L_nM state must have at least two open shells. Therefore, the lowest state for this type is the triplet state. In other words, the bonding in the $L_nM(C)(H)$ product is between the triplet L_nM state and two doublet radicals (the C radical and the H radical). Similarly to the bonding in a water molecule, from the valence-bond point of view, it is represented as bonding between a triplet oxygen atom and two doublet hydrogen atoms [69].

As schematically illustrated in **Figure 1**, the singlet-triplet excitation energy plays a decisive role in the VBSCD model [64–68]. The singlet-triplet excitation energy (i.e., the energy between the I_R and the I_P) corresponds to the energy gap, S, in the VBSCD model. In terms of the reactants, I_R is the ground state and I_P is in an excited state whose energy is greater than I_R. When the reaction is in progress, the energy of I_R increases and that of I_P decreases. The transition state occurs at a point along the reaction coordinate where the energy curves for I_R and I_P cross (see the dotted curve in **Figure 1**). Finally, in terms of the products, I_R assumes the excited-state configuration and I_P a ground state. These two configurations cross. This is the simplest description of the ground state energy profiles for the chemical reactions of the related molecular systems [64–68].

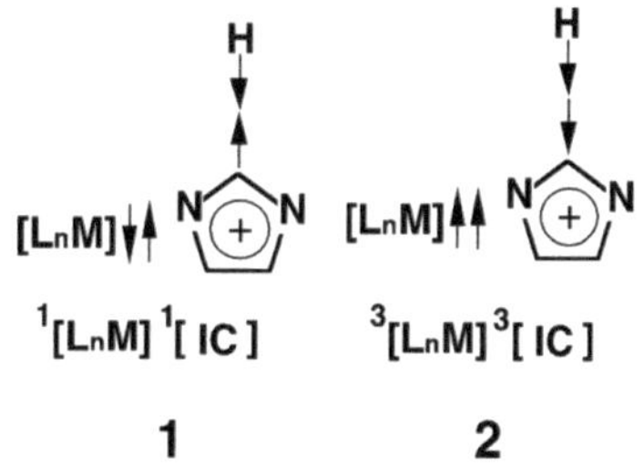

Scheme 2.

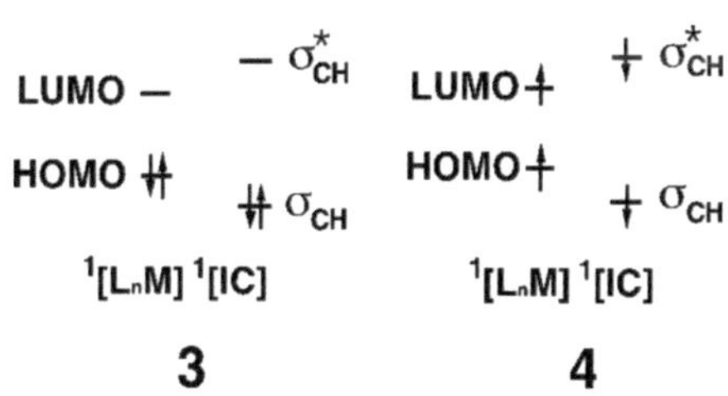

Scheme 3.

Figure 1 shows that the energy of point **2** (left in **Figure 1**), the anchor point for $^3[L_nM]^3[IC]$ in the reactant geometry, is governed by the singlet-triplet energy gap for both L_nM and C—H; i.e., ΔE_{st} (= $E_{triplet} - E_{singlet}$ for L_nM) + $\Delta E_{\sigma\sigma^*}$(= $E_{triplet} - E_{singlet}$ for C—H). In other words, the smaller the value of $\Delta E_{st} + \Delta E_{\sigma\sigma^*}$, the lower is the activation barrier and the more exothermic is the reaction [64–68]. If a reactant, L_nM, has a singlet ground state with a small triplet excitation energy, there is a greater probability that a triplet L_nM contributes to the singlet reaction and the reactions occur readily. Both the order of the singlet and triplet states and the magnitude of the singlet-triplet energy separation also determine the existence and the height of the energy barrier.

Since CH_2 and 16-electron CpML and 14-electron L_2M are isolobal [70], each has two valence orbitals with the same symmetry patterns (**5**), in which each fragment has one orbital of a′ and a″ symmetry.

In this qualitative theoretical treatment, the transition-metal fragment L_2M and CpML has an empty electrophilic orbital (i.e., a', as shown in **5**) that interacts with a filled hydrocarbon fragment orbital. This facilitates a concerted 1,2-hydrogen migration. In other words, the net molecular result of the insertion of the L_2M and CpML complexes into a C—H σ bond of an IC is that a new M—C σ bond and a new M—H σ bond are formed and the C—H σ bond of an IC is broken. This analysis is used to interpret the results in the following section.

4. Results and discussion

4.1. The geometries of the triscarbene-nickel-hydrido complex and the triscarbene-palladium-hydrido complex

The geometrical structures of the triscarbene-nickel-hydrido complex (**Pro-Ni**) and the triscarbene-palladium-hydrido complex (**Pro-Pd**) are firstly determined theoretically. The optimized geometries for these two species are computed at the M06-L/Def2-SVP level of theory. As

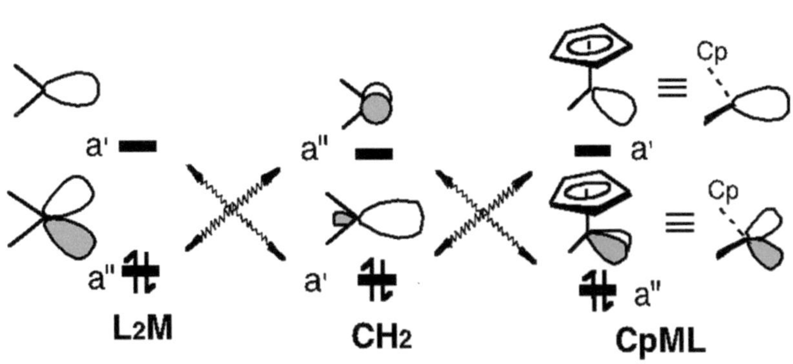

5

Scheme 4.

shown in **Figure 2**, the M06-L calculations show that the computed M—C bond lengths for both molecules (average 1.924 and 2.089 Å at M06-L) compare favorably with the average M—C bond lengths that are determined from X-ray data (1.907 and 2.057 Å) [43]. Similarly, the average values for the ∠C—M—C and ∠C—M—H angles for these two structures are calculated to be 98.26° and 82.58° (Ni) and 97.41° and 82.23° (Pd), which agrees reasonably well with the experimental data (97.85, 82.00, 95.94, and 84.00°, respectively) [43], as shown in **Figure 2**. Given the agreement between the M06-L method using the Def2-SVP basis set and the available experimental data [43], it is expected that the same relative accuracy is applicable to any discussion of their reactivities and the reaction mechanisms, for which experimental data are still not available.

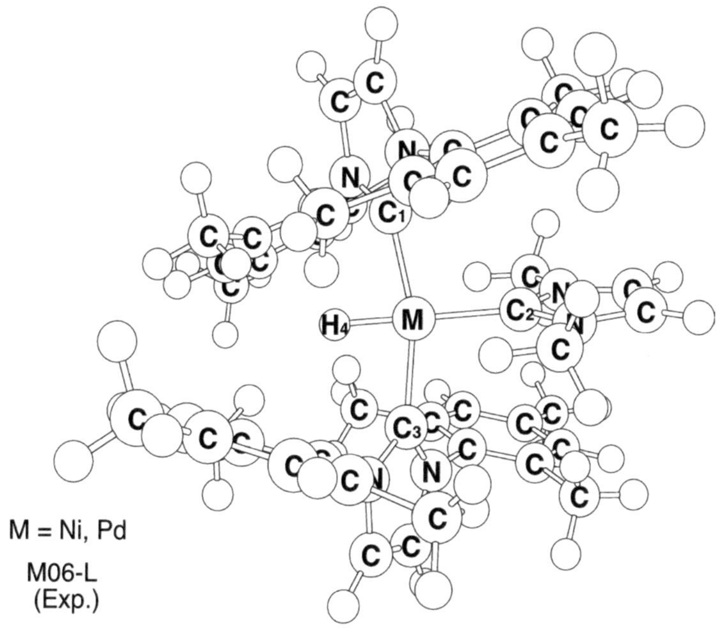

M					
Ni	M-C_1	M-C_2	M-C_3	M-H_4	C_1-M-C_3
	1.911	1.958	1.904	1.394	163.4
	(1.895)	(1.936)	(1.891)	(1.380)	(164.2)
	C_1-M-C_2	C_2-M-C_3	C_3-M-H_4	C_1-M-H_4	C_2-M-H_4
	98.29	98.24	83.76	81.39	179.0
	(97.30)	(98.40)	(83.00)	(81.00)	(179.0)
Pd	M-C_1	M-C_2	M-C_3	M-H_4	C_1-M-C_3
	2.057	2.144	2.067	1.572	167.2
	(2.031)	(2.111)	(2.030)	(1.570)	(168.0)
	C_1-M-C_2	C_2-M-C_3	C_3-M-H_4	C_1-M-H_4	C_2-M-H_4
	97.85	96.97	82.05	85.40	178.3
	(96.20)	(95.78)	(82.00)	(86.00)	(177.0)

Figure 2. Selected geometrical parameters (in Å and deg) for the triscarbene-nickel-hydrido complex (**Pro-Ni**) and the triscarbene-palladium-hydrido complex (**Pro-Pd**), calculated at the M06-L/Def2-SVP level of theory and a comparison with the experimental values [43]. Hydrogens are omitted for clarity.

4.2. The geometries and energetics of the L_2M + 1,2-dimethylimidazolium cation

The results for four regions on the potential energy surfaces for L_2M (M = Ni, Pd, Pt; L = 1,3-aryl-NHC, aryl = 2,4,6-trimethylphenyl) and 1,2-dimethylimidazolium cation (IC) are shown: 14-electron L_2M plus free IC (**Rea**), a precursor complex (**Pcx**), the transition state (**TS**), and the oxidative addition product (**Pro**). The fully optimized geometries for the key points, calculated at the M06-L/Def2-SVP level, are shown in **Figure 3**. The important geometrical parameters and relative energies and the potential energy profiles at the same level of theory are listed in **Table 1** and **Figure 4**, respectively. Four points are noteworthy.

1. For the optimized structures, see **Figure 3**. For the relative free energies, see **Figure 4**.
2. The C_3—H_4 bond distance in IC (reactant) is calculated to be 1.090 Å.
    1. The reactants, **Rea-Ni, Rea-Pd**, and **Rea-Pt**, are computed as both low-spin (singlet) and high-spin (triplet state) complexes. The M06-L computations demonstrate that these transition metal complexes all adopt the singlet ground state. The computations also show that the singlet-triplet triplet free energy splitting (ΔE_{st}; kcal/mol) for these fragments are in the order: **Rea-Ni** (23.7) < **Rea-Pd** (50.1) < **Rea-Pt** (63.9). These values are much greater than those for other previously studied L_2M complexes that have various ancillary ligands [51–56]. Therefore, it is possible that the oxidative addition reactions (Eq. (1)) that are studied in this work proceed on the singlet surface. The singlet surface is therefore the focus of this study, from this point.
    2. The optimized transition state structures (**TS-Ni, TS-Pd**, and **TS-Pt**) and arrows that indicate the main atomic motion in the transition state eigenvector are shown in **Figure 3**. These model computations show that the oxidative addition reactions that are studied using these model reactants all proceed in a concerted fashion via a three-center transition state, as shown in **Figure 3**, and all reactions are exothermic. It is noted that for the oxidative addition reactions involving the group 10 transition metals that are studied in this work, the free energies for the transition states are all less than those for the corresponding reactants. It is theoretically predicted that these oxidative addition reactions proceed readily, even at room temperature. Further supporting evidence comes from the fact that the oxidative additions between **Rea-Ni** and **Rea-Pd** species and an imidazolium cation have been experimentally proven to be easy [43].
    3. According to the theoretical analysis of the VBSCD model that is discussed in Section 3, the smaller the value of ΔE_{st} for L_2M, the lower is the barrier height and the more exothermic is the reaction and the faster is the oxidative addition reaction. The model evidence confirms this prediction. For the M06-L calculations for the model systems that have group 10 transition metals, a plot of the activation barrier ($\Delta E^{\ddagger}$) versus the ΔE_{st} is shown, for which the best fit is $\Delta E^{\ddagger} = 0.518\Delta E_{st} - 11.2$. The linear correlation between ΔE_{st} and the Gibbs free energy (ΔG), which is also calculated at the same level of theory, is $\Delta G = 0.566\,\Delta E_{st} - 67.5$. The theoretical results definitely show that for the facile oxidative addition of C—H bonds, an understanding of the ΔE_{st} of the coordinatively unsaturated 14-electron L_2M is crucial, since it can be used to predict the reactivity of the reactants.

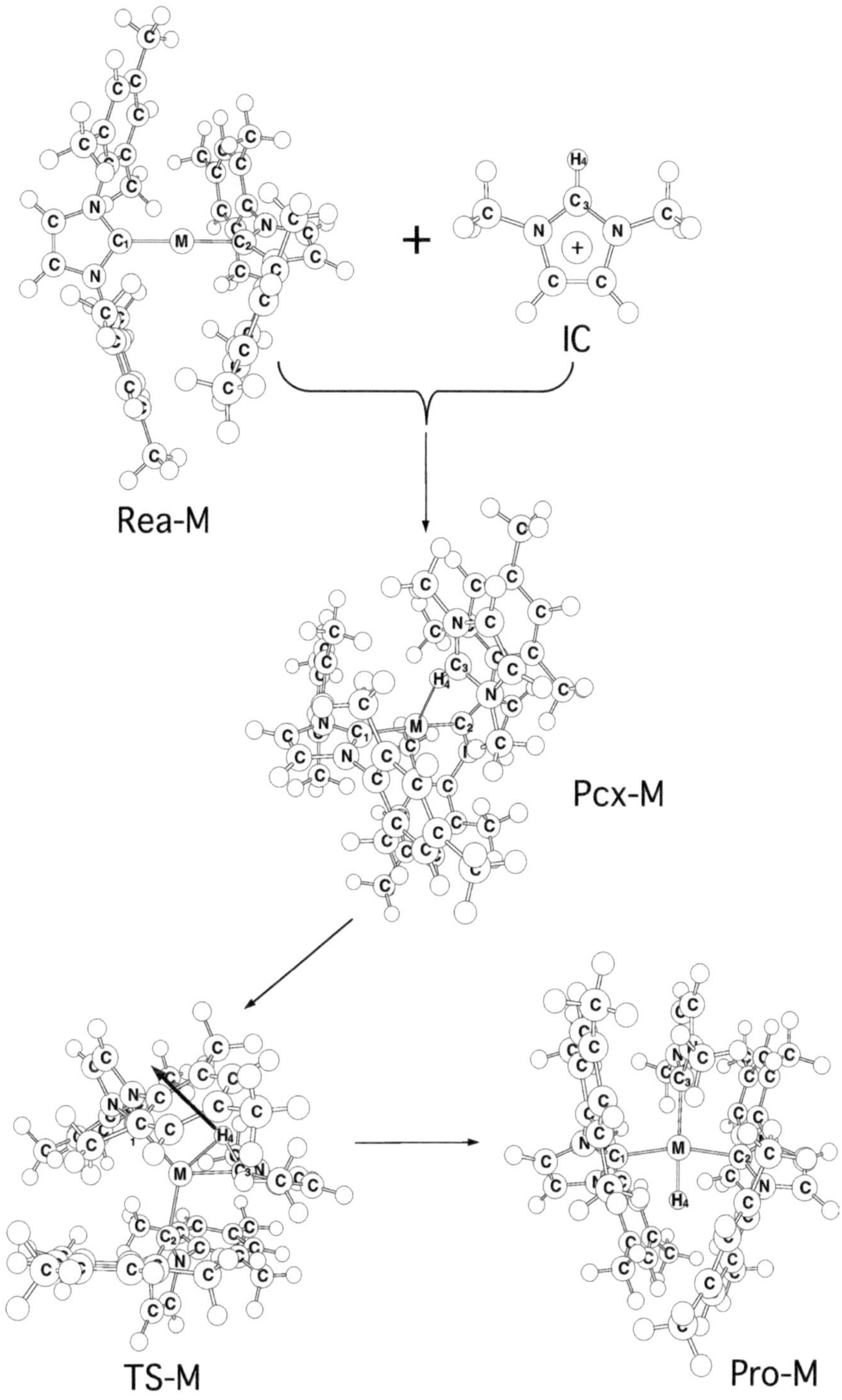

Figure 3. M06-L/Def2-SVP optimized geometries for the stationary points for the oxidative addition reactions of **Rea-M** (M = Ni, Pd, and Pt) molecules. For selected geometrical parameters and relative energies for each species, see **Table 1**. The bold arrows denote the main atomic motions in the transition state eigenvector. Some hydrogens are omitted for clarity.

	Geometrical structures					Energies	
Systems	**M-C$_1$**	**M-C$_2$**	**M-C$_3$**	**C$_3$-H$_4$**	**M-H$_4$**	**ΔE**	**ΔG**
Rea-Ni	1.836	1.836	–	–	–	0.0	0.0
Pcx-Ni	1.938	1.961	–	1.107	2.056	–50.24	–33.65
TS-Ni	1.981	1.947	1.886	1.482	1.527	–49.38	–32.26
Pro-Ni	1.911	1.904	–	1.985	1.394	–69.31	–54.36
Rea-Pd	2.015	2.015	–	–	–	0.0	0.0
Pcx-Pd	2.050	2.053	–	1.115	2.105	–39.69	–23.67
TS-Pd	2.159	2.138	2.039	1.437	1.679	–25.99	–8.417
Pro-Pd	2.057	2.067	–	2.144	1.572	–55.36	–38.23
Rea-Pt	2.003	2.003	–	–	–	0.0	0.0
Pcx-Pt	2.022	2.019	–	1.119	2.134	–41.35	–22.46
TS-Pt	2.128	2.064	2.067	1.333	1.797	–16.49	–1.384
Pro-Pt	2.054	2.060	–	2.139	1.621	–47.52	–31.98

Table 1. Selected geometrical parameters (bond distances in Å), relative energies ΔE (zero-point corrected; kcal mol^{-1}) and relative Gibbs free energies ΔG (kcal mol^{-1}) at 298 K at the M06-L/Def2-SCP level of theory for the optimized stationary points on the oxidative addition reactions (Eq. (1)) [1–30].

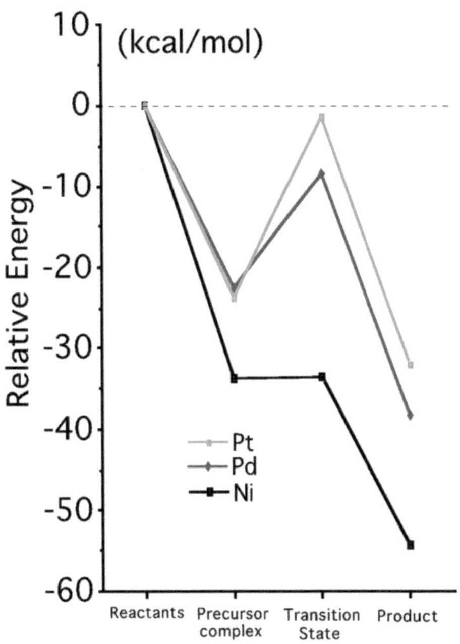

Figure 4. The reaction energy profile (in kcal/mol) for the oxidative addition reactions: L$_2$M + 1,2-dimethylimidazolium cation (M = Ni, Pd, and Pt; L = 1,3-aryl-NHC, aryl = 2,4,6-trimethylphenyl). All of the energies are calculated at the M06-L/Def2-SVP level. See also **Table 1** and **Figure 3**.

http://dx.doi.org/10.5772/67567

4.3. The geometries and energetics of the CpM′+ 1,2-dimethylimidazolium cation

Similarly to the study of the L_2M system, the M06-L/Def2-SVP level is also used to study the mechanisms for the oxidative addition reactions for CpM′L (M′ = Co, Rh, Ir; L = 1,3-aryl-NHC, aryl = 2,4,6-trimethylphenyl), as shown in **Figure 5**. The relative M06-L energies and the key geometrical parameters for the stationary points are also listed in **Table 2**. The corresponding potential energy profiles are given in **Figure 6**. Three interesting conclusions can be drawn from these figures and the table.

1. For the optimized structures, see **Figure 5**. For the relative free energies, see **Figure 6**.
2. The C_2—H_3 bond distance in IC (reactant) is calculated to be 1.090 Å.

    1. The M06-L calculations in **Table 2** show that the ground states for the CpCoL and CpIrL fragments are triplets, but the CpRhL complex is a singlet. The M06-L results also show that the ΔE_{st} value for the CpCoR and CpIrR fragments are respectively computed to be− 8.3 and − 0.88 kcal/mol and the ΔE_{st} of CpRhR was predicted to be 10.2 kcal/mol. It is worthy to note that whenever a reactant contains a heavy atom that is not necessarily directly included in the reaction, a strong spin-orbit coupling (SOC) can occur [71–75]. That is, because of the presence of the heavy atom, a triplet reactant can cause a spin-inversion process. It transfers to the singlet reactant and then forces the singlet reaction. Since these theoretical calculations show that both the CpCoR and the CpIrR species have a small value for ΔE_{st} and a heavier transition metal is involved, the SOC is expected to be substantial for the oxidative additions and these would eventually proceed in the singlet chemical reactions.

    2. **Figure 6** shows that, similar to the case for L_2M molecules, the energy of the transition state for Co, Rh, and Ir is less than that for the reactants, which demonstrates that the CpM′L (M′ = Co, Rh, and Ir) complexes readily overcome the energy barrier and then undergo oxidative addition into the C—H bond of IC in a concerted fashion, even at room temperature. The model computations show that the oxidative addition of a CpM′L fragment that has a group 9 metal (M′) decreases in the order: CpCoL > CpIrL > CpRhL. For the reverse process (right to left in **Figure 6**), the barriers to reductive elimination for the Co, Rh, and Ir systems have much higher energies than those for the corresponding oxidative addition. The theoretical evidence demonstrates that these CpM′L complexes undergo oxidative additions more easily than reductive eliminations, as noted in the introduction. That is to say, because the attached NHC groups readily donate electrons, the electrons in the metal, M′, are abundant and the CpM′L readily allows oxidative additions with the incoming molecules. The theoretical studies suggest that the CpM′L molecules prefer to undergo oxidative addition reactions with the imidazolium cation, even at room temperature.

    3. According to the VBSCD model, the smaller the ΔE_{st} value for CpM′L (if $\Delta E_{\sigma\sigma^*}$ is a constant), the lower is the barrier height, the more exothermic is the reaction and the faster is the oxidative addition reaction [64–68]. The M06-L/Def2-SVP results support this prediction. The M06-L calculations show that value for ΔE_{st} (kcal/mol) increases in the order: CpCoL (−8.3) < CpIrL (−0.77) < CpRhL (+8.0). As shown in **Table 2**, both the

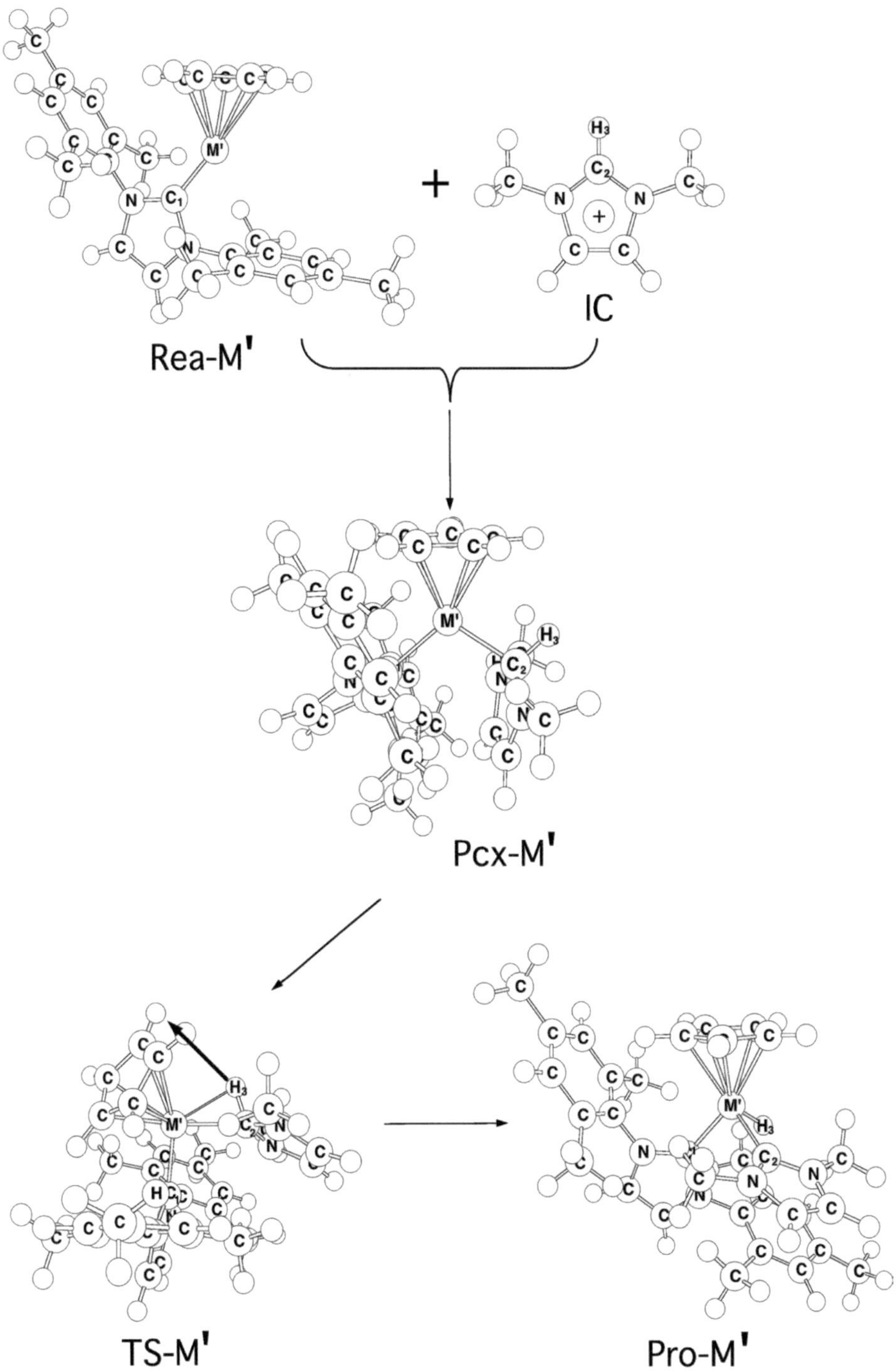

Figure 5. M06-L/Def2-SVP optimized geometries for the stationary points for the oxidative addition reactions of **Rea-M′** (M′ = Co, Rh, and Ir) molecules. For the selected geometrical parameters and relative energies for each species, see **Table 2**. The bold arrows denote the main atomic motions in the transition state eigenvector. Some hydrogens are omitted for clarity.

	Geometrical structures				Energies	
Systems	**M′-C$_1$**	**M′-C$_2$**	**C$_2$-H$_3$**	**M′-H$_3$**	**ΔE**	**ΔG**
Rea-Co	1.858	–	–	–	0.0	0.0
Pcx-Co	1.981	1.964	1.101	2.569	−20.16	−8.462
TS-Co	1.967	2.044	1.130	2.043	−17.45	+0.542
Pro-Co	1.947	1.920	2.337	1.472	−79.40	−62.38
Rea-Rh	1.953	–	–	–	0.0	0.0
Pcx-Rh	2.055	2.080	1.101	2.634	−31.35	−17.19
TS-Rh	2.075	2.158	1.147	2.002	−22.41	−6.193
Pro-Rh	2.046	2.017	2.411	1.569	−55.43	−39.08
Rea-Ir	1.923	–	–	–	0.0	0.0
Pcx-Ir	2.055	2.084	1.104	2.665	−39.30	−21.79
TS-Ir	2.037	2.147	1.146	2.156	−28.16	−10.98
Pro-Ir	2.043	2.029	2.448	1.594	−59.43	−46.20

Table 2. Key geometrical parameters (bond distances in Å), relative energies ΔE (zero-point corrected; kcal mol^{-1}) and relative Gibbs free energies ΔG (kcal mol^{-1}) at 298 K, calculated at the M06-L/Def2-SCP level of theory, for the optimized stationary points on the oxidative addition reactions (Eq. (2)) [1–30].

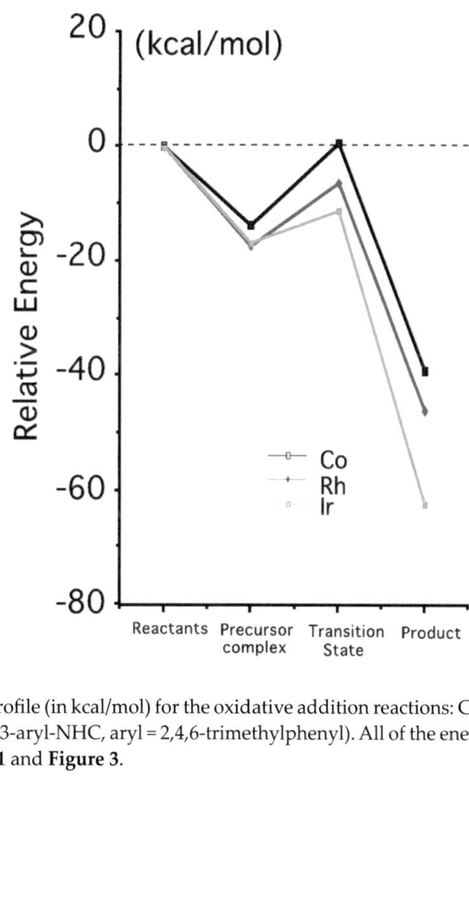

Figure 6. The reaction energy profile (in kcal/mol) for the oxidative addition reactions: CpML + 1,2-dimethylimidazolium cation (M = Co, Rh, and Ir; L = 1,3-aryl-NHC, aryl = 2,4,6-trimethylphenyl). All of the energies are calculated at the M06-L/Def2-SVP level. See also **Table 1** and **Figure 3**.

activation energies ($\Delta E^{\ddagger}$) and the Gibbs free energies (ΔG) follow the same order as the ΔE_{st} value (kcal/mol): Co (+9.28, −62.4) < Ir (+10.8, −46.2 kcal/mol) < Rh (+11.0, −39.1). In order to determine a good model for the facile oxidative addition of 16-electron CpM′L to a C-H bond of an imidazolium cation, an understanding of the ΔE_{st} of the coordinatively unsaturated CpM′L is important.

5. Conclusions

In summary, the theoretical evidence demonstrates that both a 14-electron L_2M complex and a 16-electron CpM′Lcomplex with a better electron-donating ligand, L (such as NHC), result in a smaller value for ΔE_{st} and allow a more facile oxidative addition to the saturated C—H bond. This theoretical study also demonstrates that, in terms of the VBSCD model, the ΔE_{st} value is a useful foundation for predicting the relative magnitude of the activation barriers and the reaction enthalpies for the activation of an imidazolium cation by L_2M and CpM′L. Although the computed magnitude of the barrier and the predicted geometry for the transition state for these reactions are dependent on the level of theory that is used, these qualitative predictions are in good agreement with the theoretical results that are presented here and with the available experimental observations.

It is hoped that this study will stimulate for further research into the subject.

Acknowledgements

The authors are grateful to the National Center for High-Performance Computing of Taiwan for generous amounts of computing time and the Ministry of Science and Technology of Taiwan for the financial support.

Author details

Hsin-Yi Liao[1]*, Jia-Syun Lu[2] and Ming-Der Su[2,3]*

*Address all correspondence to: midesu@mail.ncyu.edu.tw and hyliao@tea.ntue.edu.tw

1 Department of Science Education, National Taipei University of Education, Taipei, Taiwan

2 Department of Applied Chemistry, National Chiayi University, Chiayi, Taiwan

3 Department of Medicinal and Applied Chemistry, Kaohsiung Medical University, Kaohsiung, Taiwan

References

[1] Herrmann WA, Köcher C. N-Heterocyclic carbenes. Angewandte Chemie International Edition in English. 1997;**36**:2162-2187

[2] McGuinness DS, Green MJ, Cavell KJ, Skelton BW, White AH. Synthesis and reaction chemistry of mixed ligand methylpalladium–carbene complexes. Journal of Organometallic Chemistry. 1998;**565**:165-178

[3] McGuinness DS, Cavell KJ, Skelton BW, White AH. Zerovalent palladium and nickel complexes of heterocyclic carbenes: Oxidative addition of organic halides, carbon-carbon coupling processes, and the Heck reaction. Organometallics. 1999;**18**:1596-1605

[4] Carmichael AJ, Earle MJ, Holbrey JD, McCormac PB, Seddon KR. The Heck reaction in ionic liquids: A multiphasic catalyst system. Organic Letters. 1999;**1**:997-1000

[5] McGuinness DS, Cavell KJ. Donor-functionalized heterocyclic carbene complexes of palladium (ii): Efficient catalyst C—C coupling reaction. Organometallics. 2000;**19**:741-748

[6] Xu L, Chen W, Xiao J. Heck reaction in ionic liquids and the in situ identification of N-heterocyclic carbene complexes of palladium. Organometallics. 2000;**19**:1123

[7] McGuinness DS, Cavell KJ. Reaction of CO with a methylpalladium heterocyclic carbene complex: Product decomposition routes—Implications for catalytic carbonylation processes. Organometallics. 2000;**19**:4918

[8] Mathews CJ, Smith PJ, Welton T. Palladium catalysed Suzuki cross-coupling reactions in ambient temperature ionic liquids. Chemical Communications. 2000;**14**:1249-1250

[9] McGuinness DS, Saendig N, Yates BF, Cavell KJ. Kinetic and density functional studies on alkyl-carbene elimination from PdII heterocylic carbene complexes: A new type of reductive elimination with clear implications for catalysis. Journal of the American Chemical Society. 2001;**123**:4029-4040

[10] McGuinness DS, Cavell KJ, Yates BF, Skelton BW, White AH. Oxidative addition of the imidazolium cation to zerovalent Ni, Pd, and Pt: A combined density functional and experimental study. Journal of the American Chemical Society. 2001;**123**:8317-8328

[11] Duin MA, Clement ND, Cavell KJ, Elsevier CJ. C–H activation of imidazolium salts by Pt(0) at ambient temperature: Synthesis of hydrido platinum bis(carbene) compounds. Journal of Chemical Communication. 2003;**3**:400-401

[12] Dubinina GG, Furutachi H, Vicic DA. Active trifluoromethylating agents from well-defined copper(I)-CF_3 complexes. Journal of the American Chemical Society. 2008;**130**8600-8601

[13] Boydston AJ, Xia Y, Kornfield JA, Gorodetskaya IA, Grubbs RH. Cyclic ruthenium-alkylidene catalysts for ring-expansion metathesis polymerization. Journal of the American Chemical Society. 2008;**130**:12775-12782

[14] Liu Q, Perreault S, Rovis T. Catalytic asymmetric intermolecular stetter reaction of glyoxamides with alkylidenemalonates. Journal of the American Chemical Society. 2008;**130**:14066-14067

[15] Park EJ, Kim SH, Chang S. Copper-catalyzed reaction of α-aryldiazoesters with terminal alkynes: A formal [3+2] cycloaddition route leading to indene derivatives. Journal of the American Chemical Society. 2008;**130**:17268-17269

[16] Nordstrøm LU, Vogt H, Madsen R. Amide synthesis from alcohols and amines by the extrusion of dihydrogen. Journal of the American Chemical Society. 2008;**130**:17672-17673

[17] Marion N, Ramón RS, Nolan SP. [(NHC)Au^{I}]-catalyzed acid-free alkyne hydration at part-per-million catalyst loadings. Journal of the American Chemical Society. 2009;**131**: 448-449

[18] Horino Y, Yamamoto T, Ueda K, Kuroda S, Toste D. Au(I)-catalyzed cycloisomerizations terminated by sp^3 C—H bond insertion. Journal of the American Chemical Society. 2009;**131**:2809-2811

[19] Lee Y, Hoveyda H. Efficient boron-copper additions to aryl-substituted alkenes promoted by NHC-based catalysts. Enantioselective Cu-catalyzed hydroboration reactions. Journal of the American Chemical Society. 2009;**131**:3160-3161

[20] Raynaud J, Absalon C, Gnanou Y, Taton D. N-Heterocyclic carbene-induced zwitterionic ring-opening polymerization of ethylene oxide and direct synthesis of α,ω-difunctionalized poly(ethylene oxide)s and poly(ethylene oxide)-b-poly(ε-caprolactone) block copolymers. Journal of the American Chemical Society. 2009;**131**:3201-3209

[21] Jana A, Schulzke C, Roesky HW. Oxidative addition of ammonia at a silicon(II) center and an unprecedented hydrogenation reaction of compounds with low-valent group 14 elements using ammonia borane. Journal of the American Chemical Society. 2009;**131**:4600-4601

[22] Häller LJL, Page MJ, Macgregor SA, Mahon MF, Whittlesey MK. Activation of an alkyl C—H bond geminal to an agostic interaction: An unusual mode of base-induced C—H activation. Journal of the American Chemical Society. 2009;**131**:4604-4605

[23] Jeong W, Shin EJ, Culkin DA, Hedrick JL, Waymouth RM. Zwitterionic polymerization: A kinetic strategy for the controlled synthesis of cyclic polylactide. Journal of the American Chemical Society. 2009;**131**:4884-4891

[24] Boydston AJ, Holcombe TW, Unruh DA, Fréchet MJ, Grubbs RH. A direct route to cyclic organic nanostructures via ring-expansion metathesis polymerization of a dendronized macromonomer. Journal of the American Chemical Society. 2009;**131**:5388-5389

[25] Aktas H, Slootweg JC, Schakel M, Ehlers AW, Lutz M, Spek AL, Lammertsma K. N-Heterocyclic carbene-functionalized ruthenium phosphinidenes: What a difference a twist makes. Journal of the American Chemical Society. 2009;**131**:6666-6667

[26] Iwai T, Fujihara T, Terao J, Tsuji Y. Iridium-catalyzed addition of acid chlorides to terminal alkynes. Journal of the American Chemical Society. 2009;**131**:6668-6669

[27] Galan BR, Pitak M, Gembicky M, Keister JB, Diver ST. Ligand-promoted carbene insertion into the aryl substituent of an n-heterocyclic carbene ligand in ruthenium-based metathesis catalysts. Journal of the American Chemical Society. 2009;**131**:6822-6832

[28] Xiong Y, Yao S, Driess M. An isolable NHC-supported silanone. Journal of the American Chemical Society. 2009;**131**:7562-7563

[29] Zeng X, Frey GD, Kinjo R, Donnadieu B, Bertrand G. Synthesis of a simplified version of stable bulky and rigid cyclic (alkyl)(amino)carbenes, and catalytic activity of the ensuing gold(I) complex in the three-component preparation of 1,2-dihydroquinoline derivatives. Journal of the American Chemical Society. 2009;**131**:8690-8696

[30] Chiang P-C, Rommel M, Bode JW. α'-Hydroxyenones as mechanistic probes and scope-expanding surrogates for α,β-unsaturated aldehydes in N-heterocyclic carbene-catalyzed reactions. Journal of the American Chemical Society. 2009;**131**:8714-8718

[31] Herrmann WA, Elison M, Fischer J, Kocher C, Artus GRJ. Metal complexes of N-heterocyclic carbenes—A new structural principle for catalysts in homogeneous catalysis. Angewandte Chemie International Edition in English. 1995;**34**:2371-2374

[32] Herrmann WA, Runte O, Artus G. Synthesis and structure of an ionic beryllium-"carbene" complex. Journal of Organometallic Chemistry. 1995;**501**:C1–C4

[33] Herrmann WA, Goossen LJ, Köcher C, Artus GRJ. Chiral heterocylic carbenes in asymmetric homogeneous catalysis. Angewandte Chemie International Edition in English. 1996;**35**:2805-2807

[34] Herrmann WA, Elison M, Fischer J, Köcher C, Artus GRJ. N-Heterocyclic carbened[+]: Generation under mild conditions and formation of group 8-10 transition metal complexes relevant to catalysis. Chemistry: A European Journal. 1996;**2**:772-780

[35] Welton T. Room-temperature ionic liquids. solvents for synthesis and catalysis. Chemical Reviews. 1999;**99**:2071-2084

[36] Wasserscheid P, Keim W. Ionische flüssigkeiten—neue lösungen für die übergangsmetallkatalyse. Angewandte Chemie. 2000;**112**:3926-3945

[37] Wasserscheid P, Keim W. Ionic liquids—new "solutions" for transition metal catalysis. Angewandte Chemie International Edition. 2000;**39**:3772-3789

[38] Sheldon R. Catalytic reactions in ionic liquids. Chemical Communications. 2001;**23**: 2399-2407

[39] Dupont J, Souza RF, Suarez PAZ. Ionic liquid (molten salt) phase organometallic catalysis. Chemical Reviews. 2002;**102**:3667-3692

[40] Mathews CJ, Smith PJ, Welton T, White AJP, Williams DJ. In situ formation of mixed phosphine-imidazolylidene palladium complexes in room-temperature ionic liquids. Organometallics. 2001;**20**:3848-3850

[41] Dullius JEL, Suarez PAZ, Einloft S, Souza RF, Dupont J, Fischer J, Cian AD. Selective catalytic hydrodimerization of 1,3-butadiene by palladium compounds dissolved in ionic liquids. Organometallics. 1998;**17**:815-819

[42] Fonseca GS, Umpierre AP, Fichtner PFP, Teixeira SR, Dupont J. The use of imidazolium ionic liquids for the formation and stabilization of Ir^0 and Rh^0 nanoparticles: Efficient catalysts for the hydrogenation of arenes. Chemistry: A European Journal. **2003;9**:3263-3269

[43] Clement ND, Cavell KJ, Jones C, Elsevier CJ. Oxidative addition of imidazolium salts to Ni^0 and Pd^0: Synthesis and structural characterization of unusually stable metal–hydride complexes. Angewandte Chemie International Edition. 2004;**43**:1277-1279

[44] Arduengo AJIII, Harlow RL, Kline M. A stable crystalline carbine. Journal of the American Chemical Society. 1991;**113**:361-363

[45] Arduengo AJIII, Dias HVR, Harlow RL, Kline M. Electronic stabilization of nucleophilic carbenes. Journal of the American Chemical Society. 1992;**114**:5530-5534

[46] Arduengo AJIII. Looking for stable carbenes: The difficulty in starting anew. Accounts of Chemical Research. 1999;**32**:913-921

[47] Regitz M. Stable carbenes—illusion or reality? Angewandte Chemie International Edition in English. 1991;**30**:674-676

[48] Enders D, Balensiefer T. Nucleophilic carbenes in asymmetric organocatalysis. Accounts of Chemical Research. 2004;**37**:534-541

[49] Johnson JS. Catalyzed reactions of acyl anion equivalents. Angewandte Chemie International Edition. 2004;**43**:1326-1328

[50] Nair V, Bindu S, Sreekumar V. N-Heterocyclic carbenes: reagents, not just ligands!. Angewandte Chemie International Edition. 2004;**43**:5130-5135

[51] Hackett M, Ibers JA, Jernakoff P, Whitesides GM. cis-[Bis(dicyclohexylphosphino)ethane] platinum(0) reacts with unactivated carbon-hydrogen bonds. Journal of the American Chemical Society. 1986;**108**:8094-8095

[52] Hackett M, Ibers JA, Whitesides GM. Activation of the carbon-hydrogen bonds of benzene by reaction with [bis(dicyclohexylphosphino)ethane]platinum(0), generated by the thermolysis of cis-[bis(dicyclohexylphosphino)ethane]hydridoneopentylplatinum(II). Journal of the American Chemical Society . 1988;**110**:1436-1448

[53] Hackett M, Whitesides GM. [Bis(dicyclohexylphosphino)ethane]platinum(0). Reactions with alkyl, (trimethylsilyl)methyl, aryl, benzyl, and alkynyl carbon-hydrogen bonds. Journal of the American Chemical Society. 1988;**110**:1449-1462

[54] Sakaki S, Biswas B, Sugimoto M. Computational evidence for a free silylium ion. Organometallics. 1998;**17**:278-280

[55] Su M-D, Chu S-Y. Theoretical study of oxidative addition and reductive elimination of 14-electron d^{10} ML_2 complexes: a ML_2+CH_4 (M=Pd, Pt; L=CO, PH_3, $L_2=PH'_2CH_2CH_2PH_2$) case study. Inorganic Chemistry. 1998;**37**:3400-3406

[56] Sakaki S, Kai S, Sugimoto M. Theoretical study on σ-bond activation of $(HO)_2B–XH_3$ by $M(PH_3)_2$ (X = C, Si, Ge, or Sn; M = Pd or Pt). Noteworthy contribution of the boryl p_π orbital to M-boryl bonding and activation of the B–X σ-bond. Organometallics. 1999;**18**:4825-4837

[57] Zhao Y, Truhlar DG. Density functionals with broad applicability in chemistry. Accounts of Chemical Research. 2008;**41**:157-167

[58] Frisch MJ, Trucks GW, Schlegel HB, Scuseria GE, Robb MA, Cheeseman JR, Scalmani G, Barone V, Mennucci B, Petersson GA, Nakatsuji H, Caricato M, Li X, Hratchian HP, Izmaylov AF, Bloino J, Zheng G, Sonnenberg JL, Hada M, Ehara M, Toyota K, Fukuda R, Hasegawa J, Ishida M, Nakajima T, Honda Y, Kitao O, Nakai H, Vreven T, Montgomery Jr JA, Peralta JE, Ogliaro F, Bearpark M, Heyd JJ, Brothers E, Kudin KN, Staroverov VN, Keith T, Kobayashi R, Normand J, Raghavachari K, Rendell A, Burant JC, Iyengar SS, Tomasi J, Cossi M, Rega N, Millam JM, Klene M, Knox JE, Cross JB, Bakken V, Adamo C, Jaramillo J, Gomperts R, Stratmann RE, Yazyev O, Austin A J, Cammi R, Pomelli C, Ochterski JW, Martin RL, Morokuma K, Zakrzewski VG, Voth GA, Salvador P, Dannenberg JJ, Dapprich S, Daniels AD, Farkas O, Foresman JB, Ortiz JV, Cioslowski J, Fox DJ. Wallingford, CT: Gaussian, Inc.; 2013

[59] Andrae D, Haeussermann U, Dolg M, Stoll H, Preuss H. Energy-adjusted ab initio pseudopotentials for the second and third row transition elements. Theoretica Chimica Acta. 1990;**77**:123-141

[60] Metz B, Stoll H, Dolg M. Small-core multiconfiguration-Dirac–Hartree–Fock-adjusted pseudopotentials for post-d main group elements: Application to PbH and PbO. Journal of Chemical Physics, 2000;**113**:2563-2569

[61] Peterson KA, Figgen D, Goll E, Stoll H, Dolg M. Systematically convergent basis sets with relativistic pseudopotentials. II. Small-core pseudopotentials and correlation consistent basis sets for the post-*d* group 16-18 elements. Journal of Chemical Physics. 2003;**119**:11113-11123

[62] Leininger T, Nicklass A, Kuechle W, Stoll H, Dolg M, Bergner A. The accuracy of the pseudopotential approximation: Non-frozen-core effects for spectroscopic constants of alkali fluorides XF (X = K, Rb, Cs). Chemical Physics Letters. 1996;**255**:274-280

[63] Kaupp M, Schleyer PV, Stoll H, Preuss HJ. Pseudopotential approaches to Ca, Sr, and Ba hydrides. Why are some alkaline earth MX2 compounds bent? Chemical Physics. 1991;**94**:1360-1366

[64] Shaik S, Schlegel HB, Wolfe S. Theoretical Aspects of Physical Organic Chemistry. USA: John Wiley & Sons Inc; 1992

[65] Pross A. Theoretical and Physical Principles of Organic Reactivity. USA: John Wiley & Sons Inc; 1995

[66] Shaik S, Hiberty PC. A Chemist's Guide to Valence Bond Theory. USA: Wiley; 2008

[67] Shaik, S. What happens to molecules as they react? A valence bond approach to reactivity. Journal of the American Chemical Society. 1981;**103**:3692-3701

[68] Shaik S, Shurki A. Valence bond diagrams and chemical reactivity. Angewandte Chemie International Edition. 1999;**38**:586-625

[69] Siegbahn PEM. Comparison of the C—H activation of methane by M(C5H5)(CO) for M = cobalt, rhodium, and iridium. Journal of the American Chemical Society. 1996;**118**:1487-1496

[70] Hoffmann R. Building bridges between inorganic and organic chemistry (Nobel lecture). Angewandte Chemie International Edition in English. 1982;**21**:711-724

[71] Su M-D. Mechanism for di-π-methane rearrangements in nonconjugated systems. Chemical Physics Letters. 1995;**237**:317-322

[72] Su M-D. Mechanism for the photorearrangements of cyclohexadienes. Journal of Organic Chemistry 1995;**60**:6621-6623

[73] Su M-D. Role of spin-orbit coupling and symmetry in triplet carbenic addition chemistry. Journal of Physical Chemistry. 1996;**100**:4339-4349

[74] Su M-D. The role of spin-orbit coupling and symmetry in photochemical rearrangements of α,β-unsaturated cyclic ketones. Chemical Physics. 1996;**205**:277-308

[75] Su M-D. The role of spin-orbit coupling and symmetry in oxadi-π-methane rearrangements and some related photochemical reactions. Journal of Organic Chemistry 1996;**61**: 3080-3087

Chapter 8

Metal Oxide Polymer Nanocomposites in Water Treatments

Francis Opoku, Ephraim M. Kiarii,

Penny P. Govender and Messai Adenew Mamo

Additional information is available at the end of the chapter

http://dx.doi.org/10.5772/67835

Abstract

Recently, several pollutants such as dyes, pharmaceuticals and phenolic compounds, which can cause toxic effects to human health, have identified in water resources. Water pollution has extensively studied and several conventional techniques, such as chemical treatment, adsorption, biological treatment, and membrane-based separation, have adopted for pollutants removal from wastewater/ water resources. However, these techniques had led to the production of soluble refractory organic compounds and health-threatening bacteria that are hard to be removed. Recently, photocatalysis has considered as one of the most viable technology for water treatment using sunlight to eliminate harmful bacteria and pollutants owing to its cost-effectiveness and high efficiency. Metal oxide and polymers have become promising materials for water treatment owing to their properties, such as surface mobility, large surface area and superb magnetic and optical properties. This book chapter discusses recent design and synthesis of visible light response polymer/metal oxide nanocomposite through several synthetic strategies for water treatment. The results show that the polymer-metal oxide nanocomposite possesses a superior photodegradation activity toward pollutants under simulated visible light. Major challenges in polymer-metal oxide nanocomposite synthesis and future research perspectives for developing alternate synthesis methodologies are also discussed.

Keywords: metal oxides, water treatment, photodegradation, polymer, nanocomposites, pollutants

1. Introduction

Water pollution caused by toxic and hardly-degradable organic and inorganic pollutants posed a severe danger to human well-being and growth. Recently, water pollution has been a significant concern, mostly in areas where people depend on groundwater and surface water for drinking and other domestic purposes [1]. For instance, almost 40% of the world's population has limited access to potable water [2]. The concerns of water shortage have been envisaged to be continued worsen due to the rise in population growth, swiftly improving life standards, rapid modernization, and industrialisation [3]. Shortage of potable water supply is due to the misuse of water resources for irrigation, industry, and domestic purposes in several parts of the world [4]. Industrial wastewaters comprise of several complex organic pollutants such as dyes, oils, detergents, and others, which are known to be carcinogenic to human health and aquatic life [5]. Domestic wastewater containing trace levels of pharmaceuticals, personal care products and others can also induce toxic effects [6]. As a result of the health implication induced by these contaminants, scientists and government authorities are making continuous efforts to address this problem. In the past, the traditional water treatment techniques such as coagulation and adsorption only remove pollutants from water but do not entirely biodegrade them into less toxic compounds [7]. Moreover, water treatment approaches, such as membrane technologies and chemical treatment usually have high operating costs and occasionally produce other poisonous secondary contaminants [8]. Since the water industry is required to produce portable drinking water, there is a need for the development of low-cost and stable approaches to address the day-to-day deterioration of water quality. Among the several approaches used in water treatment, the advanced oxidation processes (AOPs), such as the sonolysis, Fenton reaction, ozonation, and photocatalysis have gained a considerable attention in the removal of pollutants, owing to their simplicity, low cost, high efficiency, easy handling, and good reproducibility [9]. The AOP consist of *in situ* production of nonselective and highly reactive chemical oxidants, such as hydrogen peroxide (H_2O_2), hydroxyl radical ($HO^{\bullet}$), superoxide radical ($O_2^{-\bullet}$), and ozone (O_3) to degrade nonbiodegradable and persistent pollutants into less toxic and biodegradable ones [10]. Recently, photocatalysis has been recognized as one of the most effective approaches for water treatment using sunlight and other light sources as a driving force [11]. More significantly, this technique can degrade several organic pollutants into less toxic molecules and easily biodegradable compounds without forming secondary pollutants. The use of polymer/metal oxide–based photocatalyst materials to decompose pollutants has been recognised as one of the most promising materials [12], owing to their high quantum dimension effect, low cost, photostability, and low toxicity, small-dimension and surface effect [13]. The chapter starts with a brief introduction on a metal oxide, the synthesis approach, fundamentals, and the characterization techniques, followed by a discussion of the current successes made in the development, modification, and applications of the different nanohybrid polymer/metal oxide–based nanocomposites toward the degradation of pollutants in water resources. Finally, the future perspectives and outlooks are also taken into consideration.

http://dx.doi.org/10.5772/67835

2. Inorganic metal oxide

Among the several groups of nanoparticles, inorganic metal oxide has been of considerable interest from both technological and scientific perspective. When metal oxides are transformed into nanometre scale, they show enhanced hybrid properties compared to the pure materials. Metal oxide adopts several structural geometries with an electronic structure, which can be either semiconductor, metallic, or insulator depending on the nature of the structure. The unique features of metal oxides make them the most miscellaneous class of materials, with optical, electronic, electrical, magnetic, catalytic, and photoelectronic properties covering virtually all aspects of solid state physics and materials science, and can find application such as electroceramics, gas sensing, catalysis, superconductors, and energy conversions [14]. However, there are few limitations in using inorganic metal oxides as an absorbent in water treatment. Reducing inorganic metal oxide to nanoscale size may increase the surface area; however, this increment can make the metal oxide unstable and subsequently turn out to be more susceptible to agglomeration owing to the existence of van der Waals interactions [15]. Due to this interaction, the metal oxide may lose their selectivity, mechanical strength, and high capacity. In addition, low quantum yield owing to the rapid recombination of photogenerated charge carriers and the wide energy gap of some metal oxide limits their application in water treatment [16]. To overcome these limitations, metal oxides are immobilized into other supports, such as polymeric materials [17]. Currently, polymeric nanoparticles are used in the elimination of contaminants from water due to large surface area, tunable surface chemistry, perfect mechanical rigidity, and pore size distribution [15].

2.1. Synthesis of inorganic nanomaterials

Currently, the synthesis, characterization, and application of inorganic nanomaterials represent a highly active area of scientific research. Nanofabrication is the design and production of materials with chemical and structural restrictions on the nanometer scale [18]. The design of systematic approaches for the synthesis of inorganic nanomaterials has been a major challenge from both industrial and fundamental viewpoints as the first requirement in any study associated with inorganic nanomaterials involves the synthesis and characterization. An in-depth knowledge and understanding of the synthetic method are crucial in order to design hybrid inorganic nanomaterials with unique properties. The general strategy for preparing inorganic nanomaterials in solution is to separate the nucleation and growth of nanocrystals [19]. The synthesis approaches may be categorized into bottom-up and top-down. The top-down approach uses large homogeneous objects and shrinks them down to the nanoscale, while the bottom-up uses the interactions between the small components, such as colloidal particles or molecules to assemble themselves into more discrete and complex nanoscale structures. The bottom-up approach has been accepted as the most promising technique to address several problems related to the top-down approaches [20]. Both techniques have been used to synthesise nanomaterials. Chemical processes, such as chemical vapor deposition, sol-gel, spray pyrolysis, and template synthesis are employed as bottom-up approaches. The properties and structure of the synthesized nanomaterials can be regulated based on the experimental conditions employed. The nature of the method used in the synthesizing the nanomaterials allows control over the doping ratio by different elements, particle size, the degree of particle agglomeration, and particle geometry. For example, the liquid-phase approaches, such as sol-gel, coprecipitation, solvothermal/hydrothermal processing, template syntheses and microemulsion have been very resourceful in synthesizing inorganic nanomaterials owing to their capacity to synthesise several ranges of nanomaterials with control morphology and particle size [21]. Those particle parameters give the synthesized inorganic nanomaterials new chemical and physical properties for different applications.

2.2. Inorganic metal oxide polymer nanocomposites

A composite involves an immobilization of two or more materials with distinct chemical and physical properties. Composite materials have a magnificent and several practical applications compared to the individual material due to their extraordinary explicit strength and stiffness, corrosion resistance, low density, high thermal insulation, and toughness [22]. The search for improving the properties of composite materials, which are of lower filler size, led to the design of nanocomposites. Nanocomposites are composite materials with nanoscale morphology such as nanotubes, lamellar nanostructure, or nanoparticles as one of the phases [23]. The properties of nanocomposites are influenced by the individual components, the morphology of the system, volume, and shape fraction of the filler as well as the nature of the interphase between the interface of the components [24]. The enhancement of these properties can be accomplished when there are suitable interaction and good dispersion between the matrix and the nanoparticle. Based on the nature of the matrices, nanocomposites have been classified as a metal, carbon, ceramic, and polymer [25]. Among these nanocomposites, the polymer-based has been recognized as the most attractive in several research areas, such as medicine, optoelectronics, engineering, and water remediation due to their distinctive properties emerging from the individual components [26]. The mixing of polymers and inorganic metal oxide has been an active field of research; in particular, the engineering of flexible nanocomposites has received much attention owing to the significant electrical, thermal, mechanical, and magnetic properties compared to the bulk polymers and the inorganic metal oxide [27]. In this nanocomposites, the polymer material provides convenient processing, structural flexibility, photoconductivity, metallic behavior, tunable electronic properties, and efficient luminescence [28], while the inorganic metal oxide offers high carrier mobility's, band gap tunability, thermal, and mechanical stability as well as dielectric and magnetic properties [29]. In addition to the distinctive features, new or improved phenomena can also occur due to the interface between the polymer and inorganic metal oxide [30]. Due to the large surface area, nanocomposites display many variations in their properties compared to the individual component of the metal oxide. The properties and microstructure of the nanocomposites are influenced by the interfacial interaction between the polymer and the inorganic metal oxide, where a wide range of covalent and hydrogen bonds may prevent phase separation [31].

2.3. Characterization techniques of metal oxide polymer nanocomposites

One of the key features in the design and fabrication of metal oxide/polymer nanocomposites is the in-depth characterization. Characterization of nanocomposites has been mostly centered on the surface analysis methods and conventional characterization techniques designed to determine the topography of surfaces, composition, morphology, crystallinity, shape, and size. Some of the techniques that have been used in the design and fabrication of nanocomposites have been illustrated [32].

Raman spectroscopy is an analytical technique which depends on inelastic scattering of monochromatic light by molecules due to the molecular excitation, such as rotation, vibration, and translation. Raman spectroscopy determines the vibrational frequencies of molecules that are Raman active and these frequencies rely on the mass and bond strength of atoms.

Fourier transform infra-red (FT-IR) spectroscopy is a significant analytical method which offers appropriate information on the functional groups and structure of a compound. Since FT-IR determines the stretching vibrations of molecules, it can be utilised in the identification of functional groups present in an unknown inorganic and organic compounds.

X-ray diffraction (XRD) is a nondestructive and versatile technique that gives information on the crystal structure, microstructure, chemical composition, lattice constants, and particle size of a material. XRD technique depends on a constructive interference of a beam of X-ray produced in a certain space of direction.

Scanning electron microscopy (SEM) is an electron microscope that scans materials surface with high-energy rays of an electron. SEM makes use of electrons rather than light to scan the surfaces. SEM determined the shape and morphology of material. In addition, morphological properties such as shape, size, and surface features as well as topological data of materials can be obtained.

Transmission electron microscopy (TEM) depends on the beam of electrons generated from an electron gun to interact with the material. TEM determined agglomeration, observation of defects, the effect of annealing, and dispersion in the matrix.

Ultraviolet (UV)-visible spectroscopy is a spectrophotometric technique, which comprises of the measurement of light photons in the UV-visible region. UV-visible spectroscopy measures the intensity of light before and after it has been passed through the material.

Nuclear magnetic resonance (NMR) spectroscopy relied on the population of magnetic nuclei in an external magnetic field to align the nuclei in a finite and predictable number of orientations. NMR gives information on the environment in which the nuclei of atoms are found.

Atomic force microscopy (AFM) is a scanning probe microscopy which uses a small probe to scan across the material to acquire information on the surface of the material. AFM measured the surface of thin films as well as high loading nanoparticles can also be detected.

Photoluminescence (PL) technique is the instantaneous emission of light from the analyzed materials following optical excitation. The effectiveness of PL signals is determined by the properties of the material and the nature of optical excitation. PL is used to obtain information on the compositional analysis of material, band gap, evaluation of several diode materials as well as defect evaluation of light-emitting materials.

Thermogravimetric analysis (TGA) technique is used to measure the weight of a material as a function of material time or temperature at a constant heat rate. TGA is based on heating a mixture of materials at a high temperature to decompose them in the gas phase. The TGA results are generally obtained as a curve with a percent weight against temperature under controlled atmosphere.

3. Water treatment techniques

In the past years, several biological, chemical, and physical techniques have been established to eliminate poisonous pollutants from water resources [33]. Despite the availability of several techniques to eliminate pollutants from water, each of these techniques has their intrinsic limitations.

3.1. The use of metal oxide polymer nanocomposite in chemical treatment of wastewater

Chemical treatment is a wastewater treatment technique where chemicals such as hydroxides, carbonates, and sulfides combine with the pollutants in the wastewater to form insoluble precipitates. Chemical treatment is used during wastewater treatment to slow down disinfection. Chemical treatment techniques such as ozonation and oxidation have been used for the removal of contaminants [34].

Oxidation is an essential technique which uses strong oxidising agents to remove pollutants from the wastewater. The two main classes of oxidation technique that have been reported are the chemical and the UV-assisted oxidation that uses hydrogen peroxide, chlorine, potassium permanganate, ozone, and Fenton's reagent [35]. Manganese chloride and magnesium chloride has been used to eliminate Levafix Brilliant Blue EBRA and the results demonstrated a significant increase in the chemical dosage required to remove the dyeing auxiliaries [36]. The discharging solutions of white mud have been used to eradicate Reactive Light Yellow K-6G, Direct Yellow R, Acid Orange II, and Reactive Bright Red K-2BP in wastewater [37]. The results showed almost 90% dye elimination within 90 sec.

Ozonation is also one of the most effective and attractive chemical methods owing to the strong oxidative action of O_3. In this process, O_3 is decomposed into free radicals follow by instantaneous reaction of the free radicals with the pollutants.

$$O_3 \rightarrow HO^{\bullet} + O_2^{(-\bullet)} + HO_3^{\bullet} + HO_4^{\bullet} \tag{1}$$

$$O^{\bullet} + O_2^{-\bullet} + HO_3^{\bullet} + HO_4^{\bullet} + \text{pollutants} \rightarrow \text{mineralised by} - \text{products} \tag{2}$$

The O_3 is generally created by passing oxygen *via* the gap between two discharging electrodes. During the oxidation process, agents such as high pH, UV, and peroxides are used along with ozone. The optimal pH may be alkaline or acidic based on the nature of pollutants presents in the wastewater. High dose ozonation (~60 mgL^{-1}) was used for dye removal [38]. They found that the color removal indicates the toxic potential of ozonation by-products.

However, highly toxic and unstable metabolites induce by the oxidation and ozonation technique may cause adverse effects on human health and aquatic life [39]. Chlorination process, which provides protection against regrowth of pathogens and bacteria, results in undesirable odors and tastes as well as in the formation of disinfection by-products [40]. Although ozonation has shown to be very effective for pollutants removal, sometimes ozonation generate by-products, which are more dangerous than the pollutants itself. In addition, ozonation process has been less attractive due to short lifetime and high operating costs. The combination of processes with polymers and metal oxides aids to cut cost while at the same time achieving effective degradation of pollutants. For example, polyaniline/hexanoic acid/ TiO_2/Fe_3O_4 nanocomposite with different ratios of Fe_3O_4 and TiO_2 was fabricated using *in situ* chemical polymerization *via* template-free approach [41]. The nanocomposite with chemical treatment displays a narrow sharp reflection loss peak with strong absorption at a lower frequency owing to the Fe_3O_4 nanoparticles outside the nanorods while the nanocomposite without chemical treatment displays a broad reflection loss peak with weak absorption at a higher frequency.

3.2. Metal oxide polymer nanocomposite in disinfection of biological contaminants

Biological contaminants from pathogens, as well as free-living microbes, such as protozoans, viruses, and bacteria, are possible in water [42]. Biological contaminants are classified as microorganisms, biological toxins, and natural organic matter [42]. The biological treatment techniques for the elimination of pollutants from water are regarded as extremely useful techniques owing to the less chemical usage, low cost, and eco-friendly. The general process of biological treatment techniques involves the transformation of biodegradable wastes into less toxic and simpler forms *via a* biological process using microorganisms such as fungi, bacteria, or algae. The treatment techniques are grouped into the aerobic or anaerobic process. The resultant products after anaerobic treatment are methane, biomass, and carbon dioxide, whereas water, biomass, and carbon dioxide are the resultant products after aerobic treatment. Biological processes such as microbiological, biodegradation, and enzymatic decomposition have also been used for dye removal from wastewaters. The effect of applying irradiation before *Pseudomonas sp.* to Reactive Red 120 dye was studied [43]. Decoloration and mineralization at a lower dose of irradiation were enhanced significantly. Moreover, 90% TOC removal and 98% decoloration were observed after 96 hours of microbial treatment. The potential of *Pseudomonas putida* toward the elimination of Orange II dye was investigated [44]. The results showed 92.8% dye elimination within 96 hours at pH and temperature of 8 and 30°C, respectively.

However, most of these biological treatment techniques are unable to eradicate a wide range of pollutants and most of these pollutants remain soluble in the effluent [45]. Recently, there have been reports on nanomaterials such as Zn, Ag, and Ti as a disinfectant to several waterborne disease-causing microbes owing to their charge capacity. The efficiency of polymer-metal oxide nanocomposites in water disinfections have been emphasized [46]. In particular, a polyaniline/TiO_2/graphene nanocomposite was fabricated *via in situ* oxidative polymerization of aniline with TiO_2 and graphene nanoparticles [47]. The as-fabricated nanocomposite demonstrated high antibacterial activity toward *Enterobacter ludwigii* and *Escherichia coli*, indicating its potential as a photocatalyst with antibacterial properties. The enhanced photocatalytic activity and antibacterial activity was as a result of the low recombination of the graphene electron scavenging property and the sensitising effect of polyaniline. An ultrafiltration membrane of poly(1-vinylpyrrolidone-*co*-acrylonitrile)-*g*-ZnO and poly(ether sulfone)-*g*-ZnO with high antibacterial performances and water flux was synthesized [48]. The results revealed that both membranes possess improved water flux, high antibacterial activities, and antifouling characteristic. The hybrid reverse osmosis membranes with aromatic polyamide thin films and TiO_2 particles was prepared through a self-assembly route [49]. The nanocomposite under irradiated UV light showed enhanced photocatalytic bactericidal efficiency compared to those under darkness and light condition. The nonskinned anatase/poly(vinylidene fluoride) microporous membrane was synthesized without any loss in anatase using the dry cast method [50]. The as-synthesized nanocomposite membrane showed a strong bactericide effect compared to the membrane with only UV light.

3.3. Metal oxide polymer nanocomposite in adsorptive technologies

Water can be purified in several ways, such as filtration, desalination, adsorption, osmosis, sedimentation, and disinfection [4], however, adsorption holds several benefits over the other techniques [51]. Adsorption is a surface occurrence where adsorbate are concentrated on the adsorbent surface and the process can be chemisorption or physisorption base on how the adsorbate get adsorb onto the surface of the adsorbent [52]. In this process, pollutants may adsorb on the adsorbent surface via various forces such as electrostatic, hydrogen bonding, and van der Wall interaction [53]. Normally, the adsorbents possess porous structures to permit fluid to pass through faster and increase the surface area. The adsorption process is an economical and simple technique for pollutants elimination from water since it does not require additional chemicals, large amounts of water and high amounts of energy [54]. In developing countries where there is a limited access to huge amounts of financial resources and power, this cheap and simple method might be a feasible alternative. The mechanistic process of adsorption permits flexibility in the development and usage of adsorbent. An isotherm is used to identify and describe the mechanism of adsorption between the adsorbate and adsorbent [55]. There are several adsorption isotherms, but the most commonly used isotherms are Langmuir and Freundlich. Adsorption relies on several factors such as temperature, contact time, pH, particle size, concentration of pollutants as well as the nature of the adsorbent, and adsorbate [56]. The elimination of pollutants from wastewater and water through adsorption are normally carried out using activated carbon, low-cost adsorbents, nanoparticulate adsorbents, and others. As a result of the large specific surface area, nanoadsorbent shows an extensively higher rate of adsorption for pollutants compared to the powdered activated carbon [57]. Recently, several inorganic and organic adsorbents, such as activated carbon, zeolites, clay minerals, biosorbents, montmorillonite, polymer-based adsorbent, trivalent, and tetravalent metal phosphates have been employed as an adsorbent in the adsorption process [58]. Among them, polymer/metal oxide nanocomposite containing polymers, such as polypyrrole, polythiophene, polyfuran, and polyethyleneimine have a strong affinity toward cations owing to the electrostatic interaction between the positively charged ions of the metal oxide and the lone pair of the polymers [59]. In addition, the presence of positively charged nitrogen atoms in polypyrrole offers a potential application in adsorption process as adsorbent [60]. A polyaniline-modified TiO_2 nanocomposite was synthesized through the *in situ* chemical polymerization of aniline in TiO_2 solution [61]. The as-synthesized nanocomposite showed an excellent regeneration and adsorption performances with maximum adsorption capacity (454.55 mg/g) with adsorption equilibrium time of 5 min. The acetate/polypyrrole/TiO_2, succinic-polypyrrole/TiO_2, tartaric/polypyrrole/TiO_2 and citric/polypyrrole/TiO_2 composites were fabricated [62] and the results indicate that the hydroxyl group significantly influenced the adsorption capacity and the surface physicochemical properties of the nanocomposite. Moreover, tartaric/polypyrrole/TiO_2, tartaric/polypyrrole/TiO_2 and citric/tartaric/TiO_2 showed an improved adsorption capacity of 3–4 times toward ARG and MB compared to that of acetate/polypyrrole/TiO_2 and succinic/polypyrrole/TiO_2. In addition, all the nanocomposites displayed an excellent adsorption capacity within 30 min and can be reused without any reduction in capacity for at least 4 times. The hybrid ternary reduced graphene oxide(rGO)/$ZnFe_2O_4$/polyaniline nanocomposite was fabricated through the *in situ* polymerization of aniline on the $ZnFe_2O_4$ and rGO surface [63]. The as prepared composite exhibited a superior adsorption performance in the sewage purification process. Moreover, the thermodynamic date confirmed that the adsorption behavior of the nanocomposites conforms to the Langmuir isotherm with a second-order kinetic model.

3.4. Application of metal oxide polymer nanocomposite in membrane technologies

Currently, membrane technologies have been more efficient in water and wastewater treatment owing to their effective removal of pollutants without producing any harmful by-products. Generally, the basic principle of membrane technology is to apply semipermeable membranes to eliminate particles, gases, fluids, and solutes. For the effective separation of pollutants from water reservoirs, the membrane must be water permeable as well as less permeable to particles or solutes. Water treatment by membrane technologies can be the effective removal of pollutants due to their feasibility, environmentally friendly and cost-effective [64]. Even though inorganic membranes are gaining more consideration, the majority of membranes are made of polymeric materials. Polymer materials such as polysulfone, cellulose nitrates and acetate, polytetrafluoroethylene, polypropylene, polyethersulfone, polyacrylonitrile, polyimide, polyvinylidene fluoride and polyvinyl alcohol are the most extensively used organic membrane materials. These materials are well-known as alternative approaches to pollutants removal owing to their mechanical stability, excellent permeability, chemical resistance, and selectivity of permeate [65]. The immobilization of metal oxide nanoparticles in polymer membrane has been effective for the photodegradation of contaminants in water treatment [66]. For example, membranes containing nano-TiO_2 effectively degrade contaminants (mostly chlorinated compounds) in the water system [67]. The use of TiO_2 immobilized on polyethene membranes has also proved to be very effective in degrading 1,2-dichlorobenzene [68]. A complete degradation of 4-nitrophenol was observed when a nanocomposite membrane consisted of polymers and alumina *via* layer-by-layer adsorption of citrate stabilised Au nanoparticles and polyelectrolytes [69]. A nanocomposite membrane coated with $CoFe_2O_4$(–rGO) and polyvinylidene fluoride (PVDF) on a carbon fiber cloth was fabricated in a PVDF casting solution [70]. The as-fabricated nanocomposite functioned as the cathode membrane to efficiently decompose the contaminants in the water system compared to the microbial fuel cell membrane bioreactor system. The commercial TiO_2 nanoparticles modified with polyaniline was synthesized using the *in situ* polymerization to enhance the property of membrane antifouling and avoid particle agglomeration [71]. The as-prepared nanocomposite membranes showed higher porosity, fewer microvoids, larger surface and finger-like pore size compared to the control polysulfone membrane. In addition, the results revealed that 1.0 wt% of the nanocomposite membrane exhibited water permeability, excellent hydrophilicity, and antifouling property with high rejection rate. The polysulfone ultrafiltration membranes with polyethylene Glycol 1000 as additives and polyaniline/titania nanocomposites were synthesized through the phase inversion technique [72]. The as-synthesized composites membranes displayed enhanced permeability, improved porosity, better hydrophilicity, excellent antifouling ability, and water uptake compared to the polysulfone membranes. Through the phase-inversion route, self-synthesized Cu_2O nanoparticle was introduced into the poly(ether sulfone) mixed-matrix membrane [73]. About 1.5 wt% of the nanocomposite showed an improved water permeability of 66.72×10^{-9} m s/k/Pa. Poly(l-lactide)/TiO_2 nanocomposite membrane was prepared by immersion precipitation method [74]. Moreover, the porosity and pore size on the nanocomposite membrane surface become denser with increasing the TiO_2 loading. The as-prepared nanocomposite membranes displayed enhance recycling and antifouling activity compared to the bulk poly(l-lactide) membrane. The hybrid polyacrylic acid/TiO_2 ultrafiltration membranes with enhanced antifouling performance were fabricated by incorporating TiO_2 particles grafted with polyacrylic acid [75]. The results showed that the nanocomposite membrane exhibited improved hydrophilicity, dispersed well in the membrane medium, excellent antifouling performance, and water flux compared to the pure polyacrylic acid membranes. A polyvinylidene fluoride ultrafiltration membrane was improved through the phase inversion route by incorporating TiO_2 particles in a polyvinylidene fluoride solution [76]. The results showed that the nanocomposite membranes display a larger and longer macrovoid which resulted in an enhanced water flux activity due to the increased surface hydrophilicity. A hybrid SiO_2/polyvinylchloride membrane with different loading of SiO_2 nanoparticles was prepared using the phase-inversion technique [77]. The membrane nanocomposite with 1.5 wt% SiO_2 nanoparticles displayed better performance toward bacterial attachment and protein absorption, higher flux recovery ratio, and better antifouling performance compared to the bare polyvinylchloride membrane. Hence, membrane nanocomposites exhibited applicable potential in water and wastewater treatment due to their excellent antifouling performance, cost-effectiveness, and better elimination efficacies of total bacteria (>93.6%), chemical oxygen demand (>82.9%), and suspended solids (>97.2%). The nanoporous poly(ether sulfone)/TiO_2 ultrafiltration membranes were fabricated through a nonsolvent-induced phase separation route [78]. The modified poly(ether sulfone) membrane revealed increased wettability, reduced pore size, and surface-free energy. In addition, the modified membrane with 0.5 wt% TiO_2 nanoparticle loading demonstrated enhanced antifouling activity with ~80% water flux recovery compared to the control membrane.

3.5. Metal oxide polymer nanocomposite in ion exchange technologies

Hard water often leaves a grey residue when used in cleaning and washing. An ion exchange technique which is similar to the reverse osmosis technique can be used to soften the water. Ion exchange technique is a water treatment technique, which successfully eliminates pollutants from aqueous solutions *via* a strong interaction between the functional group on the charged contaminants and the ion exchange resins [79]. Thus, this technique comprises of exchange of ions to form strong bond between the resins and solutes to achieve efficient separation. Generally, ion exchange membrane is categorized into cation and anion exchangers based on the form of ionic group attached to the membrane medium. The most common used anion exchange are weak base-type, which are type I ion exchange resins (-$N(CH_3)_3$), and type II ion exchange resins (-$N(CH_3)_2C_2H_4OH$), while cation exchangers are the weak acidic carboxylate groups (-COO-) and strong acid-type groups (-SO_3) [80]. The modification of commercially existing ion exchangers and the design of appropriate organic polymeric and inorganic metal oxide membranes with biocide and catalytic performance have been of great interest [81]. Beside polymeric membranes, ion exchange membranes can also be fabricated from other materials such as phosphate salts, bentonite, and zeolites [27]. However, these membranes are ineffective compared to the polymeric membranes owing to their high cost, too large pores and relative bad electrochemical properties [82]. In addition, ion exchange membranes fabricated from polymeric materials exhibits excellent conductivity and high chemical stability [83]. Thus, the combination of polymer materials with inorganic metal oxides offer a new form of ion exchangers with high stability, excellent reproducibility, high ion exchange capacity, and mechanical stability as well as good selectivity toward metal ions [84, 85]. Recently, several excellent metal oxide/polymer ion exchange membranes have been fabricated and effectively used in water remediation process. For instance, nanocomposite material formed by the immobilization of multivalent metal acid salts into conducting polymers, such as polypyrrole, polyaniline, or polythiophene, offers a hybrid ion exchange membrane with high reproducibility, stability, granulometric, and mechanical properties as well as excellent selectivity for heavy metals and ion-exchange capacity [86]. These results indicated that the hybrid organic-inorganic ion exchangers were highly selective toward Cd(II). A hybrid polypyrrole polyantimonic acid nanocomposite with good reproducibility, selectivity toward certain heavy metals, excellent ion-exchange capacity, thermal, and chemical stability was fabricated by mixing hydrated antimony oxide with polypyrrole [83]. The as-fabricated nanocomposite was extremely selective toward Hg(II). A crystalline acrylamide stannic silicomolybdate nanocomposite ion exchanger was prepared at pH 0.63 [87]. The authors revealed that the nanocomposite demonstrated a superior exchange capacity of 1.64 meq/g compared to the pure stannic silicomolybdate (0.40 meq/g). Based on the distribution studies results, several significant binary separations such as Cd(II)-Pb(II), Cd(II)-Cu(II), Al(III)-Pb(II), Al(III)-Cu(II), Zn(II)-Pb(II), and Zn(II)-Cu(II) were observed on the acrylamide stannic silicomolybdate column. A hybrid poly-o-toluidine/$Ce_3(PO_4)_4$ and poly-o-toluidine/$Sn(WO_4)_2$ nanocomposite with high stability, good reproducibility, high ion exchangeability, and good selectivity for heavy metals was fabricated by incorporating orthotoluidine (poly-*o*-toluidine) into $Ce_3(PO_4)_4$ precipitate [88, 89]. The distribution studies demonstrated that the nanocomposites were extremely selective toward Cd(II). Using a fibrous-type polypyrrole/$Th_3(PO_4)_4$ cation-exchanger nanocomposite, the separation of Pb(II) from aqueous solution was explored [90]. The nanocomposite was fabricated by immobilizing polypyrrole into $Th_3(PO_4)_4$ precipitate. Based on the distribution studies, the nanocomposite exhibited excellent selectivity for Pb(II) on the $Th_3(PO_4)_4$ column. A hybrid poly(methyl methacrylate)/$Zr_3(PO_4)_4$ cation-exchanger nanocomposites were synthesized by immobilizing poly(methyl methacrylate) into $Zr_3(PO_4)_4$ precipitate [91]. Based on the sorption studies, the nanocomposite was highly selective to Pb(II). A novel cellulose acetate/Zr(IV) molybdophosphate (ZMP) cation exchanger was fabricated by incorporating organic polymers into multivalent metal acid salts [92]. The as-fabricated nanocomposites displayed a superior selectivity toward Cr(III) on the ZMP column. A novel polymeric-inorganic cation-exchanger nanocomposite was fabricated through a sol-gel route by immobilizing polyaniline into zirconium titanium phosphate [93]. The nanocomposite showed high ion-exchange capacity (4.52 meq/g), good thermal and chemical stability compared to the bulk polyaniline and zirconium titanium phosphate. Moreover, the distribution studies of the nanocomposite revealed highly selectivity to Pb(II) and Hg(II).

3.6. Metal oxide polymer nanocomposite in photocatalytic degradation of pollutants

Recently, personal care and pharmaceutical products used in drugs, cosmetics, agricultural practice, and veterinary medicine have been considered as emerging contaminants [94]. Moreover, organic dyes normally used in printing, photographic, and textile industries are often wasted during the dying process and discharge into the wastewater effluents. The existence of even low concentrations of pollutants in the wastewater streams extremely affects the nature of water, which makes hard to be oxidised or biodegraded. The photodegradation of pollutants in wastewater and water systems using photocatalysis process has been an effective approach compared to the conventional approaches without high energy consumption as well as the formation of highly toxic and poor biodegradable products [95]. Photocatalysis is an AOP employed in wastewater and water treatment process, such as degradation of highly toxic pollutants as well as the oxidative elimination of microbial pathogens and micropollutants [96]. Photocatalysis process uses semiconductors such as metal oxides, nitrides, and sulfides. Normally, a metal oxide is photoactivated by the incoming photon from the light source. As a renewable and safe energy source as well as abundant and clean, natural sunlight has been the ideal source of energy for the activation process [97]. The sun distributes almost four orders of magnitude of its energy to the earth surface annually larger than the energy consumed by humans. When the metal oxide is irradiated by sunlight, electrons and holes are generated according to Eq. (3) only if the energy of the incident photons is equal to or greater than (≥) the metal oxide band gap energy (**Figure 1**).

$$\text{Metal oxide} + h\nu(\text{UV}) \rightarrow \text{Metal oxide}(e^-(\text{CB}) + h^+(\text{VB})) \quad (3)$$

The photocatalytic reaction is initiated when photogenerated electrons are transferred from the filled valence band (VB) of the metal oxide to the empty conduction band (CB), leaving positive holes in the VB (h_{VB}^+). The photogenerated electron then migrates to the metal oxide surface where separation and redox reaction occur. Both reduction and oxidation processes can occur on the surface of the photoexcited metal oxide only if the process is thermodynamically feasible. The photoinduced holes at the VB then react with the adsorbed water molecules to produce $OH^\bullet$ radical (Eq. (4)).

$$H_2O(\text{ads}) + h^+(\text{VB}) \rightarrow HO^\bullet(\text{ads}) + H^+(\text{ads}) \quad (4)$$

Instantaneously, electrons in the CB (e_{CB}^-) also react with the adsorbed oxygen molecule to produce $O_2^{-\bullet}$ radical as shown in Eq. (5).

$$O_2 + e^-(\text{CB}) \rightarrow O_2^{-\bullet}(\text{ads}) \quad (5)$$

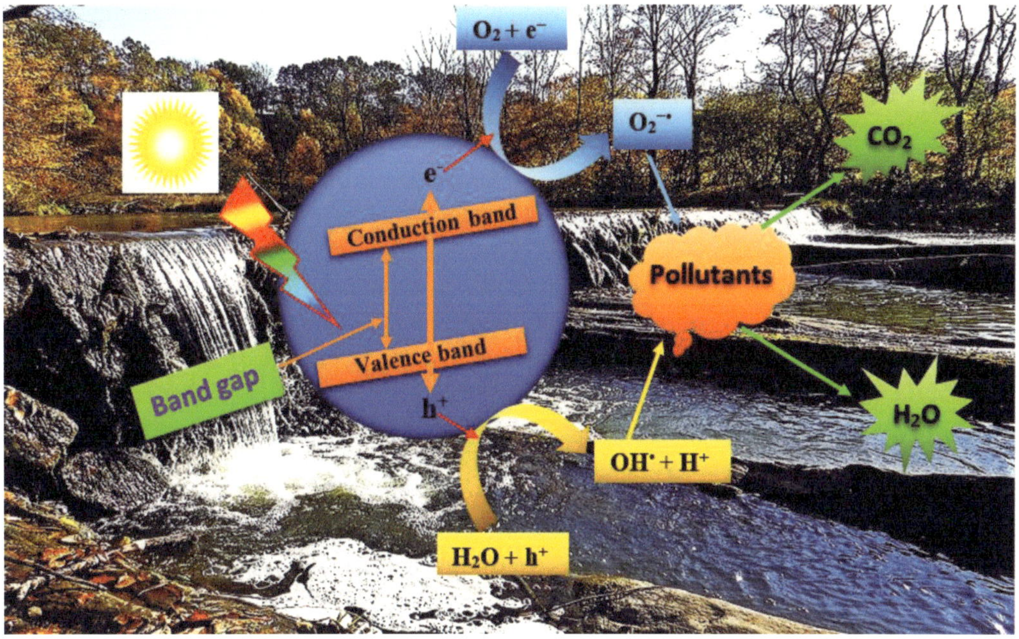

Figure 1. Schematic of photocatalysis process toward the photodegradation of pollutants.

The $HO^{\bullet}$ and $O_2^{-\bullet}$ radicals formed are extremely powerful oxidizing and reducing agent to attack the adsorbed pollutants (Eq. (6)), causing them to mineralize depending on their stability level and structure.

$$\text{Pollutants} + (HO^{\bullet}, O_2^{-\bullet}) \rightarrow \text{Degradation products} \quad (6)$$

Sometimes, the $O_2^{-\bullet}$ radical may not take part in further oxidation process and gets protonated to generate a hydroperoxyl radicals ($HO_2^{\bullet}$) and subsequently into H_2O_2, which further decomposes into extremely reactive $HO^{\bullet}$ radicals (see Eqs. (7)–(9)).

$$O_2^{-\bullet}(\text{ads}) + H^+ \rightarrow HOO^{\bullet}(\text{ads}) \quad (7)$$

$$HOO^{\bullet}(\text{ads}) + H^+ \rightarrow H_2O_2(\text{ads}) \quad (8)$$

$$H_2O_2(\text{ads}) \rightarrow 2\,HO^{\bullet}(\text{ads}) \quad (9)$$

Due to the ability of some of the metal oxides to easily absorb some of the visible light, another mechanism of photodegradation of pollutants was considered. This mechanism consist of the pollutants excitation under simulated visible light irradiation from the ground state (pollutant) to the triplet excited state (pollutant*) [98]. As a result of the migration of electron into the CB of the metal oxide, these excited state pollutants species are transformed into a semioxidized cation radical (pollutants$^{+\bullet}$). The trapped electrons combine with the dissolved oxygen molecules to generate $O_2^{-\bullet}$ radical anions, which can further undergo a series of reactions to generate an $HO^{\bullet}$ radicals which can then oxidize the adsorbed pollutants.

Since the report by Fujishima [99], numerous metal oxide–based photocatalysts has attracted substantial consideration in the degradation of highly toxic and nonbiodegradable compounds

[100] owing to its exceptional optical properties, nontoxicity, low cost, and high stability toward photo and chemical corrosion [101]. However, the major limitation of some of the metal oxide–based photocatalyst materials for potential applications includes the fast recombination of the charge carriers due to the low quantum yield and wide band gap limitation in harvesting a wider portion of the solar energy [102]. Hence, the design of metal oxide–based photocatalyst materials for the degradation of contaminants, which are of high photostability, visible light active, and photoactivity are of considerable interest to provide a fundamental insight into the underlying mechanism of photocatalysis as well as designing a more efficient and easily tunable metal oxide–based photocatalyst materials. To address these restrictions, metal oxides are normally immobilized with polymers. The incorporation of metal oxide into a polymeric material with appropriate energy levels enhances the charge migration between the inorganic metal oxide and polymer to reduce the recombination of the charge carriers [103]. Recently, nanocomposites of several conductive polymers and metal oxide nanoparticles have been fabricated with improved photocatalytic responses in the visible region. Polymeric materials are currently been applied in water treatment due to their pore size distribution, perfect mechanical rigidity, tunable, and large surface area [104]. For example, some organic nanofiber membranes, such as cellulose, polyacrylonitrile, polyvinylpyrrolidone, polytetrafluoroethylene, and polyvinyl acetate are considered as an excellent catalyst supports owing to their high porosity, high permeability, large surface area and good flexibility [105]. The improvements in the fabrication approach employed for designing materials, the characterization and computational techniques have facilitated the fine-tuning of compositional and structural characteristics of materials. The novel porous polytetrafluoroethylene nanofiber membrane with Fe_2O_3 was prepared through a three-step method [105]. Under simulated UV light irradiation, the porous nanocomposite displayed superior photocatalytic degradation toward Acid Red dye with a degradation efficiency of >99% after 80 min. TiO_2/polypropylene nanocomposite has been reported to show superb photocatalytic performance toward pollutants degradation [106]. The photocatalytic performance of a transition metal coordination polymer(TMCP)/polyoxometalate nanocomposite was enhanced by immobilizing polypyrrole into the TMCP surface *via* a superficial *in situ* chemical oxidation polymerization method [107]. Under simulated visible light irradiation, the as-prepared nanocomposite displayed better photocatalytic performance toward the degradation of Rhodamine B (RhB) dye. Transition metal ions were incorporated into TiO_2/fly-ash cenospheres with poly(o-phenylenediamine) through ion imprinting technology [108]. The results demonstrated that the as-prepared nanocomposite effectively photodegrades the tetracycline with a photodegradation rate of 71.7%. A silica nanohybrid with different Ru(II)-polypyridyl nanocomposites exhibited improved photocatalytic degradation with respect to Rhodamine 6G dye compared to the functionalized silica nanohybrid [109]. Hydrothermal and electrospinning route was used to prepare polyvinylidene fluoride/titanium dioxide nanocomposite with different compositions of anatase, brookite, and rutile and the nanocomposite displayed a relatively high photocatalytic degradation toward phenol [110]. The two types of polysulfone/carbon-covered alumina/TiO_2 membranes were successfully prepared and their activity toward RhB dye under visible light irradiation was tested [111]. The results displayed that the nanocomposite without a fabric membrane degraded 78.7% of RhB, while the nanocomposite with fabric membrane degraded 82.4% of RhB after 300 min. Polysulfone-based polyaniline-TiO_2 with a hollow fiber membrane was synthesized through a dry-wet spinning

process [112]. The results showed that the polysulfone hollow fibres containing 1.0 wt% of the as-synthesized nanocomposite exhibited a maximum rejection rate of 96.5 and 81.5% for Reactive Orange 16 and Reactive Black 5, respectively. A conductive polypyrrole-polyaniline/ TiO_2 nanocomposite was fabricated *via* an *in situ* oxidative copolymerization [113]. The as-fabricated nanocomposite showed a superior photocatalytic degradation of 4-nitrophenol and this improvement was attributed to the conjugated structure, conductivity and the synergy effect among the polymers and TiO_2. A polysulfone-sulfated/TiO_2 nanofiltration membrane showed good efficacy for methylene blue (MB) dye removal with a maximum rejection of 90.4% [114]. A novel and highly efficient α-Fe_2O_3/polypyrrole nanocomposite was successful designed and fabricated *via* a simple and mild one-step chemical route [115]. The as-fabricated nanocomposite showed a significant photocatalytic degradation toward MB dye under simulated UV irradiation and ambient temperature compared to the bulk Fe_2O_3. The improved photodegradation performance was due to the crystalline nature and mesoporous structure of the nanocomposites as well as the synergetic effect between α-Fe_2O_3 and polypyrrole, which improve the charge separation and recombination rate. The method of preparation and photodegradation activity of polymer/metal oxide nanocomposites are presented in **Table 1**.

Nanocomposites	Method of preparation	Pollutants	References
PVDF/(Ag, Pt)/rGO/TiO_2	Phase inversion process	Methyl Orange (MO)	[116]
Au/polyaniline/TiO_2	One-step chemical redox	RhB	[117]
MoO_3/polyimide	One-pot homopolymerization	MO	[118]
polyimide/ZnO	Electrospinning and direct ion-exchange process	MB	[119]
Fe_3O_4/polypyrrole/silver	Wet chemical	MO and Orange II	[120]
polypyrrole/TiO_2	Reverse microemulsion polymerization *In situ* Fractal growth	MO RhB RhB	[121] [122] [123]
ZnO/polypyrrole/rGO	simple and fast electrochemical	MB	[124]
Polypyrrole/polyvinyl alcohol/TiO_2	*In situ* polymerization	RhB	[125]

Table 1. Method of preparation and photocatalytic degradation of metal oxide modified with polymers.

4. Conclusion and outlook

Metal oxide polymer nanocomposites offer great prospects to revolutionize water treatment owing to their unique properties. In this chapter, we have comprehensively reviewed the various synthesis and characterization technique of nanomaterials, unique properties of nanocomposites and progress in water treatment using polymer/inorganic metal oxide nanocomposites. The fundamentals and unique features of metal oxide, as well as the underlying mechanism of the visible light response metal oxide–based photocatalyst, have been deliberated. The polymer/inorganic metal oxide nanocomposite has been extensively used for the photodegradation of pollutants, ion exchanger, adsorbent, membrane, and photocatalytic disinfection in water treatment. For instance, polymer-based nanocomposites have several intrinsic importance such as high mechanical strength, long-term stability, and low-cost fabrication process, while the inorganic metal oxide possesses superior optical, electronic, magnetic, and catalytic properties. Thus, the hybridization of polymers and metal oxide could enhance several properties of the resultant nanocomposites. In conclusion, the polymer/metal oxide nanocomposite reviewed in this book chapter possesses superior photodegradation activity toward pollutants under simulated light irradiation owing to the large expose area of the nanocomposite, effective migration and separation of charge carriers as well as the strong electrostatic interaction between the catalyst and the pollutants. Moreover, the nanocomposite, exhibit high antibacterial activity, strong affinity toward ions owing to the strong electrostatic interactions between the positively charged ions of the inorganic metal oxide and the lone pair of the organic polymers, excellent regeneration, and adsorption performances, better antifouling property, excellent hydrophilicity due to porosity, larger surface pore size, less microvoids as well as good ion-exchange capacity. Though this book chapter is not a comprehensive, but can give a basic idea about polymer nanocomposites in the various water treatment approaches.

Though considerable development in water treatment using metal oxide and polymer nanocomposite has been accomplished, studies in this area are still at the primary stage and additional advancement is essential. The design and fabrication of materials for water treatment process, particular visible light-response polymer/metal oxide–based photocatalyst materials are significant, but several reports briefly consider only the technical hurdles, high operating cost, and environmental risks. In future, to enhance the practicability of visible light response polymer/metal oxide–based photocatalyst materials in water treatment, several key issues, such as improving the photostability and efficiency of the photocatalyst materials need to be addressed. Moreover, the usage of photocatalytic materials in the photodegradation of pollutants is still a major issue considering the degradation and dissolution of the pollutants, which can hinder their photocatalytic performances. Hence, care should be taken when designing a functional polymer/metal oxide–based material with suitable physicochemical properties. Developing a suitable immobilization approach with cost-effective solid-liquid separation is also important. Currently, an exhausting catalyst during the photocatalysis process will endanger the regeneration of the catalysts and impacts severe effects on the environment as a result of catalysts outflow. To acquire an improved photodegradation efficiency, incorporating different techniques is essential. The accumulation of an enormous amount of theoretical study is also significant to prove an in-depth understanding of the preparation, performances, optimisation, and properties of polymer/metal oxide–based photocatalysts for water treatment.

Acknowledgements

The authors would like to acknowledge the financial contributions from the Faculty of Science: the University of Johannesburg-South Africa, Centre for Nanomaterials and Science Research: Department of Applied Chemistry and the National Research Foundation (TTK14052167682).

http://dx.doi.org/10.5772/67835

Author details

Francis Opoku, Ephraim M. Kiarii, Penny P. Govender and Messai Adenew Mamo*

*Address all correspondence to: messaim@uj.ac.za

Department of Applied Chemistry, University of Johannesburg, Johannesburg, South Africa

References

[1] Zaporozec, A. Ground-water pollution and its sources. GeoJournal 5, pp. 457-471, 1981.

[2] Suriyaraj, S. and Selvakumar, R. Advances in nanomaterial based approaches for enhanced fluoride and nitrate removal from contaminated water. RSC Advances 6, pp. 10565-10583, 2016.

[3] World Health Organization. Guidelines for Drinking-water Quality. Geneva: World health organization, 2011.

[4] Shannon, M. A., Bohn, P. W., Elimelech, M., Georgiadis, J. G., Marinas, B. J. and Mayes, A. M. Science and technology for water purification in the coming decades. Nature 452, pp. 301-310, 2008.

[5] Yang, M. A current global view of environmental and occupational cancers. Journal of Environmental Science and Health, Part C 29, pp. 223-249, 2011.

[6] Bottoni, P., Caroli, S. and Caracciolo, A. B. Pharmaceuticals as priority water contaminants. Toxicological & Environmental Chemistry 92, pp. 549-565, 2010.

[7] Padmanabhan, P., Sreekumar, K., Thiyagarajan, T., Satpute, R., Bhanumurthy, K., Sengupta, P., Dey, G. and Warrier, K. Nano-crystalline titanium dioxide formed by reactive plasma synthesis. Vacuum 80, pp. 1252-1255, 2006.

[8] Gaya, U. I. and Abdullah, A. H. Heterogeneous photocatalytic degradation of organic contaminants over titanium dioxide: a review of fundamentals, progress and problems. Journal of Photochemistry and Photobiology C: Photochemistry Reviews 9, pp. 1-12, 2008.

[9] Bremner, D. H., Molina, R., Martinez, F., Melero, J. and Segura, Y. Degradation of phenolic aqueous solutions by high frequency sono-Fenton systems (Fe_2O_3/SBA-15-H_2O_2). Applied Catalysis B: Environmental 90, pp. 380-388, 2009.

[10] Chong, M. N., Jin, B., Chow, C. W. and Saint, C. Recent developments in photocatalytic water treatment technology: a review. Water Research 44, pp. 2997-3027, 2010.

[11] Yang, L.-Y., Dong, S.-Y., Sun, J.-H., Feng, J.-L., Wu, Q.-H. and Sun, S.-P. Microwave-assisted preparation, characterization and photocatalytic properties of a dumbbell-shaped ZnO photocatalyst. Journal of Hazardous Materials 179, pp. 438-443, 2010.

[12] Chen, X. and Mao, S. S. Titanium dioxide nanomaterials: synthesis, properties, modifications, and applications. Chemical Reviews 107, pp. 2891-2959, 2007.

[13] Ball, P. and Garwin, L. Science at the atomic scale. Nature 355, pp. 761-766, 1992.

[14] Wu, Y., Yan, H., Huang, M., Messer, B., Song, J. H. and Yang, P. Inorganic semiconductor nanowires: rational growth, assembly, and novel properties. Chemistry-A European Journal 8, pp. 1260-1268, 2002.

[15] Pan, B., Pan, B., Zhang, W., Lv, L., Zhang, Q. and Zheng, S. Development of polymeric and polymer-based hybrid adsorbents for pollutants removal from waters. Chemical Engineering Journal 151, pp. 19-29, 2009.

[16] Palmisano, G., García-López, E., Marcì, G., Loddo, V., Yurdakal, S., Augugliaro, V. and Palmisano, L. Advances in selective conversions by heterogeneous photocatalysis. Chemical Communications 46, pp. 7074-7089, 2010.

[17] Hua, M., Zhang, S., Pan, B., Zhang, W., Lv, L. and Zhang, Q. Heavy metal removal from water/wastewater by nanosized metal oxides: a review. Journal of Hazardous Materials 211, pp. 317-331, 2012.

[18] Whitesides, G. M., Kriebel, J. K. and Mayers, B. T. Self-assembly and nanostructured materials, in Nanoscale Assembly: Chemical Techniques, Nanostructure Science and Technology, Springer, Berlin, Germany, 2005, pp. 217-239.

[19] Sugimoto, T. Fine Particles: Synthesis, Characterization, and Mechanisms of Growth. CRC Press, Boca Raton, 2000.

[20] Lu, T., Li, Y. and Chen, T. Techniques for fabrication and construction of three-dimensional scaffolds for tissue engineering. International Journal of Nanomedicine 8, pp. 337-350, 2013.

[21] Cushing, B. L., Kolesnichenko, V. L. and O'Connor, C. J. Recent advances in the liquid-phase syntheses of inorganic nanoparticles. Chemical Reviews 104, pp. 3893-3946, 2004.

[22] Horrocks, A. R. and Price, D. Fire Retardant Materials. Abington Cambridge, England: Woodhead Publishing, 2001.

[23] Friedrich, K., Fakirov, S. and Zhang, Z. Polymer Composites: From Nano- to Macro-Scale. New York/USA: Springer Science & Business Media, Inc., 2005.

[24] Osman, M. A., Rupp, J. E. and Suter, U. W. Effect of non-ionic surfactants on the exfoliation and properties of polyethylene-layered silicate nanocomposites. Polymer 46, pp. 8202-8209, 2005.

[25] Ratna, D. Epoxy Composites: Impact Resistance and Flame Retardancy. London, United Kingdom: Ismithers Rapra Publishing, 2007.

[26] Kickelbick, G. Concepts for the incorporation of inorganic building blocks into organic polymers on a nanoscale. Progress in Polymer Science 28, pp. 83-114, 2003.

[27] Xu, T. Ion exchange membranes: state of their development and perspective. Journal of Membrane Science 263, pp. 1-29, 2005.

[28] Fischer, H. Polymer nanocomposites: from fundamental research to specific applications. Materials Science and Engineering: C 23, pp. 763-772, 2003.

[29] Krasia-Christoforou, T. Organic–inorganic polymer hybrids: synthetic strategies and applications, in Hybrid and Hierarchical Composite Materials, C.-S. Kim, C. Randow, and T. Sano, Editors, Springer International Publishing, Cham, 2015, pp. 11-63.

[30] Sakata, T., Hashimoto, K. and Hiramoto, M. New aspects of electron transfer on semiconductor surface: dye-sensitization system. Journal of Physical Chemistry 94, pp. 3040-3045, 1990.

[31] Sanchez, C., Alonso, B., Chapusot, F., Ribot, F. and Audebert, P. Molecular design of hybrid organic-inorganic materials with electronic properties. Journal of Sol-Gel Science and Technology 2, pp. 161-166, 1994.

[32] Laurent, S., Forge, D., Port, M., Roch, A., Robic C., Vander Elst, L. and Muller, R. N. Magnetic iron oxide nanoparticles: synthesis, stabilization, vectorization, physicochemical characterizations, and biological applications. Chemical Reviews 108, pp. 2064-2110, 2008.

[33] Nandi, B. K., Goswami, A. and Purkait, M. K. Removal of cationic dyes from aqueous solutions by kaolin: kinetic and equilibrium studies. Applied Clay Science 42, pp. 583-590, 2009.

[34] Tünay, O., Kabdasli, I., Eremektar, G. and Orhon, D. Color removal from textile wastewaters. Water Science and Technology 34, pp. 9-16, 1996.

[35] Tsai, T., Kao, C., Yeh, T. and Lee, M. Chemical oxidation of chlorinated solvents in contaminated groundwater: review. Practice Periodical of Hazardous, Toxic, and Radioactive Waste Management 12, pp. 116-126, 2008.

[36] Bouyakoub, A., Lartiges, B., Ouhib, R., Kacha, S., El Samrani, A., Ghanbaja, J. and Barres, O. $MnCl_2$ and $MgCl_2$ for the removal of reactive dye Levafix Brilliant Blue EBRA from synthetic textile wastewaters: an adsorption/aggregation mechanism. Journal of Hazardous Materials 187, pp. 264-273, 2011.

[37] Zhang, G., Li, X., Li, Y., Wu, T., Sun, D. and Lu, F. Removal of anionic dyes from aqueous solution by leaching solutions of white mud. Desalination 274, pp. 255-261, 2011.

[38] Mezzanotte, V., Fornaroli, R., Canobbio, S., Zoia, L. and Orlandi, M. Colour removal and carbonyl by-production in high dose ozonation for effluent polishing. Chemosphere 91, pp. 629-634, 2013.

[39] Reddy, E. P., Davydov, L. and Smirniotis, P. TiO_2-loaded zeolites and mesoporous materials in the sonophotocatalytic decomposition of aqueous organic pollutants: the role of the support. Applied Catalysis B: Environmental 42, pp. 1-11, 2003.

[40] Gopal, K., Tripathy, S. S., Bersillon, J. L. and Dubey, S. P. Chlorination byproducts, their toxicodynamics and removal from drinking water. Journal of Hazardous Materials 140, pp. 1-6, 2007.

[41] Koh, Y. N., Sambasevam, K. P., Yahya, R. and Phang, S. W. Improvement of microwave absorption for PAni/HA/TiO_2/Fe_3O_4 nanocomposite after chemical treatment. Polymer Composites 34, pp. 1186-1194, 2013.

[42] Dugan, N. R. and Williams, D. J. Cyanobacteria passage through drinking water filters during perturbation episodes as a function of cell morphology, coagulant and initial filter loading rate. Harmful algae 5, pp. 26-35, 2006.

[43] Paul, J., Kadam, A. A., Govindwar, S. P., Kumar, P. and Varshney, L. An insight into the influence of low dose irradiation pretreatment on the microbial decolouration and degradation of reactive red-120 dye. Chemosphere 90, pp. 1348-1358, 2013.

[44] Garg, S. K., Tripathi, M., Singh, S. K. and Tiwari, J. K. Biodecolorization of textile dye effluent by Pseudomonas putida SKG-1 (MTCC 10510) under the conditions optimized for monoazo dye orange II color removal in simulated minimal salt medium. International Biodeterioration & Biodegradation 74, pp. 24-35, 2012.

[45] Servos, M., Bennie, D., Burnison, B., Jurkovic, A., McInnis, R., Neheli, T., Schnell, A., Seto, P., Smyth, S. and Ternes, T. Distribution of estrogens, 17β-estradiol and estrone, in Canadian municipal wastewater treatment plants. Science of the Total Environment 336, pp. 155-170, 2005.

[46] Jain, P. and Pradeep, T. Potential of silver nanoparticle-coated polyurethane foam as an antibacterial water filter. Biotechnology and Bioengineering 90, pp. 59-63, 2005.

[47] Kumar, R., Ansari, M. O., Parveen, N., Oves, M., Barakat, M., Alshahri, A., Khan, M. Y. and Cho, M. H. Facile route to a conducting ternary polyaniline@TiO_2/GN nanocomposite for environmentally benign applications: photocatalytic degradation of pollutants and biological activity. RSC Advances 6, pp. 111308-111317, 2016.

[48] Jo, Y. J., Choi, E. Y., Choi, N. W. and Kim, C. K. Antibacterial and hydrophilic characteristics of poly(ether sulfone) composite membranes containing zinc oxide nanoparticles grafted with hydrophilic polymers. Industrial & Engineering Chemistry Research 55, pp. 7801-7809, 2016.

[49] Kwak, S.-Y., Kim, S. H. and Kim, S. S. Hybrid organic/inorganic reverse osmosis (RO) membrane for bactericidal anti-fouling: preparation and characterization of TiO_2 nanoparticle self-assembled aromatic polyamide thin-film-composite (TFC) membrane. Environmental Science & Technology 35, pp. 2388-2394, 2001.

[50] Alaoui, O. T., Nguyen, Q. T., Schaetzel, P. and Mbareck, C. Dye and bacteria photodegradations with anatase-loaded microporous poly (vinylidene fluoride) membranes. Catalysis Science & Technology 1, pp. 1412-1422, 2011.

[51] Dubey, S. P., Gopal, K. and Bersillon, J. Utility of adsorbents in the purification of drinking water: a review of characterization, efficiency and safety evaluation of various adsorbents. Journal of Environmental Biology 30, pp. 327-332, 2009.

[52] Gupta, V. Application of low-cost adsorbents for dye removal—A review. Journal of Environmental Management 90, pp. 2313-2342, 2009.

[53] Ahmad, A., Mohd-Setapar, S. H., Chuong, C. S., Khatoon, A., Wani, W. A., Kumar, R. and Rafatullah, M. Recent advances in new generation dye removal technologies: novel search for approaches to reprocess wastewater. RSC Advances 5, pp. 30801-30818, 2015.

[54] Kumar, V., Talreja, N., Deva, D., Sankararamakrishnan, N., Sharma, A. and Verma, N. Development of bi-metal doped micro-and nano multi-functional polymeric adsorbents for the removal of fluoride and arsenic (V) from wastewater. Desalination 282, pp. 27-38, 2011.

[55] Gilles, C., MacEwan, T., Nakhwa, S. and Smith, D. A system of classification of solution adsorption isotherms, and its use in diagnosis of adsorption mechanisms and measurement of specific surface areas of solids. Journal Chemistry Society 2, pp. 3973-3993, 1960.

[56] Dąbrowski, A. Adsorption-from theory to practice. Advances in Colloid and Interface Science 93, pp. 135-224, 2001.

[57] Gehrke, I., Geiser, A. and Somborn-Schulz, A. Innovations in nanotechnology for water treatment. Nanotechnology, Science and Applications 8, pp. 1-17, 2015.

[58] Mahmud, H. N. M. E., Huq, A. O. and Binti Yahya, R. The removal of heavy metal ions from wastewater/aqueous solution using polypyrrole-based adsorbents: a review. RSC Advances 6, pp. 14778-14791, 2016.

[59] Abdi, M. M., Kassim, A., Mahmud, H. E., Yunus, W. M. M., Talib, Z. A. and Sadrolhosseini, A. R. Physical, optical, and electrical properties of a new conducting polymer. Journal of Materials Science 44, pp. 3682-3686, 2009.

[60] Mahmud, H., Hosseini, S. and Yahya, R. B. Removal of nickel ions from aqueous solution by polypyrrole conducting polymer. Key Engineering Materials 594, pp. 793-797, 2014.

[61] Wang, N., Li, J., Lv, W., Feng, J. and Yan, W. Synthesis of polyaniline/TiO_2 composite with excellent adsorption performance on acid red G. RSC Advances 5, pp. 21132-21141, 2015.

[62] Feng, J., Chen, J., Wang, N., Li, J., Shi, J. and Yan, W. Enhanced adsorption capacity of polypyrrole/TiO_2 composite modified by carboxylic acid with hydroxyl group. RSC Advances 6, pp. 42572-42580, 2016.

[63] Feng, J., Hou, Y., Wang, X., Quan, W., Zhang, J., Wang, Y. and Li, L. In-depth study on adsorption and photocatalytic performance of novel reduced graphene oxide-$ZnFe_2O_4$-polyaniline composites. Journal of Alloys and Compounds 681, pp. 157-166, 2016.

[64] Owen, G., Bandi, M., Howell, J. and Churchouse, S. Economic assessment of membrane processes for water and waste water treatment. Journal of Membrane Science 102, pp. 77-91, 1995.

[65] Mulder, M. The use of membrane processes in environmental problems. An introduction, in Membrane Processes in Separation and Purification, J.G. Crespo and K.W. Böddeker, Editors, Springer Netherlands, 1994, pp. 229-262.

[66] Xu, J., Bachas, L. and Bhattacharyya, D. Synthesis of nanostructured bimetallic particles in polyligand-functionalized membranes for remediation applications, in Nanotechnology Applications for Clean Water: Solutions for Improving Water Quality, R. Sustich, J. Duncan, and N. Savage, Editors, William Andrew Publishing, Boston, 2014, pp. 369-393.

[67] Wu, L. and Ritchie, S. Enhanced dechlorination of trichloroethylene by membrane-supported Pd-coated iron nanoparticles. Environmental Progress 27, pp. 218-224, 2008.

[68] Molinari, R., Palmisano, L., Drioli, E. and Schiavello, M. Studies on various reactor configurations for coupling photocatalysis and membrane processes in water purification. Journal of Membrane Science 206, pp. 399-415, 2002.

[69] Dotzauer, D. M., Dai, J., Sun, L. and Bruening, M. L. Catalytic membranes prepared using layer-by-layer adsorption of polyelectrolyte/metal nanoparticle films in porous supports. Nano Letters 6, pp. 2268-2272, 2006.

[70] Li, Y., Sun, J., Liu, L. and Yang, F. A composite cathode membrane with $CoFe_2O_4$–rGO/PVDF on carbon fiber cloth: synthesis and performance in a photocatalysis-assisted MFC-MBR system. Environmental Science: Nano, pp. 1-11, 2017.

[71] Teli, S. B., Molina, S., Sotto, A., Calvo, E. G. A. and Abajob, J. D. Fouling resistant polysulfone–PANI/TiO_2 ultrafiltration nanocomposite membranes. Industrial & Engineering Chemistry Research 52, pp. 9470-9479, 2013.

[72] Pereira, V. R., Isloor, A. M., Ahmed, A. A. and Ismail, A. F. Preparation, characterization and the effect of PANI coated TiO_2 nanocomposites on the performance of polysulfone ultrafiltration membranes. New Journal of Chemistry 39, pp. 703-712, 2015.

[73] Krishnamurthy, P. H., Yogarathinam, L. T., Gangasalam, A. and Ismail, A. F. Influence of copper oxide nanomaterials in a poly(ether sulfone) membrane for improved humic acid and oil–water separation. Journal of Applied Polymer Science 133, pp. 1-10, 2016.

[74] Jiang, B., Wang, B., Zhang, L., Sun, Y., Xiao, X., Hao, L. and Yang, N. Enhancing antifouling performance of poly(l-lactide) membranes by TiO_2 nanoparticles. Journal of Applied Polymer Science 133, pp. 1-8, 2016.

[75] Yuan, S., Wang, J., Wang, X., Long, S., Zhang, G. and Yang, J. Poly(arylene sulfide sulfone) hybrid ultrafiltration membrane with TiO_2-g-PAA nanoparticles: preparation and antifouling performance. Polymer Engineering & Science 55, pp. 2829-2837, 2015.

[76] Moghadam, M. T., Lesage, G., Mohammadi, T., Mericq, J.-P., Mendret, J., Heran, M., Faur, C., Brosillon, S., Hemmati, M. and Naeimpoor, F. Improved antifouling properties of TiO_2/PVDF nanocomposite membranes in UV-coupled ultrafiltration. Journal of Applied Polymer Science 132, pp. 1-13, 2015.

[77] Yu, Z., Liu, X., Zhao, F., Liang, X. and Tian, Y. Fabrication of a low-cost nano-SiO_2/PVC composite ultrafiltration membrane and its antifouling performance. Journal of Applied Polymer Science 132, pp. 1-11, 2015.

[78] Low, Z.-X., Wang, Z., Leong, S., Razmjou, A., Dumée, L. F., Zhang, X. and Wang, H. Enhancement of the antifouling properties and filtration performance of poly(ethersulfone) ultrafiltration membranes by incorporation of nanoporous titania nanoparticles. Industrial & Engineering Chemistry Research 54, pp. 11188-11198, 2015.

[79] Greluk, M. and Hubicki, Z. Evaluation of polystyrene anion exchange resin for removal of reactive dyes from aqueous solutions. Chemical Engineering Research and Design 91, pp. 1343-1351, 2013.

[80] Rivas, B. L., Sánchez, J. and Urbano, B. F. Polymers and nanocomposites: synthesis and metal ion pollutant uptake. Polymer International 65, pp. 255-267, 2016.

[81] Chesnut, D. J., Hagrman, D., Zapf, P. J., Hammond, R. P., LaDuca, R., Haushalter, R. C. and Zubieta, J. Organic/inorganic composite materials: the roles of organoamine ligands in the design of inorganic solids. Coordination Chemistry Reviews 190, pp. 737-769, 1999.

[82] Ohya, H., Masaoka, K., Aihara, M. and Negishi, Y. Properties of new inorganic membranes prepared by metal alkoxide methods. Part III: New inorganic lithium permselective ion exchange membrane. Journal of Membrane Science 146, pp. 9-13, 1998.

[83] Khan, A. A. and Alam, M. M. New and novel organic–inorganic type crystalline 'polypyrrolel/polyantimonic acid'composite system: preparation, characterization and analytical applications as a cation-exchange material and Hg (II) ion-selective membrane electrode. Analytica Chimica Acta 504, pp. 253-264, 2004.

[84] Alberti, G., Costantino, U., Allulli, S. and Massucci, M. Crystalline insoluble acid salts of tetravalent metals-XIV forward and reverse sodium-potassium ion exchange isotherms on crystalline zirconium phosphate. Journal of Inorganic and Nuclear Chemistry 35, pp. 1339-1346, 1973.

[85] Niwas, R., Khan, A. A. and Varshney, K. G. Synthesis and ion exchange behaviour of polyaniline Sn(IV) arsenophosphate: a polymeric inorganic ion exchanger. Colloids and Surfaces A: Physicochemical and Engineering Aspects 150, pp. 7-14, 1999.

[86] Khan, A. A. and Alam, M. M. Synthesis, characterization and analytical applications of a new and novel 'organic–inorganic' composite material as a cation exchanger and Cd(II) ion-selective membrane electrode: polyaniline Sn(IV) tungstoarsenate. Reactive and Functional Polymers 55, pp. 277-290, 2003.

[87] Khan, A. M., Ganai, S. A. and Nabi, S. A. Synthesis of a crystalline organic–inorganic composite exchanger, acrylamide stannic silicomolybdate: binary and quantitative separation of metal ions. Colloids and Surfaces A: Physicochemical and Engineering Aspects 337, pp. 141-145, 2009.

[88] Khan, A. A. and Akhtar, T. Synthesis, characterization and analytical application of nano-composite cation-exchange material, poly-o-toluidine Ce(IV) phosphate: Its application in making Cd(II) ion selective membrane electrode. Solid State Sciences 13, pp. 559-568, 2011.

[89] Khan, A. A. and Shaheen, S. Synthesis and characterization of a novel hybrid nano composite cation exchanger poly-o-toluidine Sn(IV) tungstate: its analytical applications as ion-selective electrode. Solid State Sciences 16, pp. 158-167, 2013.

[90] Khan, A. A., Inamuddin and Alam, M. M. Determination and separation of Pb^{2+} from aqueous solutions using a fibrous type organic–inorganic hybrid cation-exchange material: polypyrrole thorium(IV) phosphate. Reactive and Functional Polymers 63, pp. 119-133, 2005.

[91] Siddiqui, W. A., Khan, S. A. and Inamuddin. Synthesis, characterization and ion-exchange properties of a new and novel 'organic–inorganic' hybrid cation-exchanger: poly(methyl methacrylate) Zr(IV) phosphate. Colloids and Surfaces A: Physicochemical and Engineering Aspects 295, pp. 193-199, 2007.

[92] Nabi, S. A. and Naushad, M. Synthesis, characterization and analytical applications of a new composite cation exchanger cellulose acetate-Zr(IV) molybdophosphate. Colloids and Surfaces A: Physicochemical and Engineering Aspects 316, pp. 217-225, 2008.

[93] Khan, A. A. and Paquiza, L. Characterization and ion-exchange behavior of thermally stable nano-composite polyaniline zirconium titanium phosphate: its analytical application in separation of toxic metals. Desalination 265, pp. 242-254, 2011.

[94] Gadge, S. T. and Bhanage, B. M. Recent developments in palladium catalysed carbonylation reactions. RSC Advances 4, pp. 10367-10389, 2014.

[95] Malato, S., Fernández-Ibáñez, P., Maldonado, M., Blanco, J. and Gernjak, W. Decontamination and disinfection of water by solar photocatalysis: recent overview and trends. Catalysis Today 147, pp. 1-59, 2009.

[96] Friedmann, D., Mendive, C. and Bahnemann, D. TiO_2 for water treatment: parameters affecting the kinetics and mechanisms of photocatalysis. Applied Catalysis B: Environmental 99, pp. 398-406, 2010.

[97] Balachandran, S., Praveen, S. G., Velmurugan, R. and Swaminathan, M. Facile fabrication of highly efficient, reusable heterostructured Ag–ZnO–CdO and its twin applications of dye degradation under natural sunlight and self-cleaning. RSC Advances 4, pp. 4353-4362, 2014.

[98] Ma, Y. and Yao, J.-n. Photodegradation of rhodamine B catalyzed by TiO_2 thin films. Journal of Photochemistry and Photobiology A: Chemistry 116, pp. 167-170, 1998.

[99] Fujishima, A. Electrochemical photolysis of water at a semiconductor electrode. Nature 238, pp. 37-38, 1972.

[100] Chen, J., Qiu, F., Xu, W., Cao, S. and Zhu, H. Recent progress in enhancing photocatalytic efficiency of TiO_2-based materials. Applied Catalysis A: General 495, pp. 131-140, 2015.

[101] Qu, X., Alvarez, P. J. J. and Li, Q. Applications of nanotechnology in water and wastewater treatment. Water Research 47, pp. 3931-3946, 2013.

[102] Kanakaraju, D., Glass, B. D. and Oelgemöller, M. Titanium dioxide photocatalysis for pharmaceutical wastewater treatment. Environmental Chemistry Letters 12, pp. 27-47, 2014.

[103] Huang, J.-H., Ibrahem, M. A. and Chu, C.-W. Interfacial engineering affects the photocatalytic activity of poly (3-hexylthiophene)-modified TiO_2. RSC Advances 3, pp. 26438-26442, 2013.

[104] Qadir, D., Mukhtar, H. and Keong, L. K. Mixed matrix membranes for water purification applications. Separation & Purification Reviews 46, pp. 62-80, 2017.

[105] Kang, W., Li, F., Zhao, Y., Qiao, C., Ju, J. and Cheng, B. Fabrication of porous Fe_2O_3/PTFE nanofiber membranes and their application as a catalyst for dye degradation. RSC Advances 6, pp. 32646-32652, 2016.

[106] Niu, F., Zhang, L.-S., Chen, C.-Q., Li, W., Li, L., Song, W.-G. and Jiang, L. Hydrophilic TiO_2 porous spheres anchored on hydrophobic polypropylene membrane for wettability induced high photodegrading activities. Nanoscale 2, pp. 1480-1484, 2010.

[107] Xu, X., Gao, X., Cui, Z., Liu, X. and Zhang, X. Loading of PPy on the surface of transition metal coordination polymer modified polyoxometalate (TMCP/POM): a feasible strategy to obtain visible light active and high quantum yield POM based photocatalyst. Dalton Transactions 43, pp. 13424-13433, 2014.

[108] Yu, X., Gao, X., Lu, Z., Liu, X., Huo, P., Liu, X., Wu, D. and Yan, Y. Preparation and photodegradation properties of transition metal ion–poly-o-phenylenediamine/TiO_2/fly-ash cenospheres by ion imprinting technology. RSC Advances 3, pp. 14807-14813, 2013.

[109] Vilvamani, N., Gupta, R. D. and Awasthi, S. K. Ru(II)–polypyridyl complex-grafted silica nanohybrids: versatile hybrid materials for Raman spectroscopy and photocatalysis. RSC Advances 5, pp. 13451-13461, 2015.

[110] Tan, J. Z., Nursam, N. M., Xia, F., Truong, Y. B., Kyratzis, I. L., Wang, X. and Caruso, R. A. Electrospun PVDF–TiO_2 with tuneable TiO_2 crystal phases: synthesis and application in photocatalytic redox reactions. Journal of Materials Chemistry A, 5, pp. 641-648, 2017.

[111] Mahlambi, M. M., Mahlangu, O. T., Vilakati, G. D. and Mamba, B. B. Visible light photodegradation of rhodamine B dye by two forms of carbon-covered alumina supported TiO_2/polysulfone membranes. Industrial & Engineering Chemistry Research 53, pp. 5709-5717, 2014.

[112] Pereira, V. R., Isloor, A. M., Zulhairun, A. K., Subramaniam, M. N., Lau, W. J. and Ismail, A. F. Preparation of polysulfone-based PANI-TiO_2 nanocomposite hollow fiber membranes for industrial dye rejection applications. RSC Advances 6, pp. 99764-99773, 2016.

[113] Deng, F., Min, L., Luo, X., Wu, S. and Luo, S. Visible-light photocatalytic degradation performances and thermal stability due to the synergetic effect of TiO_2 with conductive copolymers of polyaniline and polypyrrole. Nanoscale 5, pp. 8703-8710, 2013.

[114] Pereira, V. R., Isloor, A. M., Bhat, U. K., Ismail, A., Obaid, A. and Fun, H.-K. Preparation and performance studies of polysulfone-sulfated nano-titania (S-TiO_2) nanofiltration membranes for dye removal. RSC Advances 5, pp. 53874-53885, 2015.

[115] Harraz, F. A., Ismail, A. A., Al-Sayari S. and Al-Hajry A. Novel α-Fe_2O_3/polypyrrole nanocomposite with enhanced photocatalytic performance. Journal of Photochemistry and Photobiology A: Chemistry 299, pp. 18-24, 2015.

[116] Ong, W., Gao, M. and Ho, G. Hybrid organic PVDF–inorganic M–rGO–TiO_2 (M = Ag, Pt) nanocomposites for multifunctional volatile organic compound sensing and photocatalytic degradation–H_2 production. Nanoscale 5, pp. 11283-11290, 2013.

[117] Zhang, H., Tao, Z., Tang, Y., Yang, M. and Wang, G. One-step modified method for a highly efficient Au-PANI@TiO_2 visible-light photocatalyst. New Journal of Chemistry 40, pp. 8587-8592, 2016.

[118] Ma, C., Zhou, J., Zhu, H., Yang, W., Liu, J., Wang, Y. and Zou, Z. Constructing a high-efficiency moo_3/polyimide hybrid photocatalyst based on strong interfacial interaction. ACS Applied Materials & Interfaces 7, pp. 14628-14637, 2015.

[119] Ding, Q., Miao, Y.-E. and Liu, T. Morphology and photocatalytic property of hierarchical polyimide/ZnO fibers prepared via a direct ion-exchange process. ACS Applied Materials & Interfaces 5, pp. 5617-5622, 2013.

[120] Cheng, Y., Gao, F., An, L., Li, X. and Wang, G. Different combinations of Fe_3O_4 microsphere, Polypyrrole and silver as core–shell nanocomposites for adsorption and photocatalytic application. Advanced Powder Technology 25, pp. 1600-1607, 2014.

[121] Sun, L., Shi, Y., Li, B., Li, X. and Wang, Y. Preparation and characterization of polypyrrole/TiO_2 nanocomposites by reverse microemulsion polymerization and its photocatalytic activity for the degradation of methyl orange under natural light. Polymer Composites 34, pp. 1076-1080, 2013.

[122] Gao, F., Hou, X., Wang, A., Chu, G., Wu, W., Chen, J. and Zou, H. Preparation of polypyrrole/TiO_2 nanocomposites with enhanced photocatalytic performance. Particuology 26, pp. 73-78, 2016.

[123] Macedo, E. R., Oliveira, P. S. and de Oliveira, H. P. Synthesis and characterization of branched polypyrrole/titanium dioxide photocatalysts. Journal of Photochemistry and Photobiology A: Chemistry 307, pp. 108-114, 2015.

[124] Pruna, A., Shao, Q., Kamruzzaman, M., Li Y., Zapien, J., Pullini, D., Mataix, D. B. and Ruotolo, A. Effect of ZnO core electrodeposition conditions on electrochemical and photocatalytic properties of polypyrrole-graphene oxide shelled nanoarrays. Applied Surface Science 392, pp. 801-809, 2017.

[125] Cao, S., Zhang, H., Song, Y., Zhang, J., Yang, H., Jiang, L. and Dan, Y. Investigation of polypyrrole/polyvinyl alcohol-titanium dioxide composite films for photocatalytic applications. Applied Surface Science 342, pp. 55-63, 2015.

Chapter 9

Nature, Sources, Resources, and Production of Thorium

Miloš René

Additional information is available at the end of the chapter

http://dx.doi.org/10.5772/intechopen.68304

Abstract

Thorium is a naturally occurring, slightly radioactive element. It is widely distributed in nature with an average concentration of 10.5 ppm Th in the upper earth's crust. In general, thorium occurs in relatively small number in Th-enriched minerals: thorite, thorianite, monazite, bastnaesite, and thorogummite. However, the main world resources of thorium are coupled with monazite and bastnaesite. Monazite-enriched placer deposits occurring mainly in India, Brazil, Australia, and the USA form the recently available resources of thorium. Other commercially interested concentrations of thorium are coupled with bastnaesite mined from carbonatite deposits, especially from Bayan Obo deposit in China. Currently, the worldwide thorium resources by major deposit types are estimated to total about 6.2 million tons of Th. Issues associated with thorium's natural radioactivity are a significant deterrent to its commercial use. The monazite concentrates are recently produced only in India, Brazil, Malaysia, Thailand, and Vietnam, with a total amount of about 7000 tons. Consequently, experimental nuclear reactors based on thorium fuel cycle are operated recently only in India. In the long term, consumption of thorium could increase substantially if its use as a nuclear fuel becomes commercialized.

Keywords: thorium, geochemistry, mineralogy, monazite, thorium fuel cycle

1. Introduction

Thorium is a naturally occurring, slightly radioactive element. It is found in small amounts in most rocks, where it is about three times abundant than uranium. Thorium is relatively enriched in acid igneous rocks, especially in granites. The most common thorium mineral is monazite. In uranium ore deposits, thorium is concentrated in thorite and thorianite. In magmatic carbonate-enriched rocks (carbonatites), thorium is associated with rare earth elements (REE) in bastnaesite. However, the most important reserves of thorium occur in placer deposits, which contain monazite. Monazite is in placer deposits mined together with other heavy minerals, such as rutile, zircon, ilmenite, and cassiterite. The principal monazite-producing countries are India, Brazil, Malaysia, and Thailand. Major end uses of thorium are refractories, lamp mantles, and aerospace alloys. Relatively restricted is using thorium in energy production. Although research into thorium-fuelled nuclear reactors continues, there exist no industrial-scale nuclear reactors using thorium. However, India continued its plan for a development of its nuclear power program based on the thorium-fuelled nuclear reactors.

2. History

Thorium was identified as an element in the mineral thorite in 1828 by the Swedish chemist Berzelius. Newly discovered element was named for Thor, the Scandinavian God of thunder and lighting, because of its use in energy. In 1885, thorium came into commercial use when it was discovered that a fabric mantle impregnated with a thorium compounds would give a steady, bright white light when heated. This discovery led to the development of the Welsh mantle, which was adopted in gas lighting and later in kerosene lamps. Thorium derived from monazite occurring in the Brazilian beach sands was produced as early as 1885. In 1911, monazite from the Indian beach sand deposits mastered world monazite markets.

During this time, the German manufacturers organized a monopoly of the thorium nitrate industry. World War I restricted German supplies of thorium compounds and enabled US production of thorium nitrate to expand. In the early 1920s, electricity began to replace gas and kerosene for general lighting purposes, and the need for thorium mantles declined. Up to the end of World War II, dominate monazite producers were India and Brazil. Since 1945 some other countries have started with their monazite production (e.g., Australia and Malaysia). During World War II started new using of thorium as a component in a high-temperature alloys.

After the war, monazite was processed largely for its nuclear fuel potential. The discovery in 1946 that ^{232}Th could be transmuted into ^{233}U increased the interest in thorium. However, the decision to develop nuclear reactors based on uranium fuels slowed development of thorium-fuelled reactors and reduced thorium demand. During the 1950s, some became new producers of thorium, namely Canada and South Africa, where uranium ores from uranium-enriched quartz-pebble conglomerates contain also some thorium. At this time distinctly increased interest in the rare earth elements (REE) and monazite was mined in the first place for its REE content. Some other thorium was also acquired from REE bearing bastnaesite, occurring in carbonate-enriched magmatic rocks (carbonatites). Much of thorium contained in residues is being stockpiled by private industry [1].

New interest about using thorium as nuclear fuels started in 1960s together with ideas in the development of Fast Breeder Reactors (FBR). Basic research of thorium fuels cycles are being undertaken by Brazil, Germany, the USA, India, Italy, Australia, Canada, China, France, USSR, Romania, and some other countries. Several experimental and prototype nuclear power reactors were successfully operated from the mid-1950s to the mid-1970s using (Th, U)O_2, (Th, U)C_2, and $LiF/BeF_2/ThF_4/UF_4$ fuel. The activity of the Nuclear Cycle Division of the IAEA in this area was supported mainly by organizing some technical committee meetings [2–5]. However, thorium fuels have not been introduced commercially because the estimated uranium resources turned out to be sufficient. On the other hand, using thorium in nuclear energy cycle has some significant precedence: (i) the intrinsic proliferation resistance of thorium fuel cycle, (ii) better thermophysical properties and chemical stability of ThO_2, as compared to UO_2, (iii) lesser long-lived minor actinides than the traditional uranium fuel cycle, (iv) superior plutonium incineration in (Th, Pu)O_2 fuel as compared to (U, Pu)O_2, and (v) attractive features of thorium related to accelerated-driven system and energy amplifier. However, there are several challenges in the form and back end of the thorium fuel cycles. Irradiated ThO_2 and spent ThO_2-based fuels are difficult to dissolve in HNO_3 because of the inertness of ThO_2. The high gamma radiation associated with the short-lived daughter products of ^{232}U, which is always associated with ^{233}U, necessitates remote reprocessing and refabricating of fuel. The protactinium formed in thorium fuel cycle also causes some problems, which need to be suitably resolved. Consequently, recently the various experimental nuclear reactors based on thorium fuel cycle are operated only in India. Some other basic research on thorium fuel cycle continued in China, France, Japan, Norway, Russia, and the USA [6].

The other thorium's commercial uses included catalysts, high-temperature ceramics, and welding electrodes. Other no energy uses of thorium are in electron tubes, special use lighting such as airport runway lighting, high-refractive glass, radiation detectors, computer memory components, photoconductive films, target material for X-ray tubes, and fuel cell elements. Its use in most of these products is generally limited because of concerns over its naturally occurring radioactivity. Consequently, no radioactive substitutes have been developed for many applications of thorium. Beryllium, aluminium, and yttrium oxides can be substituted for thorium oxide as a refractory. Yttrium compounds have replaced thorium compounds in incandescent lamp mantles. Magnesium alloys containing Zn, Al, REE, Y, and Zr can substitute for magnesium-thorium alloys in aerospace applications. Research is being conducted to find a replacement for thorium in lamp mantles. These substitutions for thorium in no energy uses are expected to increase because of growing public concern and governmental regulations on radioactive materials [7].

3. Geochemistry of thorium and thorium minerals

Thorium is widely distributed in nature with an average concentration of 10.5 ppm Th in the upper earth's crust, while the middle crust has an average of 6.5 ppm Th and the lower crust an average of 1.2 ppm Th [7]. Thorium is relatively depleted in mafic igneous rocks (basalts) where the concentration averages about 1 ppm Th, although alkali varieties enriched in Na and K relative to Ca range up to 5 ppm Th. Granitic rocks show a distinct increase over mafic igneous rocks, averaging 20–30 ppm Th. Thorium together with REE could be accumulated during fractional crystallization of alkali igneous rocks. Under some circumstances by this fractionation, a separate carbonate-enriched melt will form, resulting in carbonatites. With carbonatites are associated some complex fluoro-carbonate minerals such as bastnaesite. During weathering, thorium remains in the refractory solid form and is mostly transported as distinct mineral grains (typically as monazite). Sandstones contain about 2 ppm Th, with beach sands containing 10 ppm Th, and limestone averages about 2 ppm. Shale contains 10–15 ppm Th, small amounts of thorium may adsorb clay particles during weathering. Twelve isotopes of thorium are known, with atomic masses from 223 to 234. However, natural thorium is present as nearly 100% ^{232}Th isotope. The other important natural isotope of thorium ^{230}Th is generally presented in uranium minerals.

In general, thorium occurs in relative small number of Th-enriched minerals: thorite ($ThSiO_4$), thorianite (ThO_2), monazite [(Ce, La, Nd,Th,U)PO_4], bastnaesite [(Ce, La)CO_3F], and thorogummite [$Th(SiO_4)_{1-x}(OH)_{4-x}$]. However, the main world resources of thorium are coupled with monazite and bastnaesite. Monazite is a primary source of light REE. Monazite concentrates, which are mined from beach sands in India, Brazil, the USA, Malaysia, Korea, and Sri Lanka, contain 3.1–14.32 wt.% ThO_2 and 40.7–65.0 wt.% REO (rare earths oxides). However, monazites from some granitic rocks could contain up to 27 wt.% ThO_2.

Other, especially, potentially based resources of thorium are coupled with carbonate-enriched magmatic rocks (carbonatites), containing bastnaesite (up to 2.8 wt.% ThO_2), parisite [$CaREE_2(CO_3)_3(F,OH)_2$] (up to 4.0 wt.% ThO_2), and synchysite [$CaREE(CO_3)_2(F, OH)$] (up to 5.0 wt.% ThO_2). Highly rare alkali-rich nephelinite syenites from the Lovozero pluton on the Kola peninsula (Russia) contain rare REE-enriched mineral loparite (Na,REE,Ca) (Ti, Nb) O_2 with up to 1.6 wt.% ThO_2 [8]. Some higher concentrations of thorium have also important apatite ore deposit on the Kola Peninsula in the Russia. Apatite containing higher concentrations of thorium occurs also in the alkalic magmatic rocks on the Vishnevyye Mountains of the Urals range in the Russia [1]. A large variety of other minerals contain minor amounts of thorium (e.g., allanite, xenotime, zircon, and uraninite).

4. Ore deposits

Thorianite, thorite, and uranothorite are the only true thorium minerals, but they are not recently recovered. Some resources of these minerals are coupled with quartz-pebble conglomerates in Canada (Elliot Lake region) and South Africa (Witwatersrand). World thorium resources in terms of the genetic types of ore deposits are displayed in **Table 1** [9].

Deposit type	Metric tons of thorium
Placers	2,182,000
Carbonatites	1,783,000
Vein-type	1,528,000
Alkaline igneous rocks	584,000
Others	135,000
Total	6,212,000

Table 1. World thorium resources in terms of the deposit types [9].

The main recently available resources of thorium are coupled with monazite-enriched placer deposits in alluvial or marine sediments occurring mainly in Australia, India, Brazil, Venezuela, the USA, and Egypt. These deposits contain variable proportions of monazite, ilmenite, rutile, xenotime, zircon, and/or cassiterite. The main minerals, which are mined on these placer deposits, are ilmenite, rutile, zircon, and cassiterite. Associated minerals, which are rarely of economic significance, can include garnet and kyanite. Monazite, when also extracted, represents only an accidental product.

Placer deposits are found where water waves have concentrated heavy mineral grains on a sea beaches. These deposits may occur in both modern and ancient shorelines. Many of the heavy mineral sand deposits are concentrated by wave action in both parallel and transgressive dunes. Monazite placers are reported from Australia, Egypt, India, Liberia, Brazil, Burma, Malaysia, Sri Lanka, and the USA.

The placer deposits in Australia are mined for their ilmenite, rutile, and zircon content. The monazite content in heavy mineral concentrates varied from 0.2 to 1.5 wt.%. Present-day shoreline deposits are evolved on the east coast in the SE Queensland. The most important placer sand deposits are coupled with the Tertiary fossil shoreline deposits in the Murray Basin, in the southwest Australia. Monazite grades in this deposits are around 1–1.5 wt.%.

Although monazite occurs associated with ilmenite and other heavy minerals in beach sands, skirting the entire Peninsular India, its economic concentration is confined to only some areas with suitable physiographic conditions. The west coast placers are essentially beach or barrier deposits with development of dunes evolved on dry months. On the other hand, the east coasts deposits consist of extensive dunes fringing the coasts. The beach sands of Chavara bar (Kerala) on the West coast contain 73 vol.% heavy minerals, 60–70 vol.% ilmenite, 4–7 vol.% garnet, 5–8 vol.% zircon, and 0.5–1.0 vol.% monazite. The east coast beach placers and dunes are low grade with 8–20 vol.% of heavy minerals. The most important placer deposit on the East coast is the Chatrapur deposit (Orissa) with about 20 vol.% heavy minerals and 0.5 vol.% monazite. In the Malaysian deposits, monazite is associated with columbite, xenotime, and cassiterite. The cassiterite placers at Trengganu contain as much as 58 vol.% monazite. In Sri Lanka, the largest placer deposit near Pulmoddai contains 3 million tons of sand with 0.4 vol.% monazite, 18 vol.% rutile, and 62 vol.% ilmenite [10].

In Brazil, monazite occurs associated with ilmenite and zircon in placer deposits evolved along the eastern and south-eastern Atlantic coast. In Burma, placer deposits occur in the southern Shan states. Weathering of quartz veins and pegmatite dykes injected into the granites derives considerable quantities of cassiterite and wolframite occurring in the placers.

In the USA, alluvial deposits of monazite are known to occur in the intermountain valleys of Idaho, the Carolina Piedmont of North and South Carolina, and the beach deposits of north-eastern Florida to south-eastern Georgia. The three monazite placer districts, namely the North and South Carolina stream deposits, Idaho stream deposits, and Florida-Georgia beaches, are the largest volume known alluvial thorium deposits in the USA. The modern and raised Pleistocene and Pliocene beach deposits of north-eastern Florida and south-eastern Georgia host low-grade but persistent concentrations of thorium. Heavy minerals constitute a small part of the beach sands. The most abundant heavy mineral in this beach deposits is ilmenite, in many places forming more than 50 vol.% of the heavy-mineral fraction. Monazite forms a minor part of the heavy-mineral fraction, usually less than 1 vol.%. The beach placer deposits of this region contain total reserves of about 14,700 tons of ThO_2, which occur in 330,000 tons of monazite. These placer deposits were mined primarily for ilmenite and rutile. Mining ceased in this area in late 1978 since increasing environmental regulations made mining operations more costly [11].

The ore bodies constituting quartz-pebble conglomerates are represented in particular by the Blind River-Elliot Lake deposits in Ontario, Canada and the Witwatersrand deposits in South Africa. These ore bodies occur mainly in pyrite-bearing oligomictic conglomerates. In the Blind River-Elliot Lake deposits, thorium together with uranium occurs mainly in a brannerite-uraninite-monazite mineral assemblage. Principal ore minerals are uraninite, brannerite, and monazite, with minor coffinite, uranothorite, xenotime, and gummite. Uraninite is partly enriched in thorium with average content of 6.5 wt.% ThO_2. The Witwatersrand reefs are not only rich in gold but also represent significant uranium deposit. The principal uranium minerals are uraninite and lesser uranothorite, brannerite, and coffinite. Uraninite from this deposit is enriched in thorium (average 3.9 wt.% Th) [12].

Carbonatites, which recently represented the most important source of REE, are also considered as potential source of thorium. Carbonatites are igneous rocks containing >50% of primary carbonate minerals. The most carbonatites are actually polygenetic and show evidence of hydrothermal and metasomatic reworking. These rocks generally contain <50 ppm Th; however, some contain higher concentrations. The majority of carbonatites occur in association with broadly coeval ultramafic and alkaline silicate rocks.

From the mid-1960s to 1985, the carbonatite-hosted Mountain Pass deposit in the USA was the world's main source of REE, producing over 20 kt REO at its zenith. Recently, almost all (~97%, or 120–130 kt REO in 2006–2010) of the world's REE supply comes from China, with 40–50% of this production contributed by the giant Fe–REE–Nb deposit at Bayan Obo [8].

Carbonatite-related deposits can be subdivided into deposits where magmatic and/or hydrothermal processes are important and those where secondary processes such as supergene enrichment and laterization predominate. The most important primary carbonatite-related deposit is the Bayan Obo deposit in Inner Mongolia, China, which represents 70% of the world's REE resources. The most abundant REE minerals at Bayan Obo are monazite and bastnaesite. Carbonatites with appreciable REE and Th mineralization have been reported also at Khibiny (Kola, Russia), Ozerny and Arshan (Siberia, Russia), Fen (Norway), Sokli (Finland), Mount Weld, Cummins Range, Mud Tank (Australia), Palabora (South Africa), Khanneshin (Afghanistan), Amba Dogar (India), Barra do Itapirapuã (Brazil), Tundulu and Kangankunde (Malawi), and Wigu Hill (Tanzania) among many others [13].

Highly potential sources of REE and Th represent intrusions of alkaline and peralkaline igneous rocks. Some well-known examples of these rock series being the Ilímaussaq intrusion in southern Greenland, Lovozero and Khibiny alkaline plutons on the Kola Peninsula (Russia), the Red Wine-Letitia alkaline province in Canada, alkaline laccolith at Poços de Caldas

(Brazil), and alkaline syenite body at Pilanesberg (South Africa) [14, 15]. All these magmatic complexes are zoned or layered, enriched with Na and K, and contain a variety of relatively rare minerals including thorite, monazite, loparite, zircon, and apatite. Thorium is present in thorite, monazite, zircon, loparite, and some other accessory minerals of REE. Local thorium contents in these rock complexes may range up to 1500 ppm Th, but overall they rarely contain in excess of 50 ppm Th [16].

On the Ilímaussaq complex is bounded highly interested REE-Y-U-Th mneralization at Kvanefjeld (Kuannersuit) coupled on agpaitic nepheline syenites. Recently performed exploration is concentrated on potential recovery of REO, Y, together with production of U and Zn as valuable by-products. The Lovozero complex in the Kola Peninsula is represented by layered intrusion of varied varieties of nepheline syenite with loparite as main economic interested mineral. Several loparite-rich units were mined since 1951 as the major source of LREE, Nb, and Ta for Soviet industry [13].

Some other potential sources of thorium represent Th-bearing granites and pegmatites, which are known in many parts of world. The thorium-enriched pegmatites most commonly occur in near granitic or syenitic bodies or in high-grade metamorphic rock series near their contacts with granitic stocks or batholiths. The principal thorium-bearing minerals in these deposits are uraninite, thorite, brannerite, uranothorite, monazite, and some other REE-, Th-, and U-bearing accessory minerals.

Most of the Th-enriched granites and pegmatites are not of commercial importance at present but provide a large reserve of thorium and uranium for the future. Thorium-bearing granitic rocks and pegmatite-bearing rock complexes occur in the Bancroft area of Ontario (Canada), Rössing (Namibia), Crockers Well, Greenbushes, Radium Hill (Australia). The Th-bearing minerals are represented by allanite, betafite, brannerite, davidite, monazite, thorite, and some others rare REE-, Th-, and U-bearing minerals. The ore zones at the Greenbushes contain low levels of thorium with average grades in range 3–25 ppm Th [17]. The uranium deposits in the Bancroft area coupled with anatectic pegmatites produced a total of 5700 tons U between 1956 and 1982. The principal ore minerals were uraninite, with up to 10 wt.% ThO_2, and uranothorite.

Uranium mineralization associated with leucogranite dykes on the Rössing deposit in Namibia was derived by the partial melting of U-rich sedimentary rocks. Uraniferous leucogranite bodies are located in high-grade metasediments of the Damara Pan-African belt. Within mineralized leucogranites, the distribution of uranium can be extremely variable. Main uranium mineral is Th-bearing uraninite with 3.3–8.0 wt.% ThO_2. The Rössing uranium deposit is mined from 1976 and total production by the end of 2007 was slightly over 90,000 tones U [18].

Relatively rare thorium-bearing vein deposits are distributed throughout the world. They are localized in shear zones, faults, breccias zones in metasedimentary and metavolcanic rocks and are often associated with alkalic rocks complexes and carbonatites. The principal Th-bearing minerals in vein deposits are thorite, thorogummite, and monazite, which are associated with some REE-minerals (allanite, bastnaesite, and xenotime). Examples of these ore deposits include the Lemhi Pass in the USA, Steenkampskraal and Vanrhynsdorp in South Africa, Eskisehir deposit in Turkey, and Nolans Bore in Australia [16, 19].

The Lemhi Pass district on the Montana-Idaho border in the USA contains numerous thorium-rich veins in the central Beaverhead Mountains. This district is thought to represent the largest concentration of thorium resources in the USA. The district contains total reserves of 64,000 tons of ThO_2 [20]. Most of mineralized veins are quartz-hematite-thorite veins, which fill fractures, shears, and brecciated zones in quartzitic rocks. Rare-earth- and Th-bearing allanite and monazite are locally abundant. The thorite veins of the Lemhi Pass district are approximately equally enriched in thorium and REE. The total REE-oxide contents range from 0.07 to 2.20 wt.%, with an average value of 0.43 wt.%. The average concentrations of Th are 0.43 wt.%. [11].

At Steenkampskraal (South Africa) from the 1950s to 1963 about 50,000 tons of monazite concentrates were extracted, which contained between 3.3 and 7.6 wt.% Th before operation of the mine was halted. New economic assessment of this deposit was completed in 2012 and currently were established resources of 86,900 wt.% of REO [9].

The Nolans Bore deposit in the Northern Territory (Australia) is coupled with mineralized shear zones evolved in variably deformed and altered granitic gneiss, pegmatite, and minor calcsilicate rocks. Massive fluorapatite dykes enriched in REE and Th form the mineralization. The thorium content of Nolans Bore fluorapatite generally ranges from 0.07 to 0.59 wt.% Th (average 0.23 wt.% Th) [21].

5. Resources

The by-product nature of the occurrence of thorium and a lack of economic interest has meant that thorium resources have seldom, if over, been accurately defined. Information on thorium resources was published in a joint report by the OECD Nuclear Energy Agency and the International Atomic Energy Agency (IAEA)—"Red Books" between 1965 and 1981, typically using the same terminology as for uranium resources at that time (e.g., reasonably assured resources and estimated additional resources I and II, the latter two categories which are recently termed inferred and prognosticated resources, respectively). No further information was published until 2003 when a global estimate of thorium resources of 4.5 million Th was presented in the 2003 Reed Book. A more comprehensive report was presented in the 2007 Red Book where resource estimates were given by deposit type and by countries and this was updated in the 2009 edition. Currently, the worldwide thorium resources by major deposit types are estimated to total about 6.2 million tons Th, including undiscovered resources (**Table 1**). In 2011 and 2013, the IAEA conducted technical meetings on thorium resources. Based on the inputs given in the meetings and details available in other open sources, identified uranium resources, regardless of resource category or cost category, have been updated for 16 countries and published in the most recent Red Book (**Table 2**) [9]. However, these identified resources (reasonably assured and inferred resources) may not have the same meaning in terms of classification as identified U resources.

The main world resources of thorium are associated with monazite placer deposits in India, Brazil, Australia, the USA, Egypt, and Venezuela. The second most important thorium resources could be mined as by-product of REO from carbonatites (China, Greenland,

Country	Metric tons of thorium
India	846,000
Brazil	632,000
Australia	595,000
The USA	595,000
Egypt	380,000
Turkey	374,000
Venezuela	300,000
Canada	172,000
Russia	155,000
South Africa	148,000
China	100,000
Norway	87,000
Greenland	86,000
Finland	60,000
Sweden	50,000
Kazakhstan	50,000
Other countries	1,725,000
World total	6,355,000

Table 2. Identified resources of thorium [9].

Norway, Finland, and Sweden). Some other thorium resources are coupled with various uranium deposits in Canada, the USA, South Africa, and Kazakhstan. Thorium in Kazakhstan could be recovered as a by-product together with REO bounded on complex U-REE-Th ores. These ores are recently processed in the SARECO plant by Stepnogorsk [9].

6. Production

The monazite concentrates arte recently produced only in five countries (**Table 3**) [22]. However, substantial but unquantifiable quantities of thorium produced China during the processing of domestic and imported mineral concentrates for production of rare-earth compounds. Issues associated with thorium's natural radioactivity are a significant deterrent to its commercial use. Limited global demand for thorium continued to create an oversupply of thorium compounds and residues. Excess thorium that was not designated for commercial use was either disposed of as a low-level radioactive waste or stored. Although research into thorium-fuelled nuclear reactor continues, there are recently no industrial-scale nuclear reactors using thorium.

Country	Metric tons
India	5500
Brazil	600
Malaysia	500
Thailand	210
Vietnam	180
Total	6990

Table 3. World production of monazite concentrates [22].

Recently, worldwide only minor amounts of ThO_2 are typically used annually. Principal uses include chemical catalysts, lighting, welding electrodes, and heat-resistant ceramics, in descending order of use.

7. Conclusion

Extraction of thorium as a by-product of REE recovery from monazite and other REE- and Th-bearing minerals (bastnäesite) seems to be the most feasible source of thorium production recently. Processing of monazite to recover REE and thorium has been done in the past in many countries. Main monazite concentrate production is currently taking place in India, Brazil, Malaysia, Thailand, and Vietnam. Substantial but unquantifiable quantities of thorium produced China during the processing of domestic and imported mineral concentrates for production of rare-earth compounds. However, issues associated with thorium's natural radioactivity are a significant deterrent to its commercial use. Consequently, recently the various experimental nuclear reactors based on thorium fuel cycle are operated only in India. In the long term, consumption of thorium could increase substantially if its use as a nuclear fuel becomes commercialized.

Acknowledgements

This work was carried out thanks to support of the long-term conceptual research organization RVO: 67985891.

Author details

Miloš René

Address all correspondence to: rene@irsm.cas.cz

Institute of Rock Structure and Mechanics, Academy of Sciences of the Czech Republic, Prague 8, Czech Republic

References

[1] Hedrick JB. Thorium. In: Mineral Facts and Problems 1985 Edition. Bull Bureau of Mines. 1985;**675**:835-846.

[2] Thorium-Based Nuclear Fuel: Current Status and Perspectives. IAEA-TECDOC-412. Vienna: IAEA; 1987. p. 161.

[3] Thorium-Based Fuel Options for the Generation of Electricity. Developments in the 1990s. IAEA-TECDOC-1155. IAEA; 2000. p. 144.

[4] Thorium Fuel Utilization: Options and Trends. IAEA-TECDOC-1319. IAEA; 2002. p. 376.

[5] Thorium Fuel Cycle—Potential Benefits and Challenges. IAEA-TECDOC-1450. Vienna: IAEA; 2005. p. 105.

[6] Thorium. U.S. Geological Survey, Mineral Commodity Summaries. U.S.Geological Survey: Washington; January 2016.

[7] Rudnick RL, Gao S. Composition of the continental crust. In: Holland HD, Turekian KK, editors. Treatise on Geochemistry 3. Amsterdam: Elsevier; 2004. pp. 1-64.

[8] Chakhmouradian AR, Wall F. Rare earth elements: Minerals, mines, magnets (and more). Elements. 2012;**8**:333-340. DOI: 10.2113/gselements.8.5.333

[9] Uranium 2014: Resources, Production and Demand. Paris: OECD; 2014. p. 508.

[10] Jayaram KMV. An overview of world thorium resources, incentives for futher exploration and forecast for thorium requirements in the near future. In: IAEA-TECDOC 412. Vienna: IAEA; 1987. pp. 8-26.

[11] Van Gosen BS, Gillerman VS, Armbrustmacher TJ. Thorium deposits of the United States—Energy resources for the future? U.S. Geological Survey Circulation. 2009;**1336**:1-29.

[12] Frimmel HE. Archaean atmospheric evolution: Evidence from the Witwatersrand gold fields, South Africa. Earth Science Reviews. 2005;**70**:1-46.

[13] Chakhmouradian AR, Zaitsev AN. Rare earth mineralization in igneous rocks: Sources and processes. Elements. 2012;**8**:347-353. DOI: 10.2113/gselements.8.5.347

[14] Wooley AR. Alkaline Rocks and Carbonatites of the World. Part 3: Africa. London: The Geological Society; 2001. p. 372.

[15] Kogarko LN, Kononova NA, Orlova MP, Wolley AR. Alkaline Rocks and Carbonatites of the World. Part Two: Former USSR. London: Chapman and Hall; 1995. p. 226.

[16] Boyle RW. Geochemical Prospecting for Thorium and Uranium Deposits. Amsterdam: Elsevier; 1982. p. 498.

[17] Partington GA, McNaught NJ, Williams IJ. A review of geology, mineralisation and geochronology of the Greenbushes pegmatite, Western Australia. Economic Geology. 1995;**90**:616-635.

[18] Cuney M, Kyser K. Hydrothermal uranium deposits related to igneous rocks. In: Cuney M, Kyser K, editors. Recent and Not-So-Recent Developments in Uranium Deposits and Implications for Exploration. Short Course Series. Mineralogical Association of Canada: Québec City; Vol. 39; 2009. pp. 117-160.

[19] Gultekin AH, Örgün Y, Suner F. Geology, mineralogyx and fluid inclusion data of the Kirilcaören fluorite-barite-REE deposit, Eskisekir, Turkey. Journal of Asian Earth Science. 2003;**21**:365-376.

[20] Staatz MH. Geology and mineral resources of the Lemhi Pass thorium district, Idaho and Montana. U.S. Geological Survey Circular. 1979;**1336**:1-29.

[21] Mernagh TP, Miezitis Y. A review of the geological processes controlling the distribution of thorium in the Earth´s crust and Australia´s thorium resources. Geoscience Australia Record. 2008;**5**:1-48.

[22] Gambogi J, Aquino KC. Thorium. In: USGS 2013 Minerals Yearbook. U.S. Geological Survey. 2013:761-764.